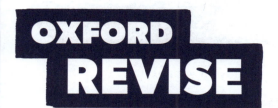

# Oxford Revise

## AQA A LEVEL
# GEOGRAPHY

## COMPLETE REVISION AND PRACTICE

Rob Bircher

Alice Griffiths

Rebecca Priest

Lucy Scovell

**OXFORD**
UNIVERSITY PRESS

# CONTENTS

 Shade in each level of the circle as you feel more confident and ready for your exam.

**How to use this book** — **viii**

**1  Water and carbon cycles** — **2**

⚙ **1.1**  Systems concepts — 2
⚙ **1.2**  The water cycle — 3
⚙ **1.3**  The drainage basin — 6
⇄ **Retrieval** — 8
⚙ **1.4**  Carbon cycle: stores and transfers — 9
⚙ **1.5**  Changes in the carbon cycle — 10
⇄ **Retrieval** — 13
⚙ **1.6**  Climate change — 14
⇄ **Retrieval** — 16
⚙ **1.7**  Case study: Tropical rainforest — 17
⚙ **1.8**  Case study: River catchment — 18
⇄ **Retrieval** — 20
✎ **Practice** — 21

**2  Hot desert systems and landscapes** — **26**

⚙ **2.1**  Hot desert systems — 26
⇄ **Retrieval** — 30
⚙ **2.2**  Desert systems and processes — 31
⚙ **2.3**  Desert landforms — 34
⇄ **Retrieval** — 38
⚙ **2.4**  Desertification — 39
⇄ **Retrieval** — 41
⚙ **2.5**  Case study: Desertification — 42
⚙ **2.6**  Case study: Hot desert environment — 43
⇄ **Retrieval** — 44
✎ **Practice** — 45

**KEY**

⚙ **Knowledge**

⇄ **Retrieval**

✎ **Practice**

## 3 Coastal systems and landscapes — 49

| | | |
|---|---|---|
| ⚙ | 3.1 Coastal systems | 49 |
| ⚙ | 3.2 Coastal processes | 51 |
| ⮂ | Retrieval | 53 |
| ⚙ | 3.3 Landforms of coastal erosion | 54 |
| ⚙ | 3.4 Landforms of coastal deposition | 56 |
| ⚙ | 3.5 Sea level change | 59 |
| ⮂ | Retrieval | 61 |
| ⚙ | 3.6 Coastal management | 62 |
| ⮂ | Retrieval | 64 |
| ⚙ | 3.7 Case Study: A coastal environment at a local scale | 65 |
| ⚙ | 3.8 Case study: A contrasting coastal landscape | 67 |
| ⮂ | Retrieval | 68 |
| ✎ | Practice | 69 |

## 4 Glacial systems and landscapes — 74

| | | |
|---|---|---|
| ⚙ | 4.1 Glacial systems | 74 |
| ⚙ | 4.2 Cold environments | 77 |
| ⮂ | Retrieval | 79 |
| ⚙ | 4.3 Glacial processes | 80 |
| ⮂ | Retrieval | 83 |
| ⚙ | 4.4 Glacial landforms | 84 |
| ⮂ | Retrieval | 88 |
| ⚙ | 4.5 Fluvioglacial and periglacial landforms | 89 |
| ⚙ | 4.6 Human impacts on fragile cold environments | 91 |
| ⚙ | 4.7 Case study: A glaciated environment | 93 |
| ⚙ | 4.8 Case study: A contrasting glaciated landscape | 94 |
| ⮂ | Retrieval | 95 |
| ✎ | Practice | 96 |

## 5 Hazards — 101

| | | |
|---|---|---|
| ⚙ | 5.1 The concept of hazard in a geographical context | 101 |
| ⮂ | Retrieval | 103 |
| ⚙ | 5.2 Plate tectonics | 104 |
| ⮂ | Retrieval | 107 |

# CONTENTS

| | | |
|---|---|---|
| ⚙ 5.3 | Volcanic hazards | ⊖ 108 |
| ⇄ **Retrieval** | | ⊖ 112 |
| ⚙ 5.4 | Seismic hazards | ⊖ 113 |
| ⇄ **Retrieval** | | ⊖ 117 |
| ⚙ 5.5 | Storm hazards | ⊖ 118 |
| ⇄ **Retrieval** | | ⊖ 121 |
| ⚙ 5.6 | Wildfires | ⊖ 122 |
| ⚙ 5.7 | Case study: A place in a hazardous setting | ⊖ 124 |
| ⚙ 5.8 | Case study: A multi-hazardous environment | ⊖ 125 |
| ⇄ **Retrieval** | | ⊖ 127 |
| ✎ **Practice** | | ⊖ 128 |

| **6 Ecosystems under stress** | | **133** |
|---|---|---|
| ⚙ 6.1 | Ecosystems and sustainability | ⊖ 133 |
| ⇄ **Retrieval** | | ⊖ 135 |
| ⚙ 6.2 | Ecosystems and processes | ⊖ 136 |
| ⚙ 6.3 | Terrestrial ecosystems | ⊖ 139 |
| ⇄ **Retrieval** | | ⊖ 140 |
| ⇄ 6.4 | Biomes | ⊖ 141 |
| ⇄ **Retrieval** | | ⊖ 146 |
| ⚙ 6.5 | Ecosystems in the British Isles over time | ⊖ 147 |
| ⇄ **Retrieval** | | ⊖ 149 |
| ⚙ 6.6 | Marine ecosystems | ⊖ 150 |
| ⇄ **Retrieval** | | ⊖ 152 |
| ⚙ 6.7 | Local ecosystems | ⊖ 153 |
| ⚙ 6.8 | Case study: A local ecosystem | ⊖ 155 |
| ⚙ 6.9 | Case study: Ecological change | ⊖ 156 |
| ⇄ **Retrieval** | | ⊖ 157 |
| ✎ **Practice** | | ⊖ 158 |

| **7 Global systems and global governance** | | **161** |
|---|---|---|
| ⚙ 7.1 | Globalisation | ⊖ 161 |
| ⇄ **Retrieval** | | ⊖ 163 |

**KEY**

⚙ **Knowledge**

⇄ **Retrieval**

✎ **Practice**

| 7.2 | Global systems | 164 |
|---|---|---|
| Retrieval | | 166 |
| 7.3 | International trade | 167 |
| Retrieval | | 170 |
| 7.4 | TNCs | 171 |
| 7.5 | Consequences of global systems | 174 |
| Retrieval | | 176 |
| 7.6 | Global governance | 177 |
| Retrieval | | 179 |
| 7.7 | The global commons | 180 |
| 7.8 | Antarctica as a global common | 182 |
| 7.9 | Global governance in Antarctica | 184 |
| Retrieval | | 186 |
| 7.10 | Globalisation critique | 187 |
| Retrieval | | 189 |
| Practice | | 190 |

## 8 Changing places — 195

| 8.1 | The concept of place | 195 |
|---|---|---|
| Retrieval | | 197 |
| 8.2 | Relationships and connections | 198 |
| 8.3 | Meaning and representation | 201 |
| Retrieval | | 206 |
| 8.4 | Quantitative and Qualitative data | 207 |
| Retrieval | | 208 |
| 8.5 | Local place study | 209 |
| 8.6 | Contrasting place study | 211 |
| Retrieval | | 213 |
| Practice | | 214 |

## 9 Contemporary urban environments — 218

| 9.1 | Urbanisation | 218 |
|---|---|---|
| 9.2 | Megacities and world cities | 222 |
| Retrieval | | 224 |

# CONTENTS

⚙ **9.3**   Urban forms   ⊖ 225

⚙ **9.4**   Urban issues   ⊖ 228

⇄ **Retrieval**   ⊖ 230

⚙ **9.5**   Urban climate   ⊖ 231

⇄ **Retrieval**   ⊖ 234

⚙ **9.6**   Urban drainage   ⊖ 235

⇄ **Retrieval**   ⊖ 239

⚙ **9.7**   Urban waste   ⊖ 240

⇄ **Retrieval**   ⊖ 243

⚙ **9.8**   Urban environmental issues   ⊖ 244

⚙ **9.9**   Sustainable urban development   ⊖ 246

⇄ **Retrieval**   ⊖ 248

⚙ **9.10**   Case study: An urban area in the UK   ⊖ 249

⚙ **9.11**   Case study: A contrasting urban area   ⊖ 250

⇄ **Retrieval**   ⊖ 251

✎ **Practice**   ⊖ 252

## 10  Population and the environment   255

⚙ **10.1**   The relationship between physical geography and population   ⊖ 255

⚙ **10.2**   Global food production and consumption   ⊖ 256

⇄ **Retrieval**   ⊖ 257

⚙ **10.3**   Climate   ⊖ 258

⚙ **10.4**   Soils   ⊖ 259

⇄ **Retrieval**   ⊖ 261

⚙ **10.5**   The environment and health   ⊖ 262

⚙ **10.6**   Biologically transmitted and non-communicable disease   ⊖ 264

⇄ **Retrieval**   ⊖ 266

⚙ **10.7**   Natural population change   ⊖ 267

⚙ **10.8**   International migration   ⊖ 270

⇄ **Retrieval**   ⊖ 271

**KEY**

⚙ Knowledge

⇄ Retrieval

✎ Practice

| | | |
|---|---|---|
| ⚙ **10.9** Population ecology | ⊖ | 272 |
| ⚙ **10.10** Global population futures | ⊖ | 274 |
| ⇄ **Retrieval** | ⊖ | 275 |
| ⚙ **10.11** Case study: Population change | ⊖ | 276 |
| ⚙ **10.12** Case study: Place and health | ⊖ | 277 |
| ⇄ **Retrieval** | ⊖ | 279 |
| ✐ **Practice** | ⊖ | 280 |

## 11 Resource security — 284

| | | |
|---|---|---|
| ⚙ **11.1** Resource development | ⊖ | 284 |
| ⇄ **Retrieval** | ⊖ | 286 |
| ⚙ **11.2** Natural resource issues | ⊖ | 287 |
| ⇄ **Retrieval** | ⊖ | 291 |
| ⚙ **11.3** Water security | ⊖ | 292 |
| ⇄ **Retrieval** | ⊖ | 297 |
| ⚙ **11.4** Energy security | ⊖ | 298 |
| ⇄ **Retrieval** | ⊖ | 305 |
| ⚙ **11.5** Mineral security | ⊖ | 306 |
| ⇄ **Retrieval** | ⊖ | 309 |
| ⚙ **11.6** Resource futures | ⊖ | 310 |
| ⚙ **11.7** Case study: Resource issues | ⊖ | 311 |
| ⚙ **11.8** Case study: How physical environment affects resources | ⊖ | 312 |
| ⇄ **Retrieval** | ⊖ | 313 |
| ✐ **Practice** | ⊖ | 314 |

## 12 Skills and fieldwork — 318

| | | |
|---|---|---|
| ⚙ **12.1** Fieldwork | ⊖ | 318 |
| ⇄ **Retrieval** | ⊖ | 321 |
| ⚙ **12.2** Cartographic skills | ⊖ | 322 |
| ⚙ **12.3** Graphical skills | ⊖ | 325 |
| ⚙ **12.4** Statistical skills | ⊖ | 329 |
| ⇄ **Retrieval** | ⊖ | 331 |
| **OS maps symbols** | ⊖ | 332 |

# HOW TO USE THIS BOOK

This book uses a three-step approach to revision: **Knowledge**, **Retrieval**, and **Practice**.
It is important that you do all three; they work together to make your revision effective.

##  Knowledge

**Knowledge** comes first. Each topic is divided into **Knowledge Organisers**. These are clear, easy-to-understand, concise summaries of the content that you need to know for your exam.

**REVISION TIP**

**Revision tips** offer you helpful advice and guidance to aid your revision and help you to understand key concepts and remember them.

**Case study**

Popular case studies that you may have learned about in class are highlighted.

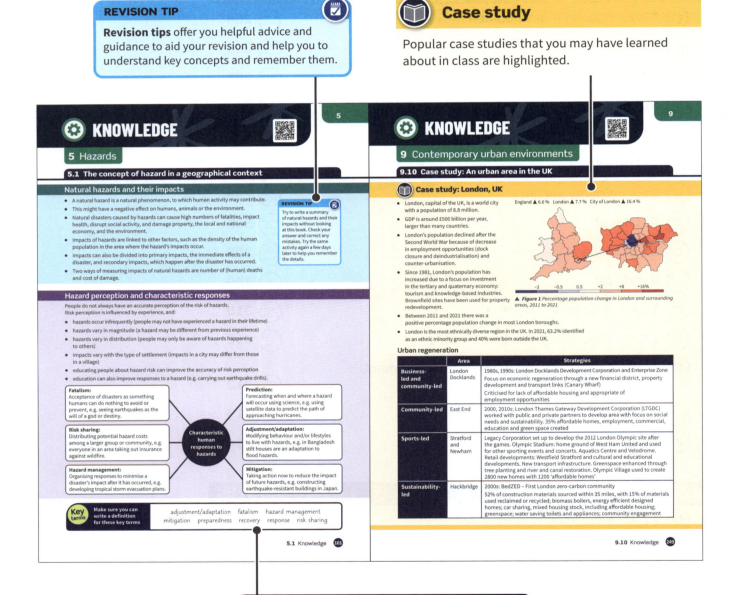

The **Key terms** box highlights the key words and phrases you need to know, remember, and be able to use confidently.

Make sure you can write a definition for these key terms

## Retrieval

The **Retrieval questions** help you learn and quickly recall the information you've acquired. These are short questions and answers about the content in the Knowledge Organisers you have just revised. Cover up the answers with some paper and write down as many answers as you can from memory. Check back to the Knowledge Organisers for any you got wrong, then cover the answers and attempt all the questions again until you can answer *all* the questions correctly.

Make sure you revisit the Retrieval questions on different days to help them stick in your memory. You need to write down the answers each time, or say them out loud, for your revision to be effective.

### Previous questions

Each Retrieval page also has some **Retrieval questions** from **previous sections**. Answer these to see if you can remember the content from the earlier sections. If you get the answers wrong, go back and do the Retrieval questions for the earlier sections again.

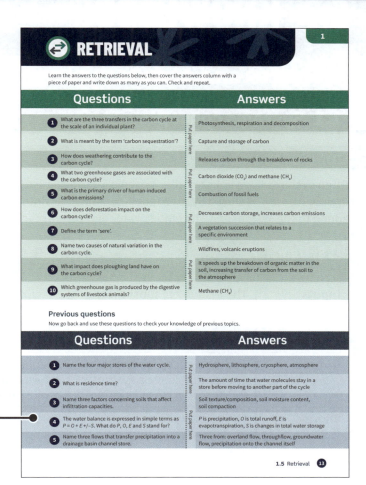

## Practice

Once you are confident with the Knowledge Organisers and Retrieval questions, you can move on to the final stage: **Practice**.

Each chapter has **exam-style questions** to help you apply all the knowledge you have learned.

### EXAM TIP

**Exam tips** show you how to interpret the questions, provide guidance on how to answer them, and give advice on how to secure as many marks as possible. Guidance is also offered on how to approach different command words.

### Answers and Glossary

You can scan the QR codes at any time to access sample answers and mark schemes for the exam-style questions, a glossary containing definitions of the key terms, as well as further revision support, or go to:
go.oup.com/OR/Alevel/A/Geog

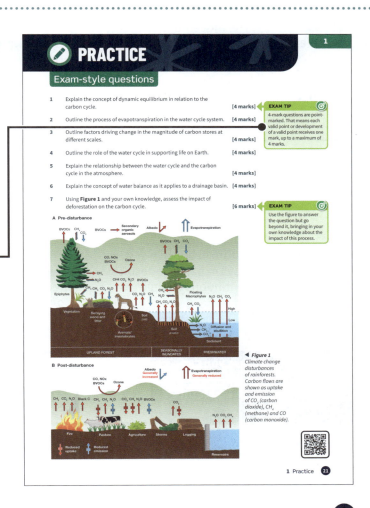

 **KNOWLEDGE**

# 1 Water and carbon cycles

## 1.1 Systems concepts

### Open systems

**Open systems** have links to other systems (e.g. local systems such as a drainage basin). A **closed system** is self-contained as nothing is lost from or added to the system (e.g. the global water and carbon cycles).

| System concept | Definition | Examples for water cycle | Examples for carbon cycle |
|---|---|---|---|
| **Inputs** | The resources, substances or energy that enter a system (or store in a closed system) | Precipitation and surface runoff are inputs to stores within the system | $CO_2$ released by respiration, combustion and volcanic eruptions are inputs to the atmosphere store |
| **Outputs** | The resources, substances or energy that leave a system (or store in a closed system) | Evaporation and transpiration return water vapour to the atmosphere store | $CO_2$ released through deforestation returns carbon to the atmosphere store |
| **Energy** | The capacity to do work or cause change within a system | Solar energy provides heat for evaporation | Solar energy is converted into chemical energy by plants during photosynthesis |
| **Stores/** components | Where energy or matter is kept for a relatively long time in a system | Atmosphere, oceans, surface water, groundwater | Atmosphere, plants, soil, oceans, fossil fuel deposits |
| **Flows/** transfers | The movement of energy or matter between stores within a system | Precipitation, evaporation, transpiration, infiltration, runoff | Photosynthesis, respiration, decomposition, combustion |
| **Positive feedback** | Amplifies or reinforces changes within a system | Evaporation leads to more water vapour in the atmosphere, resulting in higher precipitation | Forest fires release $CO_2$ into the atmosphere, which contributes to global warming |
| **Negative feedback** | Counteracts or reduces changes within a system, maintaining stability | Increased surface temperatures lead to increased evaporation and more clouds that reflect radiation; surface temperatures cool | Increased carbon dioxide levels stimulate plant growth, leading to greater carbon absorption through photosynthesis |
| **Dynamic equilibrium** | A state in which a system stays in balance despite continuous change | The rate of water entering and leaving stores/components remains relatively constant over time | The amount of carbon entering and leaving stores is balanced |

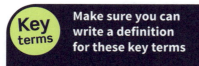

**Key terms** | Make sure you can write a definition for these key terms

> closed system   dynamic equilibrium   energy   flow/transfer
> input   negative feedback   open system   output
> positive feedback   store

## 1.2 The water cycle

### The size and distribution of major stores

- Hydrosphere: the oceans, seas, lakes and rivers. 97.5% of the Earth's water.

- Lithosphere: groundwater is stored in the pores of soil and rocks of the Earth's crust and upper mantle and requires permeable/porous geology. Groundwater is the largest source of fresh water available to humans. 0.75% of the Earth's water.

- Cryosphere: frozen water stores, including ice caps, glaciers and permafrost mainly located at the Earth's poles and in high mountain ranges. 1.7% of the Earth's water.

- Atmosphere: stored as water vapour and condensation and depends on temperature and atmospheric pressure. 0.001% of the Earth's water.

- Only 2.5% of the Earth's water is freshwater, and this is stored in the cryosphere (68.7%), the lithosphere (30.1%) and in surface water (1.2%).

- Biosphere: all living things (not a major store)

### Residence time and cloud formation

**Cloud formation**

1. Evaporation – water from various sources is heated by the Sun and turns to water vapour

2. As moist air rises it cools, either due to vertical expansion or meeting cooler air masses

3. Saturation point (or dew point) is reached when the cooling air can no longer hold all its moisture

4. The water vapour starts to condense around condensation nuclei, e.g. dust, pollen or pollutants. The condensed water droplets cluster to create different cloud formations.

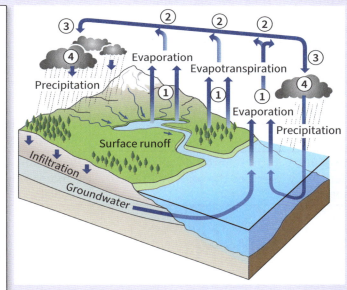

▲ **Figure 1** The water cycle

**Residence time**

Residence time is affected by location, climate conditions, geology and human activity.

Atmosphere: around 9 days

Rivers: 2–6 months; Lakes: years to decades

Ice caps and glaciers: 10–1000 years

Oceans: around 4000 years

Shallow groundwater: a few weeks to 100 years; Deep groundwater: up to 10,000 years

### Factors affecting flow/transfer processes

| | |
|---|---|
| Precipitation | Air temperature (warmer air can hold more water vapour), relative humidity, topography, atmospheric stability, seasonal changes, wind patterns |
| Evaporation | Mainly temperature, but also humidity, wind speed, surface area of water. Evaporation increases in warm, dry or windy conditions; decreases in cold, humid or calm conditions. |
| **Evapotranspiration (EVT)** | As evaporation, adding the extent of vegetation cover and soil moisture. In drought conditions, when soil moisture is low, EVT will also be low. |
| Interception | Vegetation type and density (a greater leaf surface area intercepts more precipitation), precipitation intensity (high intensity saturates interception capacity, increasing stemflow), seasonal variation |
| **Stemflow** | Vegetation characteristics (branching structure affects stemflow), leaf area index (LAI) and canopy density (higher values promote more stemflow) |
| Channel flow | Discharge, channel slope and shape (wider and deeper channels have faster flow rates than narrow, shallow channels), channel roughness (increased friction = slower flow rates) and seasonal changes |

## 1 Water and carbon cycles

### 1.2 The water cycle

## Factors affecting infiltration and overland flow

| Factors | Infiltration | Overland flow |
|---|---|---|
| Soil texture and composition | Coarse-textured sandy soils have higher capacity because of larger spaces between soil particles (pore spaces). | Fine-grained clays have lower **infiltration** capacity because soil particles are close together. |
| Soil moisture content | Drier soils have higher infiltration capacities. | Gaps between soil particles are already full of water, so infiltration is reduced. |
| Soil compaction | Uncompacted soils have higher infiltration capacity because pore spaces have not been compressed. | Compacted soils with reduced pore spaces may have limited capacity to absorb water, increasing runoff. |
| Slope gradient | Gentler slopes allow more time for infiltration as water moves more slowly downslope. | Steeper slopes mean more runoff due to gravity affecting downslope movement. |
| Vegetation cover and density | Vegetation reduces rainfall intensity at ground level, giving more time for water to infiltrate the soil. Roots improve soil structure. | Lack of/reduced vegetation cover increases the chance of overland flow as interception is reduced and rainfall intensity is not reduced at ground level. |
| Rainfall intensity and duration | Lower intensity and shorter duration rainfall is less likely to exceed the soil's infiltration capacity. | Higher intensity and longer duration rainfall is more likely to exceed the soil's infiltration capacity. |

> **REVISION TIP**
>
> You need to know how these factors affect change at the local scale of the hill slope and drainage basin.

## Changes at the local scale: the hill slope

Increased flow and erosive power during wet season; EVT also increases

Vegetated hill slope slows infiltration

Shallow water increases friction with the river bed and slows flow; steeper sections increase flow

Bare rock surfaces increase runoff

**Human activities**
- Channelisation, dredging, construction of reservoirs, weirs and bridges can all alter, regulate or redirect channel flow, impacting flow rates and channel characteristics.
- Urbanisation affects infiltration rates (impervious surfaces) and alters drainage systems.
- Agriculture affects infiltration rates by reducing vegetation cover, compacting soul with machinery, irrigation practices and lowering slope gradients (grading).

◀ **Figure 2** *Processes affecting transfers of water on a hillslope*

# Changes at the global scale: the cryosphere

Processes of accumulation and ablation drive changes in the magnitude of water stored in glaciers, ice sheets and permafrost.

▲ **Figure 3** *Melting Glacier in Iceland*

- Glaciers accumulate snow and ice through snowfall, and lose mass through melting, sublimation, calving (breaking off of icebergs), and evaporation. Positive mass balance occurs when accumulation exceeds ablation (loss), leading to glacier growth, while negative mass balance results in glacier retreat.

- Ice sheet accumulation and ablation operates in similar ways to glaciers and impacts sea levels and global hydrological cycles.

- Sea ice undergoes seasonal cycles of expansion and retreat. During winter, sea ice coverage increases, and during summer, it melts and recedes. The melting of sea ice influences ocean circulation patterns and habitats for marine organisms.

- As temperatures increase due to climate change, permafrost thaws, releasing stored water. This process can result in changes to the hydrology of high-latitude and mountainous regions, affecting surface water runoff, groundwater recharge, and ecosystems that rely on permafrost for stability.

## How have cryosphere stores changed over time?

- The Last Glacial Maximum (LGM) was 21,500 years ago, when ice sheets covered much of Europe and North America. Global temperatures were 4–5°C lower than current averages.

- The end of the last ice age came 11,700 years ago. Water loss from the cryosphere caused a 60–100 m rise in sea level.

- Between 2000 and 2019, the world's glaciers lost a total of 267 billion tonnes of ice per year.

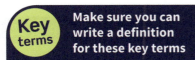

**Key terms** Make sure you can write a definition for these key terms

evapotranspiration    infiltration    stemflow

# 1 Water and carbon cycles

## 1.3 The drainage basin

## Drainage basins as open systems

- The drainage basin is an open system with inputs of precipitation and outputs of evapotranspiration and runoff. Groundwater can also be an input or output between drainage basins.

- Stores within the drainage basin include groundwater, soil moisture, vegetation (interception store), river and stream channels and surface storage.

- Flows within the drainage basin system include stemflow, throughfall (precipitation that is not intercepted by plants), overland flow, throughflow (water passing through the soil) and groundwater flow.

### Water balance

The **water balance** is the balance between inputs and outputs over a period of time.

Precipitation (P) = total runoff (O) + evapotranspiration (E) +/− changes in total water storage (S)

$$P = O + E +/- S$$

Total runoff (or streamflow) is the proportion (in %) of the total precipitation that reaches streams and rivers.

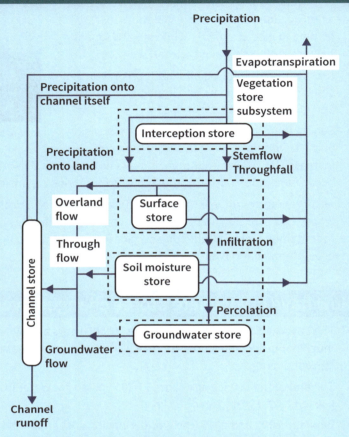

▲ **Figure 1** *Inputs, transfers, stores and outputs from a drainage basin system*

## Seasonal variations in runoff

A river regime shows changes to the water balance of a drainage basin over a longer period of time.

| UK summer | UK winter |
|---|---|
| Warmer temperatures, vegetation in full leaf and lower precipitation reduces runoff. | Lower temperatures, loss of leaf cover and higher precipitation increases runoff. |
| • EVT can exceed precipitation, especially when hot temperatures increase evaporation | • Precipitation can exceed EVT, especially where vegetation is deciduous |
| • Stores become depleted; soil dries out; throughflow declines | • Channel flow increases and total runoff increases |
| • In droughts, vegetation may die, reducing interception and EVT | • Winter flooding can occur due to saturation-excess overland flow |
| • Channel flow decreases; rivers may be fed by groundwater flows only | • Stores fill; soil becomes saturated |
| | • Lower temperatures reduce EVT |

# Causes of changes in the water cycle in a drainage basin

## Storm events

A flood hydrograph (or storm hydrograph) shows changes to river **discharge** (volume of water) following a storm event. Hydrographs can show variations in runoff between drainage basins with different characteristics. Different factors affect how quickly a river's flow responds to a storm event.

**1** The lag time is the gap between peak rainfall and peak discharge.

**2** Stormflow is overland flow and throughflow.

**3** Baseflow is groundwater flow – it shows the normal flow of the river.

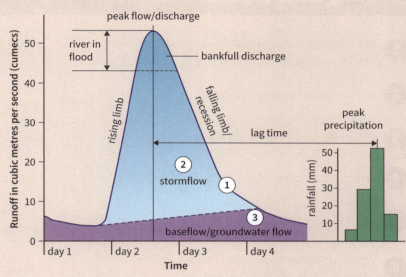

▶ **Figure 2** *Flood hydrograph*

When water reaches the river quickly, the hydrograph is 'flashy' with a short lag time and a high peak. Contributing factors can include:

- small drainage basin size
- high drainage density
- impermeable rock type
- steep slopes
- saturated soils
- low vegetation cover (or deciduous trees in winter)
- intense, heavy rainfall.

Other drainage basins may take longer to respond so the hydrograph has a long lag time and a lower peak. Contributing factors can include:

- large drainage basin
- low drainage density
- permeable rock type
- gentle slopes
- dry soils
- dense vegetation cover (broad-leaved plants)
- light precipitation.

## Human impacts

- Urbanisation replaces permeable vegetation with impermeable surfaces. Infiltration decreases, overland flow increases, drainage density increases (drains, sewers and gutters) heightening the risk of flooding. Soil water and groundwater stores are reduced. Reservoirs delay flows and increase evaporation.

- Deforestation removes vegetation cover from the soil, which reduces interception and infiltration. Soil compaction may increase especially with grazing, farm machinery, irrigation, or grading (lowering gradients). Overland flow increases; soil and groundwater stores are reduced.

- Drainage and irrigation reduces the soil store by increasing water flow through the drainage basin.

Soil moisture is decreased and the water table is lowered. Irrigation transfers water to the land, from rivers or from groundwater stores. It can increase EVT, especially in hot countries.

- Water abstraction removes water from rivers and groundwater. Groundwater can take a long time to recharge, so water abstraction can lower the water table and severely deplete groundwater stores. If the water table drops lower than sea level, salt water can flow into aquifers.

- Channelisation, dredging, weirs and bridges can alter channel flow, regulating or redirecting flow, impacting flow rates and channel characteristics.

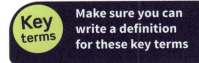

**Key terms**

**Make sure you can write a definition for these key terms**

discharge    water balance

# RETRIEVAL

Learn the answers to the questions below, then cover the answers column with a piece of paper and write down as many as you can. Check and repeat.

## Questions

| | |
|---|---|
| 1 | Is a drainage basin an open system or a closed system? |
| 2 | Define dynamic equilibrium. |
| 3 | Name the four major stores of the water cycle. |
| 4 | Which store holds 97.5% of the Earth's water? |
| 5 | What is residence time? |
| 6 | Which store has the longest residence time on average, of up to 10,000 years? |
| 7 | Which store has the shortest residence time on average, of 9 days? |
| 8 | What is the saturation point (or dew point)? |
| 9 | Name three factors concerning soils that affect infiltration capacities. |
| 10 | Positive mass balance for glaciers occurs when accumulation exceeds ablation. What is ablation? |
| 11 | The water balance is expressed in simple terms as $P = O + E +/-S$. What do $P$, $O$, $E$ and $S$ stand for? |
| 12 | Name three flows that transfer precipitation into a drainage basin channel store. |
| 13 | Define total runoff. |
| 14 | What does the lag time on a flood hydrograph show? |
| 15 | Which of the following drainage basin characteristics would be likely to produce a 'flashy' hydrograph – dry soils, large drainage basin, high drainage density? |
| 16 | In a drought, vegetation may die, reducing EVT. Is this an example of negative or positive feedback? |
| 17 | Name three human impacts which can affect the water balance of a drainage basin. |

## Answers

*Put paper here*

| |
|---|
| Open system |
| A state in which a system stays in balance despite continuous change |
| Hydrosphere, lithosphere, cryosphere, atmosphere |
| Hydrosphere |
| The amount of time that water molecules stay in a store before moving to another part of the cycle |
| Deep groundwater (in the lithosphere) |
| Atmosphere |
| The point at which cooling air can no longer hold all the moisture it is carrying |
| Soil texture/composition, soil moisture content, soil compaction |
| Loss of ice and snow |
| $P$ is precipitation, $O$ is total runoff, $E$ is evapotranspiration, $S$ is changes in total water storage |
| Three from: overland flow / throughflow / groundwater flow / precipitation onto the channel itself |
| The proportion of the total precipitation that reaches streams and rivers |
| The gap (in time) between peak rainfall and peak discharge |
| High drainage density |
| Negative, because by reducing EVT it reduces changes in the water balance |
| Three from: urbanisation / deforestation / drainage / irrigation / abstraction |

# ⚙ KNOWLEDGE

## 1 Water and carbon cycles

### 1.4 Carbon cycle: stores and transfers

## Carbon cycle stores

| Carbon store | Sub-store | Amount (GtC/1 billion metric tonnes) | Residence time |
|---|---|---|---|
| Lithosphere | Sedimentary rocks | 100,000,000 | 150–400 million years |
| | Fossil fuels | 4130 | |
| | Sea floor sediments | 6000 | |
| | Soils | 1580 | From a few months to 1300 years |
| Cryosphere | Permafrost soils | 1600 | 10,000 years |
| Hydrosphere | Surface ocean | 970 | 25 years |
| | Deep ocean | 39,000 | 3000 years |
| Biosphere | Living organisms | 613 | 13 years |
| Pedosphere | Soil organisms | 750 | 4 years |

## Carbon cycle transfers

### Plant scale

- $CO_2$ is absorbed from the atmosphere and converted into glucose by photosynthesis.
- Carbon is stored in plant tissues, entering the soil through the decomposition of dead tissue (e.g. leaves) or through the stem/trunk.
- $CO_2$ enters the atmosphere by the respiration of the plant and the decomposers.

#### Sere scale

A **sere** is a **vegetation succession** that relates to a specific environment. The final stage of the succession is called the **climax stage**.

### Continental scale

$CO_2$ in the atmosphere combines with rainfall to form weak carbonic acid.

↓

Chemical weathering of carbon-rich rocks means carbon ions are washed into rivers and oceans.

↓

Some marineorganisms use carbon to build carbonate shells. When they die, their shells build up in marine sediments.

↓

The intense heat at subduction zones of continental–ocean plate boundaries releases $CO_2$ from carbon-rich sediments.

↓

The outgassing of $CO_2$ from volcanic eruptions releases carbon from the mantle into the atmosphere.

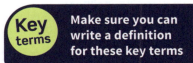

**Make sure you can write a definition for these key terms**

climax stage    sere    vegetation succession

# 1 Water and carbon cycles

## 1.5 Changes in the carbon cycle

### Factors driving change

Changes in the amount (magnitude) of carbon in carbon cycle stores involve transfer processes. Different factors, at different scales, affect the amount of change.

| Transfer process | Process description | Factors affecting change |
|---|---|---|
| **Photosynthesis** | Green plants and algae in the sea absorb light energy from the Sun, which converts $CO_2$ from the air, plus water from the soil, into glucose. Oxygen is the waste product. | Higher light intensity increases photosynthesis, as does the amount of $CO_2$ in the air. Extreme temperatures (high and low) inhibit photosynthesis, as do lack of water and lack of key nutrients. Tropical rainforests have year-round intense sunlight, daily rainfall, warm temperatures, so high levels of photosynthesis. Temperate forests see increased rates through spring into summer, and low rates during autumn and winter. Photosynthesis is lowest in polar regions. |
| **Respiration** | Almost all living organisms convert glucose into energy, releasing $CO_2$ as a waste product. | Respiration increases with metabolic activity, body size and temperature. Areas with high biodiversity have higher $CO_2$ production, including biodiverse ocean ecosystems. Plants will respire more as global temperatures increase. |
| **Decomposition** | Decomposers such as bacteria and fungi break down dead tissue, releasing carbon (mainly $CO_2$). | Decomposition tends to increase with warmer conditions. Microbial activity requires moisture, so extremely dry conditions can limit decomposition. Aerobic decomposition is faster than anaerobic (without oxygen); anaerobic produces methane. |
| **Combustion** | When organic material is burned in the presence of oxygen, energy is released, plus $CO_2$ and water. | Since the Industrial Revolution, combustion of fossil fuels has released an estimated 2.3 trillion tonnes of $CO_2$ into the atmosphere. Coal releases the most $CO_2$ when combusted and natural gas the least. |
| Weathering | Carbon in the atmosphere dissolves in rainwater to form weak carbonic acid; this reacts with carbonate minerals in rocks to release soluble carbon ions. | Warm and humid climates have higher rates of chemical weathering. Geology is an important factor: this process works with rocks containing calcium carbonate such as limestone and chalk. Where these rocks are part of the underlying geology, surface and groundwater can have high levels of dissolved carbon (calcium bicarbonate). Weathering is faster on exposed rock surfaces. |
| **Carbon sequestration** | Carbon dioxide diffuses from the atmosphere into ocean water. Warmer temperatures reduce the solubility of $CO_2$ in water. | Deep ocean currents transport carbon from the surface to deep ocean layers of water, where it can stay for thousands of years. This process is affected by the strength of ocean circulation. Marine biomass takes up carbon from the atmosphere. When it dies and sinks to the ocean floor, part of the carbon is sequestered in marine sediments and eventually turns into rocks. This process is affected by factors including nutrient availability, oxygen levels and the strength of ocean currents. |

# The impact of the carbon cycle

| Atmosphere | Land | Oceans |
|---|---|---|
| Carbon dioxide ($CO_2$) and methane ($CH_4$) are key greenhouse gases. By absorbing and reflecting radiated heat from the Earth's surface, they keep the surface warm enough to sustain life (16°C warmer than would otherwise be the case). | Plants, via photosynthesis, absorb $CO_2$ and convert it into organic matter, which is the source of energy for all non-primary producers.<br><br>This organic matter also forms the basis of soils which, like vegetation, are a significant carbon sink. | Oceans absorb 5–15 GtC of carbon from the atmosphere per year (a carbon sink).<br><br>Dissolved $CO_2$ combines with water to form carbonic acid, leading to ocean acidification.<br><br>Carbon is converted by some marine organisms into shells made of calcium carbonate. |

# Natural variation in the carbon cycle over time

## Wildfires

Wildfires can create conditions for saplings to grow, increasing carbon sequestration from the atmosphere. Some carbon from charred vegetation becomes part of the soil. However, forest fires release massive amounts of carbon from biomass into the atmosphere – in 2021, 1.7 GtCs of carbon were released from forest fires in North America and Eurasia. Where climate change is making droughts more common or severe in forested areas, instead of **carbon sinks**, forests are becoming **carbon sources**. This positive feedback loop increases the amount of carbon in the atmosphere.

> **REVISION TIP**
>
> Make an audio recording of yourself explaining natural variation in the carbon cycle over time to help you remember the key points.

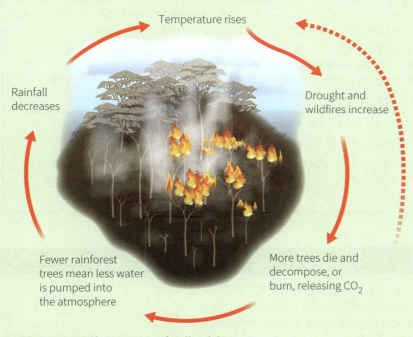

Temperature rises

Rainfall decreases

Drought and wildfires increase

Fewer rainforest trees mean less water is pumped into the atmosphere

More trees die and decompose, or burn, releasing $CO_2$

▲ **Figure 1** Wildfires positive feedback loop

## Volcanic activity

Volcanic activity currently emits at least 130 million tonnes of carbon into the atmosphere per year (some of this from submarine eruptions). Previous eras have seen much higher levels of volcanic activity; in the Palaeozoic era volcanoes emitted at least 240 million tonnes a year, contributing to natural global warming.

## 1  Water and carbon cycles

### 1.5  Changes in the carbon cycle

## Human impact

| | |
|---|---|
| **Combustion of fossil fuels** | • Around 90% of current carbon emissions from human activities come from the burning of fossil fuels. Most fossil fuels are from lithosphere stores that are 70–100 million years old.<br><br>• Some 2.3 trillion tonnes of $CO_2$ have been released from the lithosphere into the atmosphere over the last 250 years; a process that naturally takes up to 400 million years.<br><br>• This has raised the level of atmospheric $CO_2$ by around 50% from 1750. |
| **Deforestation** | • Forests are carbon sinks, absorbing roughly twice as much carbon as they emit through respiration (globally 16 GtC a year).<br><br>• Deforestation releases an average of 8.1 GtC per year, reducing the role of forests as a global carbon sink and risking them becoming a carbon source.<br><br>• Burning also releases the carbon stored in forest vegetation.<br><br>• Deforestation is focused in tropical areas, which sequester more carbon than boreal or temperate forests. |
| **Farming practices** | • Most deforestation is to clear land for farming – a carbon source.<br><br>• Ploughing and other soil treatments speeds up the breakdown of organic material in the soil, increasing transfers from the soil to the atmosphere.<br><br>• Grass-eating animals like cows and sheep eat carbon-storing plants for energy; digestion releases $CH_4$ as a waste product, so the more livestock, the more emissions. Emissions are increased when livestock manure is used as fertiliser.<br><br>• Rice cultivation also releases methane. |
| **Land use changes: urbanisation** | • 97% of all carbon emissions from human activities comes from cities.<br><br>• This is because of the demand for power, utilities and transportation.<br><br>• Urbanisation reduces carbon-absorbing vegetation cover and replaces it with substances, such as concrete, the production of which involves high emissions of $CO_2$. |

## The carbon budget

The **carbon budget** records data on how much carbon is stored and being transferred between stores. At the global scale, a carbon budget is used to set targets for reducing flows to the atmosphere from human activities, for example by increasing carbon sequestration. At the local scale of a tree, the carbon budget can be used to calculate net removal of $CO_2$ from the atmosphere or to calculate net primary productivity (creation of new biomass).

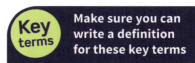 **Key terms** — Make sure you can write a definition for these key terms

carbon budget    carbon sequestration    carbon sink    carbon source    combustion    decomposition    photosynthesis    respiration

# RETRIEVAL

Learn the answers to the questions below, then cover the answers column with a piece of paper and write down as many as you can. Check and repeat.

## Questions | Answers

| | Questions | Answers |
|---|---|---|
| 1 | What are the three transfers in the carbon cycle at the scale of an individual plant? | Photosynthesis, respiration and decomposition |
| 2 | What is meant by the term 'carbon sequestration'? | Capture and storage of carbon |
| 3 | How does weathering contribute to the carbon cycle? | Releases carbon through the breakdown of rocks |
| 4 | What two greenhouse gases are associated with the carbon cycle? | Carbon dioxide ($CO_2$) and methane ($CH_4$) |
| 5 | What is the primary driver of human-induced carbon emissions? | Combustion of fossil fuels |
| 6 | How does deforestation impact on the carbon cycle? | Decreases carbon storage, increases carbon emissions |
| 7 | Define the term 'sere'. | A vegetation succession that relates to a specific environment |
| 8 | Name two causes of natural variation in the carbon cycle. | Wildfires, volcanic eruptions |
| 9 | What impact does ploughing land have on the carbon cycle? | It speeds up the breakdown of organic matter in the soil, increasing transfer of carbon from the soil to the atmosphere |
| 10 | Which greenhouse gas is produced by the digestive systems of livestock animals? | Methane ($CH_4$) |

*Put paper here*

## Previous questions

Now go back and use these questions to check your knowledge of previous topics.

## Questions | Answers

| | Questions | Answers |
|---|---|---|
| 1 | Name the four major stores of the water cycle. | Hydrosphere, lithosphere, cryosphere, atmosphere |
| 2 | What is residence time? | The amount of time that water molecules stay in a store before moving to another part of the cycle |
| 3 | Name three factors concerning soils that affect infiltration capacities. | Soil texture/composition, soil moisture content, soil compaction |
| 4 | The water balance is expressed in simple terms as $P = O + E +/- S$. What do $P$, $O$, $E$ and $S$ stand for? | $P$ is precipitation, $O$ is total runoff, $E$ is evapotranspiration, $S$ is changes in total water storage |
| 5 | Name three flows that transfer precipitation into a drainage basin channel store. | Three from: overland flow, throughflow, groundwater flow, precipitation onto the channel itself |

*Put paper here*

# 1 Water and carbon cycles

## 1.6 Climate change

## Water, carbon, climate and life on Earth

**Water and carbon:**
- Water and carbon are linked by water's ability to absorb and transfer $CO_2$.
- Photosynthesis requires both water and $CO_2$.
- Water is required for the reactions involved in decomposition.
- Dissolved $CO_2$ in rainwater is key to chemical weathering.
- Dissolved carbon ions (from weathering) are transferred by water to the oceans.

**Water and carbon are both essential for life on Earth**

**Water:**
- Water is required for the chemical reactions necessary for metabolism and nutrient transport in living organisms.
- Water's high specific heat capacity helps regulate Earth's temperature and reduces impacts of global warming.
- The water cycle refills the stores of fresh water used by terrestrial ecosystems and supports a huge variety of habitats.

**Carbon:**
- Carbon is the basic building block for all organic compounds, including carbohydrates, proteins, nucleic acids.
- Plants and algae convert carbon into glucose, producing the oxygen used by (almost all) life on Earth.
- The carbon cycle maintains soil fertility and generates the greenhouse effect which keeps the Earth warm enough for life.

## Water and carbon cycles feedback loops and climate change

Feedback loops in the water and carbon cycles can impact climate change. Here are three examples:

1. Fresh snow and ice reflect a lot of sunlight (a high **albedo**). The rise in global temperatures causes ice caps and glaciers to melt. Less ice cover means less sunlight is reflected, and more heat is absorbed by the oceans and land, increasing global temperatures, causing further ice melting – a positive feedback loop.

2. The carbon in permafrost decomposes as it thaws and is released into the atmosphere as $CO_2$ and $CH_4$, increasing the enhanced greenhouse effect and the amount of permafrost melting – a positive feedback loop.

3. Warmer oceans increase their ability to absorb $CO_2$ from the atmosphere – a negative feedback loop. However, more dissolved $CO_2$ increases seawater acidity, which can be harmful to corals and shellfish that build shells or skeletons from calcium carbonates, reducing their capacity to act as carbon sink, reducing the ocean's role in sequestering carbon – a positive feedback loop.

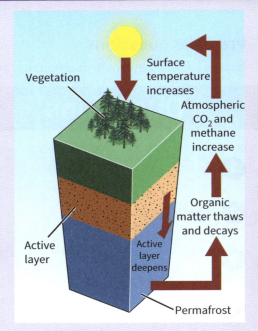

▲ **Figure 1** *Melting permafrost, a carbon cycle positive feedback*

# Interventions in carbon cycle transfers

**Afforestation** and **reforestation**: planting new forests and restoring existing ones.

**Carbon farming:** increasing long-term carbon storage in soil, including agroforestry, cover crops, no-till planting (no ploughing) and rotational grazing.

Blue carbon protection and restoration: mangroves, seagrass meadows and tidal marsh ('blue carbon') ecosystems have higher carbon sequestration rates than terrestrial forests.

**Human interventions in the carbon cycle to try to mitigate the impacts of climate change**

Carbon capture and storage (CCS): captures $CO_2$ emitted from industrial processes, and stores it deep underground.

Enhanced rock weathering: spreading finely crushed high-silica rocks, like basalt, over land. This reacts with $CO_2$ in the air to form calcium carbonate, which moves to the soil store and into groundwater.

Direct air capture (DAC): removing $CO_2$ from the atmosphere, and storing underground or using it (e.g. for carbonating fizzy drinks).

Bioenergy with carbon capture and storage (BECCS): capturing and storing the $CO_2$ produced when bioenergy is extracted from biomass.

Scale: reduction in $CO_2$ is tiny compared to the excess amount of $CO_2$ in the atmosphere.

Time: take to develop, by which time climate change may be irreversible (e.g. due to feedback loops).

**Limitations of these human interventions**

Sustainability: some interventions require a lot of energy (DAC); some have significant environmental impacts (BECCS, growing biofuels); maintenance of artificial carbon stores.

Land use: afforestation or crop production require large areas of land, which could impact food security.

Cost: developing interventions is expensive.

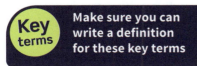

▲ **Figure 2** *Emissions from human interventions continue to increase*

**Key terms** — Make sure you can write a definition for these key terms

afforestation    albedo    carbon farming    reforestation

# RETRIEVAL

Learn the answers to the questions below, then cover the answers column with a piece of paper and write down as many as you can. Check and repeat.

## Questions

## Answers

1. State two ways in which the water cycle and its stores are important for supporting life on Earth.

Water's high specific heat capacity helps regulate the Earth's temperature; the water cycle refills the stores of fresh water used by terrestrial ecosystems

2. Give one example of how the water cycle and carbon cycle are linked.

One from: photosynthesis requires both water and carbon dioxide / water is required for the reactions involved in decomposition / dissolved $CO_2$ in rainwater for chemical weathering / dissolved carbon ions (from weathering) are transferred by water to the oceans / carbon also dissolves directly into the ocean

3. 'Increased forest fires release more $CO_2$ into the atmosphere, which contributes to global warming.' Is this an example of positive or negative feedback?

Positive feedback; it amplifies or reinforces changes within a system

4. 'Increased atmospheric $CO_2$ levels stimulate plant growth, leading to greater carbon absorption through photosynthesis.' Is this an example of positive or negative feedback?

Negative feedback; it counteracts or reduces changes within a system, maintaining stability

5. What is albedo a measure of?

The reflectivity of a surface, particularly in relation to solar radiation

6. Why does permafrost thawing lead to increased greenhouse gas emissions?

Permanently frozen soil contains organic material that has not decomposed, so when the soil thaws, decomposition releases $CO_2$ and $CH_4$ from the soil store

7. Name three examples of human interventions in the carbon cycle that are designed to mitigate the impacts of climate change.

Three from: afforestation and reforestation / carbon farming / blue carbon protection and restoration / direct air capture (DAC) / bioenergy with carbon capture and storage (BECCS) / enhanced rock weathering / carbon capture and storage (CCS)

Put paper here

## Previous questions

Now go back and use these questions to check your knowledge of previous topics.

## Questions

## Answers

1. Name three human impacts which can affect the water balance of a drainage basin.

Three from: urbanisation, deforestation, drainage, irrigation, abstraction

2. What is meant by the term 'carbon sequestration'?

Capture and storage of carbon

4. How does deforestation impact on the carbon cycle?

Decreases carbon storage, increases carbon emissions

5. Name two causes of natural variation in the carbon cycle.

Wildfires, volcanic eruptions

Put paper here

# KNOWLEDGE

## 1 Water and carbon cycles

### 1.7 Case study: Tropical rainforest

 **Case Study: Amazonia – water and carbon cycle dynamics**

**REVISION TIP**

You should revise the tropical rainforest case study that you have studied, which looks at the water and carbon cycles and the impacts of human activity and environmental change.

- Amazonia is the world's largest and most biodiverse tropical rainforest with an estimated 410 billion trees, covering 3 million km².

- Amazonia has high rates of precipitation (Manaus has over 2300 mm of rainfall per year), which vary seasonally.

- Amazonia's location in the tropics (between 5° north and 10° south of the equator) means it has warm temperatures all year (Manaus ranges from 26°C in January to 33°C in September).

- The dense canopy intercepts around 75% of rainfall, with much of this evaporating from leaf surfaces. Rainfall reaches the forest floor through stemflow and drip flow.

- These factors give Amazonia one of the highest EVT rates in the world, between 730 mm and 1460 mm per year (depending on location). Each of the 65 m tall trees can emit 1000 litres of water into the atmosphere each day. As a result, the rainforest produces almost half of its own rainfall.

- Amazonia's soils are nutrient-poor. Although the warm, wet conditions are excellent for decomposition, a lot of leaf litter is washed away by the heavy rainfall and nutrients leach out of the soil. Remaining nutrients are quickly taken up by biomass.

- Photosynthesis rates are very high at the canopy, much lower at the forest floor due to light being blocked by the canopy. Estimates are that Amazonia's trees and soil store 200 GtC, absorbing around 2.2 GtC per year.

- Carbon emissions are also very large – an estimated 1 GtC from decomposition of dead trees and leaf litter. The huge biomass respires an estimated 1.5 GtC per year.

**Feedback loops**

High rates of EVT result in more precipitation, increasing vegetation growth and EVT – positive feedback.

Dense canopy has a low albedo, increasing warming, providing improved conditions for plant growth, leading to more growth in the canopy – positive feedback.

## Human influences

- Deforestation is a major influence. Since 2000, Amazonia has lost 20.3% of its original forest – 832,000 km², 70% of which has been to clear land for cattle ranching. Livestock farming has significantly increased methane emissions from the region.

- The loss of trees is making the region drier, and hotter. Temperatures are predicted to rise by 3°C by 2050. Without tree cover, soils are drying out.

- There has been an increase in droughts in the region; severe droughts occurred in 2005, 2010 and 2015–16.

- Wildfires have become more common as Amazonia experiences droughts. It is estimated that forest fires now produce around 1.5 GtC per year.

- As a result, it is now estimated that Amazonia emits more $CO_2$ than it sequesters. Between 2010 and 2020, emissions have increased by 20%.

## 1 Water and carbon cycles

### 1.8 Case study: River catchment

 **Case Study: Pickering Beck**

**REVISION TIP**

You should revise the river catchment case study that you have studied, including the impacts of human activities on the water cycle – you may have done some fieldwork.

### Flooding events

The town of Pickering in North Yorkshire, UK has a long history of flooding from Pickering Beck, a tributary of the River Derwent. Following four flash floods in 1999, 2000, 2002 and 2007 (the last of which caused £7 million of damage to homes and businesses), UK government agencies invested in the 'Slowing the Flow' scheme. The aim was to reduce the risk of flooding in Pickering from 25% to 4%.

| Catchment details | |
| --- | --- |
| River length | 18 miles |
| Catchment size | 69 km² |
| Geology | sedimentary rocks (sandstone and limestone) |
| Relief | deep, steep-sided river valley |
| Land use | mix of forest, heather moorland, pasture for livestock and arable crops |
| Precipitation | average of 740 mm rainfall per year |
| Total runoff percentage | estimated at 50% average, rising to 70% during storm events |

▲ **Figure 1** Pickering Beck in 2019

# Storm hydrograph

- Pickering Beck has a flashy hydrograph for the June 2007 flood event, principally because of the steep sides to the river valley.
- Human influences also increased the risk of flash flooding.

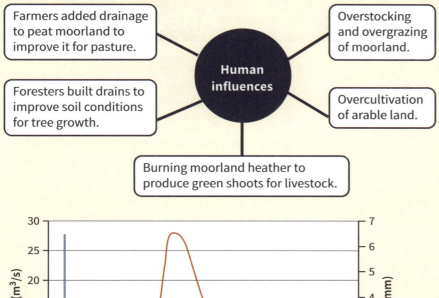

Farmers added drainage to peat moorland to improve it for pasture.

Overstocking and overgrazing of moorland.

**Human influences**

Foresters built drains to improve soil conditions for tree growth.

Overcultivation of arable land.

Burning moorland heather to produce green shoots for livestock.

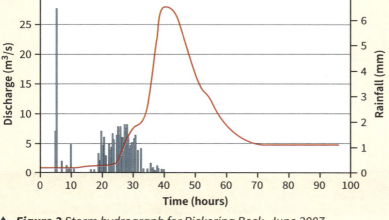

▲ *Figure 2* Storm hydrograph for Pickering Beck, June 2007

## 'Slowing the Flow' scheme (2009)

- 128 'leaky dams' constructed – large woody debris dams that block fast river flow but allow slower-moving water through.
- 187 dams made of heather bales added to smaller streams.
- A new floodplain created north of Pickering, with a 2.5 m high clay bund designed to hold back 120,000 m³ of flood water.
- 30 hectares of woodland planted alongside the river, plus 15 ha of farm woodland.
- Moorland drains responsible for rapid runoff identified and blocked.
- Buffer zones set up on moorland where heather burning is banned.
- Farmers were educated on techniques to improve soil infiltration.
- Scheme cost: £4 million (now costs £20,000 a year for maintenance). A hard engineering solution (concrete flood walls) was quoted at £20 million in 2007.

In 2015, the catchment received 50 mm of rain over 36 hours. Pickering did not flood, unlike nearby York. Analysis showed that the 'Slowing the Flow' measures had reduced the flood risk by 20%.

**Results of the scheme**

In 2019, two beavers were introduced to the river catchment, with the expectation that their dams will contribute to reducing flood risk.

# RETRIEVAL

Learn the answers to the questions below, then cover the answers column with a piece of paper and write down as many as you can. Check and repeat.

## Questions

## Answers

**1** Suggest two reasons why EVT is high in tropical rainforests.

Two from: high annual precipitation (e.g. over 2000 mm per year) / warm temperatures (e.g. over 25°C all year) / dense vegetation cover all year / leaves typically have large surface area

**2** With reference to the water cycle, account for the low nutrient levels in tropical rainforest soils.

Leaf litter is washed away by the heavy rainfall and nutrients leach out of the soil due to increased water in the soil

**3** How have human activities increased the risk of flooding in Pickering Beck?

Adding drainage to upland moorland and forestry areas in the catchment; overstocking grazing land

**4** How have human activities reduced the risk of flooding in Pickering Beck?

Building 128 'leaky' dams and 187 heather bale dams to reduce flow speeds following storm events; constructing a new floodplain and bund; planting woodland along the river; blocking moorland drainage

*Put paper here*

## Previous questions

Now go back and use these questions to check your knowledge of previous topics.

## Questions

## Answers

**1** Define dynamic equilibrium.

A state in which a system stays in balance despite continuous change

**2** What are the three transfers in the carbon cycle at the scale of an individual plant?

Photosynthesis, respiration and decomposition

**3** What two greenhouse gases are associated with the carbon cycle?

Carbon dioxide ($CO_2$) and methane ($CH_4$)

**4** State two ways in which the water cycle and its stores are important for supporting life on Earth.

Water's high specific heat capacity helps regulate the Earth's temperature; the water cycle refills the stores of fresh water used by terrestrial ecosystems

**5** 'Increased forest fires release more $CO_2$ into the atmosphere, which contributes to global warming.' Is this an example of positive or negative feedback?

Positive feedback; it amplifies or reinforces changes within a system

**6** Why does permafrost thawing lead to increased greenhouse gas emissions?

Permanently frozen soil contains organic material that has not decomposed, so when the soil thaws, decomposition releases $CO_2$ and $CH_4$ from the soil store

**7** Name three examples of human interventions in the carbon cycle that are designed to mitigate the impacts of climate change.

Three from: afforestation and reforestation; carbon farming; blue carbon protection and restoration; direct air capture (DAC); bioenergy with carbon capture and storage (BECCS); enhanced rock weathering; carbon capture and storage (CCS)

*Put paper here*

# ✏ PRACTICE

## Exam-style questions

1   Explain the concept of dynamic equilibrium in relation to the
    carbon cycle.                                                    **[4 marks]**

**EXAM TIP** 🎯

4-mark questions are point-marked. That means each valid point or development of a valid point receives one mark, up to a maximum of 4 marks.

2   Outline the process of evapotranspiration in the water cycle system.   **[4 marks]**

3   Outline factors driving change in the magnitude of carbon stores at
    different scales.                                                **[4 marks]**

4   Outline the role of the water cycle in supporting life on Earth.   **[4 marks]**

5   Explain the relationship between the water cycle and the carbon
    cycle in the atmosphere.                                         **[4 marks]**

6   Explain the concept of water balance as it applies to a drainage basin.   **[4 marks]**

7   Using **Figure 1** and your own knowledge, assess the impact of
    deforestation on the carbon cycle.                              **[6 marks]**

**EXAM TIP** 🎯

Use the figure to answer the question but go beyond it, bringing in your own knowledge about the impact of this process.

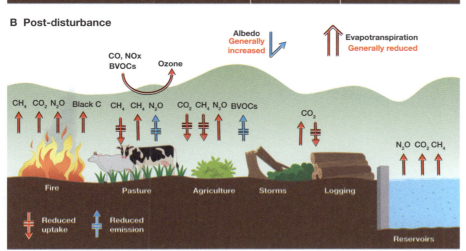

▶ **Figure 1**
*Climate change disturbances of rainforests. Carbon flows are shown as uptake and emission of $CO_2$ (carbon dioxide), $CH_4$ (methane) and CO (carbon monoxide).*

8    Assess the usefulness of **Figure 2** in depicting the pattern of soil moisture across the UK.

**[6 marks]**

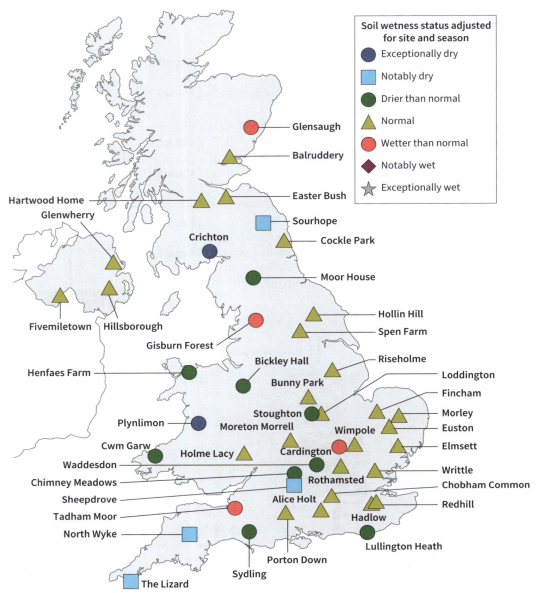

▲ **Figure 2** *Soil moisture data for selected UK Centre for Ecology and Hydrology sites, May 2023*

**9** Analyse the data shown in **Figure 3**. **[6 marks]**

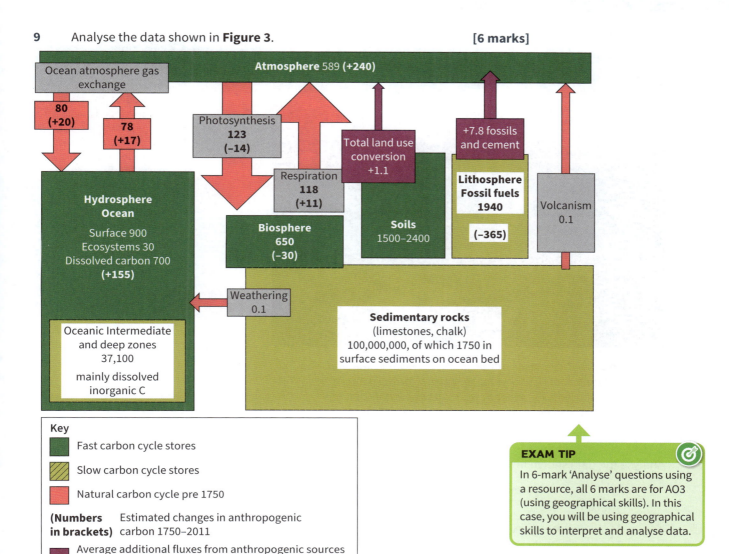

Key

■ Fast carbon cycle stores

▨ Slow carbon cycle stores

■ Natural carbon cycle pre 1750

**(Numbers in brackets)** Estimated changes in anthropogenic carbon 1750–2011

■ Average additional fluxes from anthropogenic sources estimated as an average 9 PgC/year 2000–2009

> **EXAM TIP**
>
> In 6-mark 'Analyse' questions using a resource, all 6 marks are for AO3 (using geographical skills). In this case, you will be using geographical skills to interpret and analyse data.

▲ **Figure 3** *International Panel on Climate Change (IPCC) estimates of the natural carbon cycle and human-induced changes since 1750*

**10** Analyse the data shown in **Figure 4**. **[6 marks]**

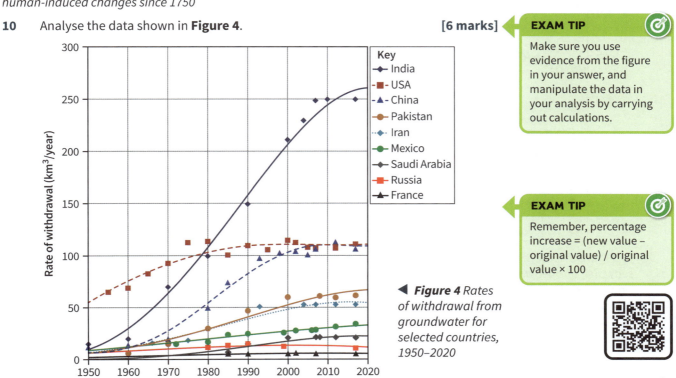

◄ **Figure 4** *Rates of withdrawal from groundwater for selected countries, 1950–2020*

> **EXAM TIP**
>
> Make sure you use evidence from the figure in your answer, and manipulate the data in your analysis by carrying out calculations.

> **EXAM TIP**
>
> Remember, percentage increase = (new value – original value) / original value × 100

11 Using **Figure 5** and your own knowledge, assess human interventions in the carbon cycle designed to mitigate the impacts of climate change. **[6 marks]**

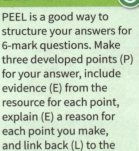

**EXAM TIP**

PEEL is a good way to structure your answers for 6-mark questions. Make three developed points (P) for your answer, include evidence (E) from the resource for each point, explain (E) a reason for each point you make, and link back (L) to the question.

▲ **Figure 5** *The Drax power station in West Yorkshire, England. Drax is an example of BECCS – bioenergy with carbon capture, use and storage. It produces power by burning sustainably-sourced compressed wood pellets, which are mainly imported from North America.*

12 Analyse the data shown in **Figure 6**. **[6 marks]**

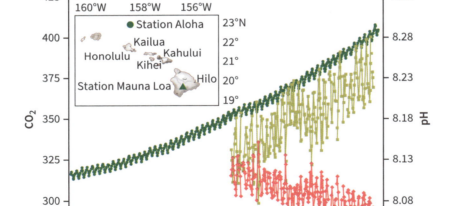

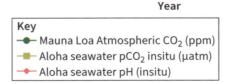

▲ **Figure 6** *A record of $CO_2$ measurements in the atmosphere and in seawater, and seawater pH, at the Mauna Loa baseline observatory, on Hawaii's Big Island. The amount of $CO_2$ dissolved in seawater is measured in $pCO_2$ (partial pressure of $CO_2$).*

13  Using **Figure 7** and your own knowledge, assess the challenges of reducing the impact of human activities on global greenhouse emissions.  **[6 marks]**

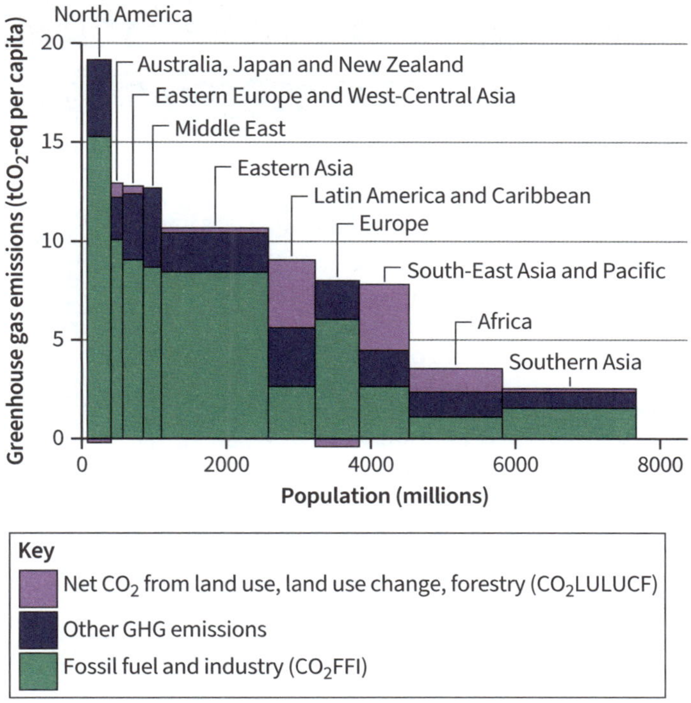

Greenhouse gas emissions (tCO$_2$-eq per capita)

North America
Australia, Japan and New Zealand
Eastern Europe and West-Central Asia
Middle East
Eastern Asia
Latin America and Caribbean
Europe
South-East Asia and Pacific
Africa
Southern Asia

Population (millions)

◀ *Figure 7* Net greenhouse gas emissions per capita, per global region, 2019

**Key**
- Net CO$_2$ from land use, land use change, forestry (CO$_2$LULUCF)
- Other GHG emissions
- Fossil fuel and industry (CO$_2$FFI)

**EXAM TIP**

Make sure your answer keeps its focus. In this case, the focus is on reducing the impact of human activities on global greenhouse emissions. There could be a danger of answers getting side-tracked into comparing the size of emissions without considering the challenges of reducing the impact of related human activities.

14  Assess the factors driving change in the magnitude of carbon stores over time in a tropical rainforest you have studied.  **[20 marks]**

**EXAM TIP**

There are 10 marks for AO1 (knowledge and understanding) and 10 marks for AO2 (applying and interpreting). Accurate and relevant knowledge and understanding (e.g. examples) gives you your AO1 marks; for AO2 you will be weighing up evidence, considering significance, making judgements, etc. as you develop your argument.

15  'Human activity needs to focus more on taking measures to reduce emissions of carbon into the atmosphere than adapting to the expected impacts of climate change.'

How far do you agree with this view?  **[20 marks]**

16  With reference to a river catchment you have studied, assess the extent to which human activities within the system have implications for **either** sustainable water supply **or** flooding.  **[20 marks]**

17  Assess the relative importance of feedback systems in driving change in the magnitude of stores in the carbon cycle over time.  **[20 marks]**

18  How far do you agree that systems in physical geography can help us understand the impact of changing inputs to the water cycle over time?  **[20 marks]**

**EXAM TIP**

A PEEL approach works well for 20-mark answers – backing up points with evidence provides AO1 and explaining points and linking them back to the question provides AO2.

19  Assess the influence of different processes driving change in the water cycle within a tropical rainforest setting you have studied.  **[20 marks]**

20  To what extent can an understanding of feedback systems help to inform our interventions in the carbon cycle designed to mitigate climate change?  **[20 marks]**

# ⚙ KNOWLEDGE

## 2 Hot desert systems and landscapes

### 2.1 Hot desert systems

## Systems concepts and desert landscapes

Open systems have links to other systems.

| System concept | Definition | Examples for hot desert systems |
|---|---|---|
| Inputs | The resources, substances or energy that enter a system | Precipitation (rainfall, fog, dew, frost), solar radiation, descending air at convergence of Hadley and Ferrel cells, water from exogenous rivers outside the desert |
| Outputs | The resources, substances or energy that leave a system | Runoff, evaporation, re-radiation of longwave radiation from the Earth's surface into the atmosphere |
| Energy | The capacity to do work or cause change within a system | **Insolation** (direct impact of heat from the sun), winds, runoff (e.g. high amounts of surface runoff and erosion as a result of infrequent but intense flashy storm events) |
| Stores/ components | Where energy or matter is kept for a relatively long time in a system | Playas, sand dunes, aquifers, rivers |
| Flows/ transfers | The movement of energy or matter between stores within a system | Wind-blown sand, surface runoff, salinisation, sediment transfer |
| Positive feedback | Amplifies or reinforces changes within a system | Removal of vegetation (e.g. by overgrazing) will reduce moisture emitted into the atmosphere by EVT, which may result in lower rainfall, which means a reduction in extent of vegetation cover |
| Negative feedback | Counteracts or reduces changes within a system, maintaining stability | High rates of weathering of a slope leads to the build-up of scree, which then protects the lower part of the slope from weathering |
| Dynamic equilibrium | A state in which a system stays in balance despite continuous change | In some desert areas, seasonal changes in wind can result in small-scale and short-term changes to sand dune profiles, but in the longer term profiles remain the same |

# The global distribution of deserts and their margins

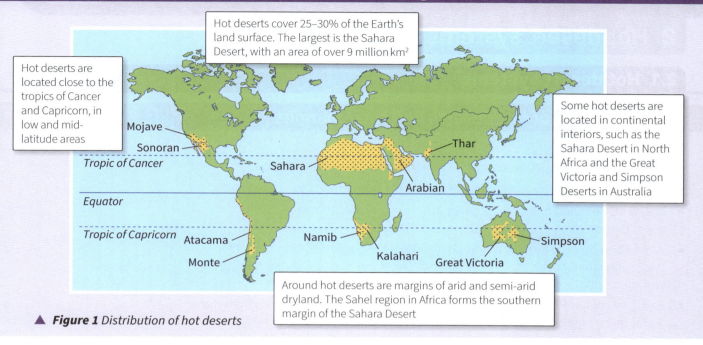

Hot deserts cover 25–30% of the Earth's land surface. The largest is the Sahara Desert, with an area of over 9 million km²

Hot deserts are located close to the tropics of Cancer and Capricorn, in low and mid-latitude areas

Some hot deserts are located in continental interiors, such as the Sahara Desert in North Africa and the Great Victoria and Simpson Deserts in Australia

Around hot deserts are margins of arid and semi-arid dryland. The Sahel region in Africa forms the southern margin of the Sahara Desert

▲ **Figure 1** Distribution of hot deserts

# Water balance and aridity index

The **water balance** is the balance between inputs and outputs over a period of time.

Precipitation (P) = total runoff (O) + evapotranspiration (E) +/− changes in total water storage (S)

$$P = O + E +/- S$$

Hot deserts have very high rates of evapotranspiration. Because inputs from precipitation are low, evapotranspiration exceeds precipitation so there is generally no runoff.

The **aridity index** (AI) is expressed as P/PET.

- This gives the ratio between the long-term average precipitation (P) and the long-term average potential evapotranspiration (PET) for an area.
- PET is a measure of the 'drying power' of the atmosphere to remove water from surfaces (e.g. the soil, plant leaves) by evaporation and through transpiration.
- An arid climate is when PET exceeds P.

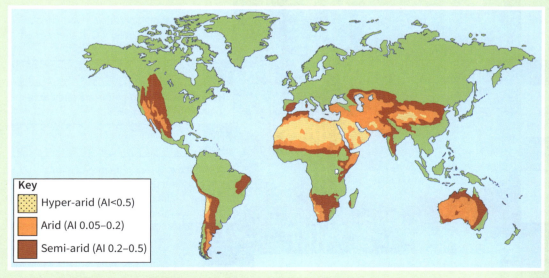

Key
- Hyper-arid (AI<0.5)
- Arid (AI 0.05–0.2)
- Semi-arid (AI 0.2–0.5)

▲ **Figure 2** Distribution of hyper-arid, semi-arid and arid regions

## 2 Hot desert systems and landscapes

### 2.1 Hot desert systems

## Characteristics of hot deserts and their margins

### Climate

- Characterised by very high summer temperatures and very low rainfall.
- Lack of cloud cover results in large diurnal temperature variations.
- Low winter temperatures, so overnight frosts and precipitation as snow are not uncommon.
- Convection currents form locally as air rises from the hot surface into the cooler upper atmosphere. This can trigger thunderstorms, producing high amounts of surface runoff.
- Rising air creates localised low pressure regions, drawing in strong winds from the surrounding high pressure.
- Downdrafts from thunderstorms can produce huge dust storms up to 100 km in width travelling at speeds of up to 100 km/h.

### Soils

- Soils contain very little organic matter because very few plants can survive the lack of precipitation, high temperatures and wide diurnal temperature range.
- Decomposition rates are very low due to lack of moisture.
- Without organic material, desert soils lack structure, are easily eroded by wind and water, and are generally infertile.
- The upper layers of desert soils suffer from **salinisation** and often have high mineral content because minerals are not leached out of the soils by precipitation.

### Vegetation

Plants that can survive arid conditions are called **xerophytes**. Adaptations include:

Succulence: many desert plants have water-storing tissues in stems, leaves or roots

Smaller leaves: or spines instead of leaves (protection) to reduce water loss from transpiration

CAM photosynthesis: stomata only open to take in $CO_2$ at night, reducing water loss as temperatures are cooler. $CO_2$ is then stored in plant vacuoles until the daytime

Thick, waxy cuticle: this covering reduces water loss from stems and leaves

**Halophytes**: desert plants are often able to grow in salinised soils

Deep root systems: these allow plants to tap into deep groundwater sources

Drought avoidance: ephemeral plants stay dormant as seeds in desert soils until sufficient rainfall causes them to grow and reproduce rapidly

Extensive root systems: shallow roots mean plants access precipitation before it evaporates

# The causes of aridity

## Atmospheric processes

Most hot deserts are located in belts of high pressure called the subtropical high: around 30° north and south of the equator.

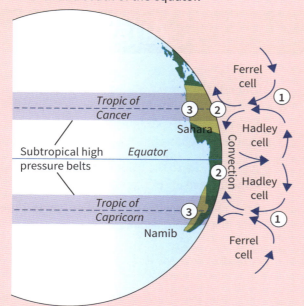

**(1)** Air moves polewards in the Hadley cell converging with the Ferrel cell; the cooling air sinks.

**(2)** Descending air becomes warmer and drier, causing high pressure conditions of cloudless skies and no rainfall.

**(3)** The high pressure causes warm, dry air to dominate the regions; any moist air is effectively kept out by the high pressure.

▲ **Figure 4** *The causes of aridity: global atmospheric circulation*

## Ocean currents

Cold ocean currents cause the air above to cool. Cooler air has reduced capacity to pick up moisture, and sinks, suppressing rain formation.

## Relief

Prevailing winds force air up the windward side of a mountain range; the air cools causing precipitation. The drier air then sinks on the leeward side creating an arid rain shadow.

## Continentality

Air near the coasts is moist because of the influence of the ocean. Further inland, this influence is absent, leading to dry conditions in continental interiors.

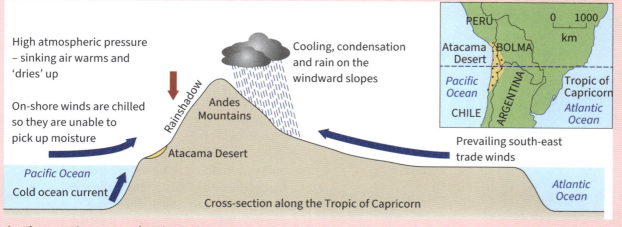

▲ **Figure 5** *The causes of aridity in the Atacama Desert, South America*

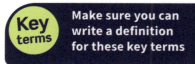
**Key terms** Make sure you can write a definition for these key terms

aridity index   halophyte   insolation   salinisation
water balance   xerophyte

# RETRIEVAL

Learn the answers to the questions below, then cover the answers column with a piece of paper and write down as many as you can. Check and repeat.

## Questions | Answers

| # | Question | Answer |
|---|----------|--------|
| 1 | Is a hot desert an open system or a closed system? | Open system |
| 2 | Define dynamic equilibrium. | A state in which a system stays in balance despite continuous change |
| 3 | Describe the global distribution of hot deserts. | Located close to the tropics of Cancer and Capricorn; some located in continental interiors |
| 4 | The aridity index (AI) is expressed as P/PET. What do P and PET stand for? | P = long-term average precipitation and PET = long-term potential evapotranspiration |
| 5 | In terms of the aridity index P/PET, when is a climate classed as arid? | When PET exceeds P |
| 6 | Explain why hot deserts have large diurnal temperature variations. | Cloudless skies (high pressure) means the land surface heats up rapidly in the day (high day time temperatures) but is then rapidly radiated at night (low night time temperatures) |
| 7 | Explain why desert soils are generally infertile. | Lack of organic matter because few plants can survive arid conditions; slow rates of decomposition because of lack of soil moisture, lack of leaching means salinisation as salts accumulate in the soil |
| 8 | What are xerophytes? | Plants that can tolerate arid conditions |
| 9 | What are halophytes? | Plants that can tolerate salty conditions |
| 10 | Explain succulence as an adaptation for arid conditions. | Ability of plants to store water in their tissues, e.g. leaves, stem, roots |
| 11 | How do ephemeral plants avoid droughts? | Their seeds remain dormant in the soil until there is sufficient rainfall for them to grow |
| 12 | How does CAM photosynthesis help plants retain water? | Plants only open their stomata to take in $CO_2$ at night; closing stomata in the daytime reduces loss of water through EVT |
| 13 | Name the two atmospheric cells associated with the formation of arid regions. | Hadley cell and Ferrel cell |
| 14 | How can relief cause aridity? | Prevailing winds force air up the windward side of a mountain range, causing the air to cool and precipitation to occur. The drier air then sinks on the leeward side creating an arid rain shadow |
| 15 | What is insolation? | Direct impact of heat from the sun |

Put paper here

# ⚙ KNOWLEDGE

## 2 Hot desert systems and landscapes

### 2.2 Desert systems and processes

## Sediment sources

**Sediment sources** include weathering, mass movement (rock falls, talus creep, soil creep), and sediment washed by surface runoff, transported by rivers and carried by wind. Some sources may originate outside the desert, for example wind-blown sediment and sediment carried into deserts by exogenous rivers.

## Sediment cells

A **sediment cell** is a systems model for understanding sediment movement. Within a sediment cell, there is erosion, transport and deposition of sediment within a long-term cycle.

- Inputs to the sediment cell are sediment sources, e.g. the erosion of weathered sediment from hillslopes.
- Transfers describe the movement of sediment within the desert system, for example by wind or water.
- Sediment sinks are where sediment is deposited, for example in dunes.

- Outputs describe sediment lost to the desert system, for example deposited in the ocean.
- Human activities can interrupt sediment cell systems, causing imbalances.

## Sediment budgets

A **sediment budget** is the balance between the input, output and storage of sediment within a desert system over a given period of time.

- Hot desert sediment budgets typically have low sediment inputs because of the very limited erosion of hillslopes by water.
- Wind erosion is often the dominant transfer, though flash floods can erode significant amounts of sediment.

- Sediment sinks are localised and dynamic (e.g. dunes) because the lack of vegetation and surface water means the dry sediment is easily redistributed.
- Outputs are mainly by wind; transport by rivers is minimal.

## Weathering processes

- Hot deserts generally have slow rates of weathering because of the absence of water, and therefore mechanical weathering is more important than chemical weathering.

- However, where water is present, high temperatures intensify chemical weathering, and salts weathered out of rocks are deposited as the water evaporates.

### Mechanical weathering in hot deserts

**Thermal fracture:** the wide diurnal temperature fluctuation causes rocks to expand and contract, causing cracks in the rock surface. Over time, repeated heating and cooling cycles cause the rocks to weaken and break apart.

**Exfoliation:** for rocks with layers of different mineral compositions, outer layers expand in the day, while inner layers remain cooler. At night, outer layers contract more than the inner layers, resulting in the peeling away of thin sheets or slabs from the rock surface.

**Block and granular disintegration:** the combined effect of thermal fracture and different minerals expanding and contracting at different rates can cause rocks to split into blocks (along joints or bedding planes in sedimentary rocks) or to crumble as mineral grains separate (in granular rocks).

## 2.2 Desert systems and processes

### The role of wind in hot deserts

Wind is a source of energy in hot desert environments (together with insolation and runoff) and plays a significant role in redistributing sediment and in aeolian landform development where vegetation is scarce and water is limited.

### Wind erosion

**Deflation:** wind removes loose surface materials such as sand, silt and clay from the desert's surface, leaving behind coarser particles and rock fragments, resulting in desert pavement

**Abrasion:** wind-borne particles act as abrasive tools, wearing down exposed rock surfaces, leading to the formation of distinctive landforms such as ventifacts

▲ *Figure 1 Digging beneath a desert pavement shows how underlying sand is protected by the pavement*

▲ *Figure 2 Igneous rock ventifacts can develop faceted sides reflecting prevailing wind directions*

### Wind transportation

Suspension – dust, fine silts and sand are picked up by the wind and can be carried for long distances out of the desert system (output)

Surface creep – coarse sand and pebbles are rolled along by the wind

Saltation – sand particles 'hop' along the surface: usually involves stronger winds and smaller particles than surface creep

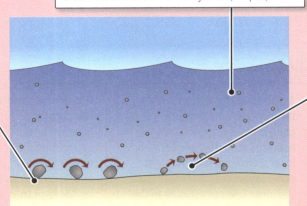

▲ *Figure 3 There are three forms of wind transportation*

Surface roughness: a smooth, flat desert surface encourages transportation as wind speed is higher, while obstacles will reduce wind speed and may result in deposition.

Wind strength: strong, high energy winds can pick up and transport more and larger sediment.

Sediment availability: hamada (bare rock) areas have little sediment available for wind transportation, regardless of wind strength.

Particle size and shape: fine, well-sorted sand grains are more easily lifted by the wind than larger, poorly-sorted particles. Round and smooth grains also have higher mobility than angular and rough grains, as they offer less resistance to wind forces.

**Factors affecting wind transportation**

Saltation threshold: the minimum wind speed required to initiate saltation. Smaller, lighter particles have lower saltation thresholds than larger particles. If wind speeds are below the saltation threshold, sediment will move by surface creep.

## Wind deposition

As wind energy drops, transported sediment is deposited. This often occurs when the wind encounters an obstacle that causes its velocity to decrease, for example on the leeward side of rock formations.

# The role of water in hot deserts

## Sources of water

| Exogenous | Endoreic | Ephemeral |
|---|---|---|
| **Exogenous** rivers come from outside the desert system. For example, the River Nile originates in the highlands of East Africa and carries water into Egypt's deserts. | **Endoreic** rivers do not leave the river system and do not flow into the sea. In hot deserts, endoreic rivers often form salt lakes or salt flats because of the high rates of evaporation. For example, the waters of the Great Salt Lake in Utah have high salinity and come from mountains within the desert. | **Ephemeral** rivers and streams flow intermittently, for example only after storm events or following spring snow melt in mountain regions. |

## Episodic flooding events

Strong convectional currents sometimes produce intense, localised storm events. Desert surfaces are often compacted through mineral cementation ('desert crust') with little or no vegetation preventing effective infiltration, so intense precipitation quickly exceeds the soil's capacity to absorb water, leading to surface runoff and flash flooding. There are two main types of flash flooding in deserts:

| Sheet flooding | Channel flash flooding |
|---|---|
| Runoff spreads out evenly over a broad area with gentle slopes and minimal natural channels to guide the flow. The water can cover large areas, creating temporary lakes or ponds. | Water flows with higher velocity and force within existing channels or dry riverbeds (wadis). Significant erosion and damage to the surrounding landscape can result. |

**Key terms**

Make sure you can write a definition for these key terms

abrasion   block and granular disintegration   deflation
endoreic   ephemeral   exfoliation   exogenous
sediment budget   sediment cell   sediment source   thermal fracture

# 2 Hot desert systems and landscapes

## 2.3 Desert landforms

### Relationship between process and time

Desert landforms are the result of interactions between geology and processes such as weathering and mass movement over time. Some landforms were formed long ago by processes no longer operating. The development of most desert landforms involves the interaction of several factors.

**REVISION TIP**

When writing about the development of a specific landform or landscape consider the relative importance of these factors or their inter-relationship.

**Factors contributing to desert landforms**

- Time: influence of (different) past climates
- Sources of energy: insolation, winds, runoff
- Geology: rock type, differential erosion, tectonics
- Geomorphological processes: weathering, mass movement, erosion, transportation, deposition
- Wind: erosion, transportation, deposition
- Sediment sources: sediment cells, sediment budgets
- Water: exogenous, endoreic and ephemeral rivers, sheet flooding, channel flash flooding

### Landforms and landscape

Collectively, landforms form a landscape – the visible features of an area. There are three main characteristic desert landscapes:

- reg – stony desert
- hamada – bare rock
- erg (sand sea) – sand dominates.

Desert landscapes have often taken a very long time to develop, due to the slow progress of weathering and erosion. Some characteristic desert landscapes may be the result of processes that are no longer acting on landforms, such as tectonics.

- The Namib Desert in south-west Africa has dunes estimated to be 55 million years old.
- Landforms in the western USA's Colorado Plateau desert are strongly influenced by tectonics: uplifted sedimentary rocks have been exposed to wind and water erosion over millions of years to form canyons and mesas.
- The Sahara Desert contains river channels formed when northern Africa was wetter than today.

### Deflation hollows (process: deflation)

A **deflation hollow** is a depression in the ground, varying from a few metres in width to several kilometres. The bottom of the hollow is typically a harder, more wind-resistant layer where lakes, oases or swamps may form.

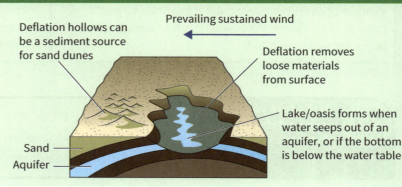

Deflation hollows can be a sediment source for sand dunes

Prevailing sustained wind

Deflation removes loose materials from surface

Lake/oasis forms when water seeps out of an aquifer, or if the bottom is below the water table

Sand
Aquifer

▶ **Figure 1** *Creation of a deflation hollow and salt lake*

## Desert pavements (deflation)

A **desert pavement** is a flat or gently-sloping desert surface covered by closely-packed, unconsolidated but interlocking stones too large to be transported. Sheet wash might also remove finer particles.

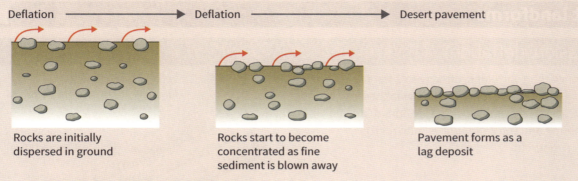

Deflation ———→ Deflation ———→ Desert pavement

Rocks are initially dispersed in ground

Rocks start to become concentrated as fine sediment is blown away

Pavement forms as a lag deposit

▲ **Figure 2** *Formation of a desert pavement*

## Ventifacts (abrasion)

Stones with sides (facets) that have been abraded by wind-driven sand. Mushroom-shaped pillars develop where abrasion is concentrated at the base of the pillar. Requires frequent strong winds. Strong, heavily sedimented winds will erode too quickly for large **ventifacts** to develop.

▲ **Figure 3** *A ventifact on a desert pavement*

## Yardangs (abrasion, deflation)

Wind-eroded ridges of bedrock or consolidated clay. **Yardangs** have a blunt windward end and a tapering leeward. Vary in length from a few centimetres, to several kilometres. They form when hard and soft bands of material are eroded, leaving the harder ridges separated by troughs.

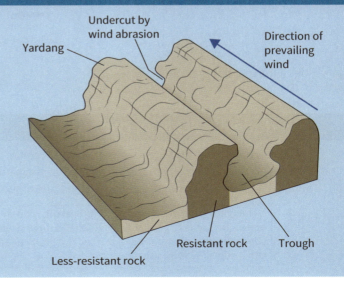

Yardang

Undercut by wind abrasion

Direction of prevailing wind

Less-resistant rock

Resistant rock

Trough

▶ **Figure 4** *Formation of a yardang*

# 2 Hot desert systems and landscapes

## 2.3 Desert landforms

### Zeugen (abrasion, deflation)

A blockier appearance than yardangs, often with a flat top wider than the base. **Zeugen** are formed when weaknesses in the horizontally layered resistant rock are exploited by mechanical weathering and the softer underlying rock is exploited and often undercut.

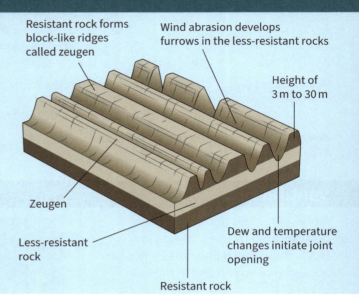

Resistant rock forms block-like ridges called zeugen

Wind abrasion develops furrows in the less-resistant rocks

Height of 3 m to 30 m

Zeugen

Less-resistant rock

Resistant rock

Dew and temperature changes initiate joint opening

▶ *Figure 5 Formation of zeugen*

### Barchan dunes (saltation, deposition)

**Barchan dunes** initially need a subtle change in land surface so that sediments collect, and an abundant supply of well-sorted sediment. Eventually the slip face reaches the angle of repose (35° for dry sand) and slumps, causing the dune to migrate downwind.

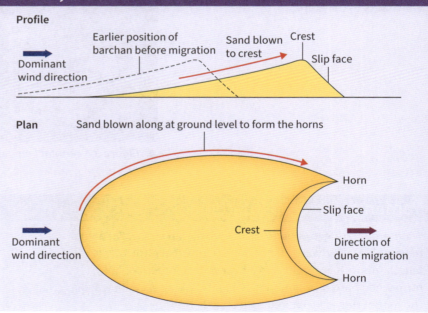

Profile

Dominant wind direction

Earlier position of barchan before migration

Sand blown to crest

Crest

Slip face

Plan

Sand blown along at ground level to form the horns

Dominant wind direction

Crest

Horn

Slip face

Direction of dune migration

Horn

▶ *Figure 6 Formation of a barchan dune*

### Seif dunes (saltation, deposition)

Large, narrow, linear sand dunes, often forming a chain. **Seif dunes** are created from barchans parallel to the prevailing wind, but need a change of wind direction to extend one of the horns.

> A change in wind direction from X (prevailing wind) to Y causes one of the horns to gradually lengthen. Continued alternating wind patterns, eventually create a seif (longitudinal) dune. Other slip faces may also develop.

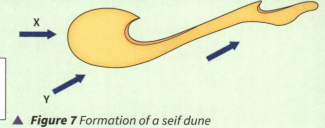

X

Y

▲ *Figure 7 Formation of a seif dune*

# Water-formed landforms

## Wadis (channel flash flooding)

Riverbeds ranging in length from a few metres to hundreds of kilometres that only contain water during periods of heavy rain. **Wadis** may be lined with vegetation because of higher soil moisture content, so are important wildlife corridors. Infrequent intense rainfall that transports huge quantities of sediment are factors in their formation.

## Bahadas (alluvial deposition)

An **alluvial fan** is a fan-shaped accumulation of sediment, which spreads out from where a channel emerges onto a plain and loses energy depositing its load. Sediment is sorted, with coarse sediment deposited at the channel mouth and finer material carried further away to the edge of the fan. A **bahada** is the merging of several alluvial fans.

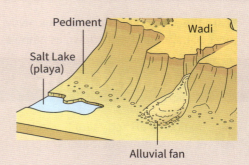

▲ **Figure 8** Water-formed landscape in a hot desert

## Pediments (fluvial erosion, sheet flooding)

Gently sloping, smooth bedrock surfaces found at the base of mountains or escarpments. Mechanical weathering weakens steep rock faces on mountain slopes and during periods of intense rainfall, the weathered mountain slopes are heavily eroded by water. Over millions of years, the mountain slope retreats, exposing a rock **pediment** at the base of the slope. The lower parts are often covered by a bahada.

## Playas (evaporation)

**Playas** are flat and typically salty dry lake beds formed by the evaporation of ephemeral lakes. Build up of the salt layer creates an extensive crusty surface.

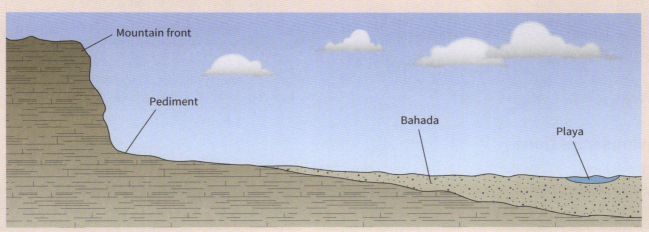

▲ **Figure 9** The idealised profile of a pediment

## Inselbergs (differential erosion)

Rocky, prominent, isolated outcrops within the landscape that are tens of millions of years old. They are a more resistant type of rock than the surroundings. Their formation involved sub-surface chemical weathering under a past humid climate. '**Inselberg**' can be used as an umbrella term to include all relic hills, including mesas and buttes.

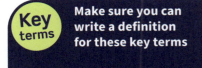

**Key terms** Make sure you can write a definition for these key terms

> alluvial fan    bahada    barchan dune    deflation hollow
> desert pavement    inselberg    pediment    playa    seif dune
> ventifact    wadi    yardang    zeugen

# RETRIEVAL

Learn the answers to the questions below, then cover the answers column with a piece of paper and write down as many as you can. Check and repeat.

## Questions / Answers

| # | Questions | Answers |
|---|-----------|---------|
| 1 | Do hot desert sediment budgets typically have low or high sediment inputs? | Low |
| 2 | Do hot deserts typically have slow or fast rates of weathering? | Slow |
| 3 | Which is typically more important in hot deserts: chemical or mechanical weathering? | Mechanical |
| 4 | Explain how thermal fracture weathering occurs. | Rocks repeatedly expand in the hot days and contract in the cold nights, leading to cracks and fractures |
| 5 | Name the type of mechanical weathering that can result in spheroidal landforms. | Exfoliation |
| 6 | Name the two types of wind erosion. | Deflation and abrasion |
| 7 | Identify two factors affecting wind transportation. | Two from: wind strength / surface roughness / sediment availability / particle size and shape |
| 8 | What is an exogenous source of water? | A source from outside the desert system |
| 9 | What are the two main types of flash flooding in deserts? | Sheet flooding and channel flash flooding |
| 10 | Identify the two main factors required for a deflation hollow to form. | Sustained strong winds (erosion by deflation), surface composed of loose particles |

*Put paper here*

## Previous questions

Now go back and use these questions to check your knowledge of previous topics.

## Questions / Answers

| # | Questions | Answers |
|---|-----------|---------|
| 1 | Is a hot desert an open system or a closed system? | Open system |
| 2 | The aridity index (AI) is expressed as P/PET. What do P and PET stand for? | P = long-term average precipitation and PET = long-term potential evapotranspiration |
| 3 | Explain why desert soils are generally infertile. | Lack of organic matter because few plants can survive arid conditions; slow rates of decomposition because of lack of soil moisture, lack of leaching means salinisation as salts accumulate in the soil |
| 4 | Name the two atmospheric cells associated with the formation of arid regions. | Hadley cell and Ferrel cell |
| 5 | What is insolation? | Direct impact of heat from the sun |

*Put paper here*

## 2 Hot desert systems and landscapes

### 2.4 Desertification

## Changing extent and distribution of hot deserts

- At the last glacial maximum 20,000 years ago, hot deserts were more extensive than today, and aridity was more widespread.

- Between 11,500 and 8000 years ago, the global climate was warmer and more humid. Today's deserts were grasslands and monsoon rains reached present-day arid areas.

- This warmer and wetter period began to end around 5000 years ago, and by around 3000 years ago deserts had developed in their present-day distribution and extent.

## Natural climate change

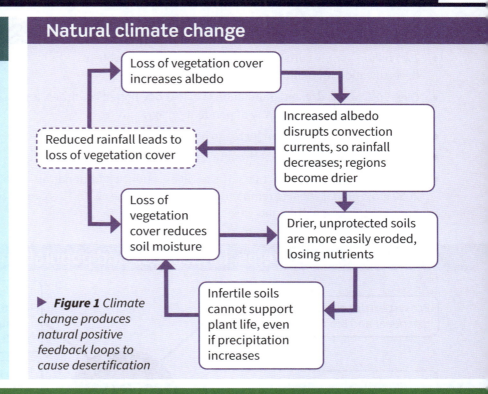

▶ **Figure 1** Climate change produces natural positive feedback loops to cause desertification

## Causes of desertification

- **Desertification** is characterised by a loss of vegetation, soil erosion and a general degrading of the land. Soils become useless for agriculture and natural habitats tend to rapidly deteriorate.

- In the past, desertification has been a natural process in which climate change sets off positive feedback loops. Human activities replicate and intensify this process.

## Human-induced climate change

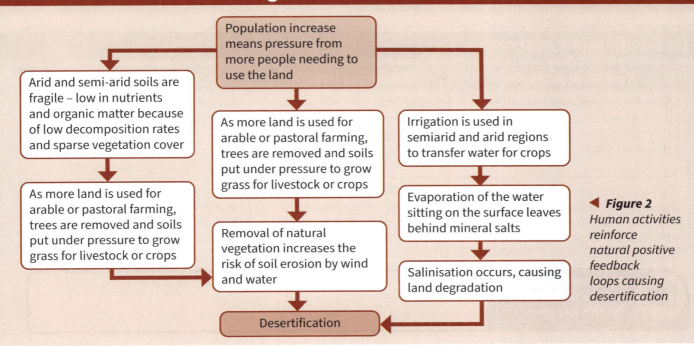

◀ **Figure 2** Human activities reinforce natural positive feedback loops causing desertification

## 2 Hot desert systems and landscapes

### 2.4 Desertification

## Desertification

- Two billion people currently live in areas most vulnerable to desertification (drylands).
- Over 90% of this dryland population is in less developed countries, and in regions where populations are mostly dependent on agriculture.
- Dryland populations are projected to increase by 43% (to 4 billion people) by 2050.
- The number and duration of droughts has increased by 29% since 2000.
- Predictions: 30% of the Earth's surface will experience additional 'aridification' by 2050 if the global average temperature increase reaches 2°C. This can be cut by two thirds if temperature increase remains below 1.5°C.

**REVISION TIP**

Learning key facts and statistics will help you discuss and compare them in detail in an exam.

## Impacts on ecosystems, landscapes and populations

Ecosystem degradation: habitats disappear and biodiversity decreases.

Soil erosion: fertile topsoil becomes more susceptible to erosion by wind and water, leaving behind infertile and compacted soil.

Drought and water scarcity: lakes, rivers and groundwater reserves become depleted, impacting both human and animal populations.

**Impacts of desertification**

Loss of livelihoods: crop failure, loss of livestock and reduced income for people relying on rain-fed agriculture lead to poverty and food insecurity.

Human displacement and migration: as land becomes unproductive, people move to look for food, water and opportunities. This can lead to conflicts over resources and humanitarian crises.

Climate change: vegetation loss reduces carbon uptake by plants, increased albedo is linked to temperature increases, and changes in convection patterns reduces rainfall.

## Desertification: responses

| Resilience strategies | Mitigation strategies | Adaptation strategies |
|---|---|---|
| Afforestation and reforestation: planting trees and restoring forests helps reduce erosion by stabilising soil and increasing soil moisture. | Sustainable land management: agroforestry and crop rotations can prevent soil degradation and improve soil productivity. | Diversification: developing alternative sources of income for farming, such as ecotourism or small-scale businesses, to reduce pressure on fragile ecosystems. |
| Soil and water conservation: terracing, contour farming and water harvesting help retain soil moisture, reduce erosion and enhance soil fertility. | Restoration of degraded land: reseeding land, introducing drought-resistant plants and controlling erosion. | Early warning systems: early warning for droughts and low soil moisture helps communities prepare for and respond to adverse climate events. |

**Key terms** Make sure you can write a definition for this key term

desertification

# RETRIEVAL

Learn the answers to the questions below, then cover the answers column with a piece of paper and write down as many as you can. Check and repeat.

## Questions | Answers

| # | Question | | Answer |
|---|----------|---|--------|
| 1 | Were conditions in current desert areas between 11,500 and 8000 years ago warmer and wetter, or cooler and drier? | Put paper here | Warmer and wetter |
| 2 | Define desertification. | | A process in which semi-arid regions become increasingly dry and lose vegetation cover |
| 3 | Why is loss of vegetation cover associated with unpredictable rainfall in arid and semi-arid regions? | Put paper here | Rainfall patterns associated with convection currents are disrupted when surface albedo increases; loss of vegetation also lowers EVT, reducing water vapour in the air |
| 4 | What is salinisation? | | The accumulation of salts at or near the surface of soils |
| 5 | Explain the connection between population increase, firewood and desertification. | Put paper here | In regions where wood is the main source of fuel for cooking and heating, increased populations means more demand for wood, which increases deforestation; the removal of vegetation cover increases soil erosion |
| 6 | Name two impacts of desertification. | Put paper here | Two from: ecosystem degradation / soil erosion / drought and water scarcity / loss of livelihoods / human displacement and migration / climate change |
| 7 | What is the aim of desertification resilience strategies? | | To strengthen the capacity of ecosystems and communities to withstand and recover from desertification |

## Previous questions

Now go back and use these questions to check your knowledge of previous topics.

## Questions | Answers

| # | Question | | Answer |
|---|----------|---|--------|
| 1 | What is an exogenous source of water? | | A source from outside the desert system |
| 2 | What are the two main types of flash flooding in deserts? | Put paper here | Sheet flooding and channel flash flooding |
| 3 | Do hot deserts typically have slow or fast rates of weathering? | | Slow |
| 4 | Name the two types of wind erosion. | | Deflation and abrasion |
| 5 | What are xerophytes? | | Plants that can tolerate arid conditions |

# KNOWLEDGE

## 2 Hot desert systems and landscapes

### 2.5 Case study: Desertification

 **Case study: Desertification in Burkina Faso**

| Location | North Africa, northern third is in the Sahel |
|---|---|
| Population | 20.1 million – growing by 3.1% per year |
| Industry | 90% of total population are subsistence farmers, with much land overgrazed |
| Annual precipitation | 250 mm, PET is 1875 mm |
| Droughts | Every 2–3 years |
| Predicted temperature increase | 3–4°C by 2080–99 (IPCC) |

> **REVISION TIP**
>
> You will have studied a case study looking at causes and impacts of desertification at a local scale, with an evaluation of resilience, mitigation and adaptation responses. You should revise the case study you have studied. This page has examples of responses in Burkina Faso.

## Soil and water conservation (SWC) strategies

Since the 1970s, government programmes have promoted SWC strategies based on traditional farming methods. These have significantly improved soil structure, crop yields, groundwater recharge and rainfall infiltration: 300,000 hectares of land has been restored to productivity, with an additional 80,000 tonnes of food produced each year. Rural poverty has been reduced by 50% and food security has been improved for three million people.

However, experts conclude that work is needed to establish the best strategy for individual situations – site-specific approaches are needed.

| SWC strategy | Description | Advantages | Evaluation |
|---|---|---|---|
| Zai planting pits | Shallow pits (30 cm wide, 15 cm deep) dug before rainy season begins. Organic material is placed in the pits and covered. A low berm is made on the downslope side. | The pits collect water, and the berms help to prevent runoff, reducing erosion. Termites break down the material, releasing nutrients into the soil, and their burrows improve soil permeability. | Yields of sorghum, maize and millet are increased by up to 100%. They work best with 300–800 mm rainfall. |
| Agroforestry | Useful tree species are grown among arable crops. Trees used for livestock fodder or for fuelwood; cashew nut trees, mango and baobab trees provide food or medicines. | Tree roots help protect soil from erosion, encourage infiltration and their shade reduces evaporation. Nitrogen-fixing trees improve soil fertility; fuelwood trees supply local fuel and reduce deforestation; fruits can be sold. | Agroforestry in Burkina Faso reduced time collecting wood for fuel from 2.5 hours to 0.5 hours per day. |
| Stone bunds | Lines of stones set out along contour lines on sloping farmland and barren land. Grasses can be planted along bunds or field margins. | The stones slow surface runoff, increase infiltration into the soil and trap sediment. Grass bands reduce wind and water erosion. Grass species are also used locally to make mats. | Stone bunds reduce soil erosion by 95% and runoff by 45%, increase infiltration by 15% and crop yields by 30%. Larger bunds raise the water table. |
| Rock dams | A series of long, low dams made of stones placed across wadis and gullies. Also placed across land that is vulnerable to flash flooding to prevent gullies from forming. | Highly erosive flash flooding is forced to spread out, which slows the flow leading to deposition. | Such large quantities of sediment are deposited that gullies can be filled in a couple of years, and then used for growing crops. |

# 2.6 Case study: Hot desert environment

 **Case study: Barchan sand dunes in the Namib Desert**

| | |
|---|---|
| **Location** | Coastal desert in south-west Africa, extending for 2000 km along the coasts of Angola, Namibia and South Africa |
| **Area** | 81,000 km² |
| **Annual precipitation** | 2–200 mm |
| **Daytime temperatures** | 10°C (coast)–45°C (inland) |
| **Aridity** | Arid climate caused by the descent of dry air of the Hadley Cell, plus the influence of the cold Benguela current along the coast |
| **Landscape** | Sand (erg) at coast (34,000 km²); stony desert (reg) and mountains inland |
| **Sediment source** | The Orange River to the south of the sand sea; other ephemeral rivers including the Kuiseb. In the Sossusvlei area, several dunes exceed 300 m in height |
| **Landforms** | South of the sand sea is an area of yardangs, which are orientated in the direction of the prevailing south-south-west winds |
| | Barchan dunes dominate in this area, though they are less common than seif dunes and other dune types across the whole of the sand sea. |

**REVISION TIP**

You will have studied a case study looking at key themes about hot desert systems and landscapes. You should revise the case study you have studied, which may have included some fieldwork on sand dunes. This page looks at fieldwork investigations into barchan sand dunes in an area of the Namib Desert in south-west Africa.

## Research questions

1. Do these barchan dunes display the same characteristics as found in other dune fields?
   - Dune height: the dunes on the Kuiseb delta have a mean height of 7.7 m, which is comparable to that found in other studies of barchan dune fields.
   - Relationship between dune height and horn width: a strong positive correlation was found between dune height and horn width (ratio of 0.85). This supports the theory that dunes reach a state of dynamic equilibrium with the local environmental conditions.

2. Does grain size influence the asymmetry of the dune horns?
   - Many of the barchans in the study area have abnormal longitudinal profiles, with one horn longer than the others.
   - Significantly coarser particle sizes were found on the more elongated horns, likely to be because coarser grains are diverted around the dune base through sand creep, so that dune horns are characterised by coarser mean grain sizes.
   - Other external factors are also likely to be involved, including shifting wind patterns and changes in sediment supply.

3. Has the rate and direction of dune movements changed significantly over the 30-year period of record keeping?
   - Aerial photos from a 30-year time series show that the average speed of dune movement changes over time, as does the direction of movement.
   - This corresponds with weather records showing changes in wind direction related to fluctuations in the strength of the South Atlantic anticyclone.

# RETRIEVAL

Learn the answers to the questions below, then cover the answers column with a piece of paper and write down as many as you can. Check and repeat.

## Questions | Answers

| | Question | Answer |
|---|---|---|
| 1 | SWC is an example of a resilience strategy. What does it stand for? | Soil and water conservation |
| 2 | What are zai planting pits? | Shallow pits (30 cm wide, 15 cm deep) dug before rainy season begins |
| 3 | Explain how rock dams help combat desertification. | Highly erosive flash flooding is forced to spread out, which slows the flow leading to deposition |
| 4 | Explain one way in which agroforestry helps build resilience against desertification. | One from: reduction in soil erosion from splash erosion / trees shade crops and soil, reducing water loss from evaporation and EVT / tree roots encourage infiltration, reducing soil erosion and increasing soil moisture |
| 5 | Explain how stone bunds help combat desertification. | The stones slow down surface runoff, increase infiltration into the soil and trap sediment |
| 6 | Suggest one reason why a study of barchans finds a significant number with one horn longer than the other. | One from: changes in wind direction / changes in sediment supply / differences in sediment coarseness |
| 7 | By what percentage has rural poverty been reduced in Burkina Faso as a result of SWC strategies? | 50% |

*Put paper here*

## Previous questions

Now go back and use these questions to check your knowledge of previous topics.

## Questions | Answers

| | Question | Answer |
|---|---|---|
| 1 | Explain why hot deserts have large diurnal temperature variations. | Cloudless skies (high pressure) means the land surface heats up rapidly in the day (high day time temperatures) but is then rapidly radiated at night (low night time temperatures) |
| 2 | How do ephemeral plants avoid droughts? | Their seeds remain dormant in the soil until there is sufficient rainfall for them to grow |
| 3 | Identify two factors affecting wind transportation. | Two from: wind strength / surface roughness / sediment availability / particle size and shape |
| 4 | Define desertification. | A process in which semi-arid regions become increasingly dry and lose vegetation cover |
| 5 | Why is loss of vegetation cover associated with unpredictable rainfall in arid and semi-arid regions? | Rainfall patterns associated with convection currents are disrupted when surface albedo increases; loss of vegetation also lowers EVT, reducing water vapour in the air |

*Put paper here*

# PRACTICE

## Exam-style questions

**1**  Outline the role of wind in the process of erosion in hot deserts.  **[4 marks]**

**2**  Outline sources of sediment in hot desert environments.  **[4 marks]**

**3**  Outline the characteristics of vegetation found in hot desert environments.  **[4 marks]**

> **EXAM TIP**
>
> 4-mark questions are point-marked. That means each valid point or development of a valid point receives one mark, up to a maximum of 4 marks.

**4**  Explain the concept of the sediment budget as it applies to hot desert systems.  **[4 marks]**

**5**  Explain the origin and development of deflation hollows.  **[4 marks]**

**6**  Explain the concept of positive feedback in relation to desertification.  **[4 marks]**

**7**  Evaluate the usefulness of **Figure 1** for an investigation into water balance in this location.  **[6 marks]**

> **EXAM TIP**
>
> Consider how effectively this graph depicts the data. It is easy to read? What are the figure's limitations and are they to do with the graph used or the dataset itself?

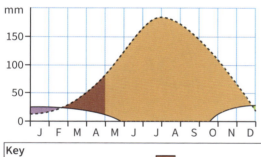

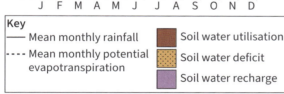

**Key**
— Mean monthly rainfall
---- Mean monthly potential evapotranspiration
Soil water utilisation
Soil water deficit
Soil water recharge

◄ *Figure 1 A water balance graph for Baghdad, Iraq*

**8**  Analyse the data shown in **Figure 2**.  **[6 marks]**

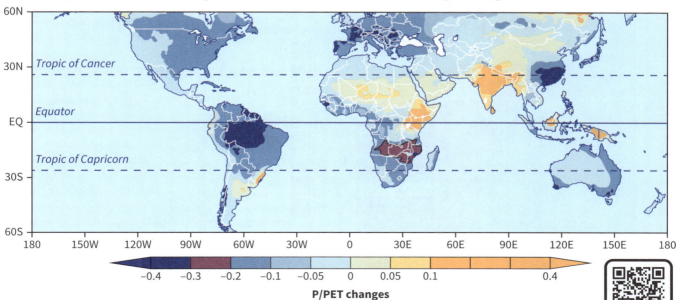

▲ **Figure 2** *Projected changes in the aridity index by 2100. The aridity index is expressed as P/PET, which gives the ratio between the long-term average precipitation (P) and the long-term average potential evapotranspiration (PET) for an area.*

9    Analyse the extent of the relationships shown in **Figure 3a** and **Figure 3b**. [6 marks]

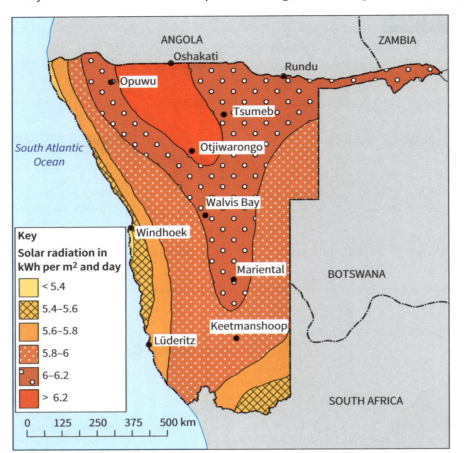

▲ **Figure 3a** *Average values of solar radiation in Namibia (kWh per m²)*

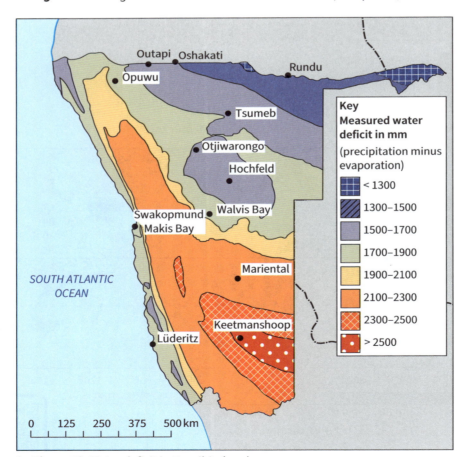

▲ **Figure 3b** *Water deficit in Namibia (mm)*

**10** Using **Figure 4** and your own knowledge, assess the role of weathering in the development of these landforms. **[6 marks]**

▲ *Figure 4 Corestone landforms in Mangystau, Kazakhstan*

Note: Mangystau is a region in the south-west of Kazakhstan. Average precipitation is 150 mm per year. Summer temperatures reach averages of 35°C, while in winter temperatures are frequently below freezing.

**11** Using **Figure 5** and your own knowledge, assess the role of wind in the development of this landscape. **[6 marks]**

▲ *Figure 5 Features of the landscape of the White Desert in Egypt*

Note: The chalk and limestone formations of the White Desert were formed 60 million years ago from sediment accumulations at the bottom of a shallow sea. Less resistant chalk layers are sandwiched in places between more resistant layers of limestone.

**12** Using **Figure 6** and your own knowledge, assess the role of water in the development of this landscape. **[6 marks]**

▲ *Figure 6 A landscape in Death Valley, a valley in the north of the Mojave Desert, USA*

Note: This aerial view shows a mountain range and a series of fan-like features on a flat valley floor. Death Valley was formed when a block of land between two faults dropped down, creating a flat valley between two mountain ranges.

**13** Using **Figure 7** and your own knowledge, assess the relative importance of factors leading to the development of these landforms. **[6 marks]**

▲ **Figure 7** *A landscape in the Namib Desert, Namibia*

Note: The landforms in this landscape are aligned approximately north-west to south-east and extend from between 16 km to 32 km in length, reaching heights between 60 metres and 240 metres. The sediment source is the Orange River, several kilometres away.

> **EXAM TIP**
>
> Examiners recognise that you will not have seen the photo of the landscape before. A note may well be provided to help you identify what you are looking at; make sure you read such notes carefully as they will contain valuable information.

**14** To what extent can desertification be seen as a characteristic process of a natural system? **[20 marks]**

> **EXAM TIP**
>
> There are 10 marks for AO1 (knowledge and understanding) and 10 marks for AO2 (applying and interpreting). Accurate and relevant knowledge and understanding (e.g. examples) gives you your AO1 marks; for AO2 you will be weighing up evidence, considering significance, making judgements, etc. as you develop your argument.

**15** For a local scale landscape you have studied, assess the impacts of climate change on desertification. **[20 marks]**

**16** 'Mitigation is the best response to desertification.'

With reference to a local scale landscape you have studied, how far do you agree? **[20 marks]**

**17** 'Historic, one-off events have a much greater influence on landscape development than ongoing processes in hot deserts.'

How far do you agree with this statement? **[20 marks]**

**18** 'Hot deserts and their margins are too fragile for development by human activity.'

How far do you agree with this view? **[20 marks]**

**19** With reference to a landscape where desertification has occurred, assess the impact of human activity upon the natural system. **[20 marks]**

**20** Assess the relative importance of weathering and erosion in the development of a hot desert landscape/environment you have studied. **[20 marks]**

# ⚙ KNOWLEDGE

## 3 Coastal systems and landscapes

### 3.1 Coastal systems

## Systems concepts and coastal landscapes

Open systems have links to other systems.

| System concept | Definition | Examples for glacial systems |
|---|---|---|
| Inputs | The resources, substances or energy that enter a system | Energy from wind and waves, sediment |
| Outputs | The resources, substances or energy that leave a system | Removal of sediment to the ocean and evaporation |
| Energy | The capacity to do work or cause change within a system | Energy from wind and waves |
| Stores/components | Where energy or matter is kept for a relatively long time in a system | Snow, ice |
| Flows/transfers | The movement of energy or matter between stores within a system | Accumulations of sand in depositional landforms – beaches, spits, offshore bars |
| Positive feedback | Amplifies or reinforces changes within a system | An increase in wave energy will lead to an increase in erosion |
| Negative feedback | Counteracts or reduces changes within a system, maintaining stability | An increase in deposition will steepen the beach profile, which encourages the formation of destructive waves that will flatten the beach |
| Dynamic equilibrium | A state in which a system stays in balance despite continuous change | Sediment accumulation happens at the same rate as it is removed |

## Sediment cells

Sediment theoretically stays within each cell with no transfer between one cell and its neighbour

**Sediment cells** are closed systems

**Sediment cells**

All have a distinctive coastal landscape, made up of different coastal landforms

Sediment cells have sources of sediment, transfers of sediment and sediment sinks (depositional landforms)

- The **sediment budget** describes the balance of sediment in a cell considering losses of sediment (deposition) and sediment gains (input from sediment sources).
- A budget in dynamic equilibrium means that the inputs and outputs are equal, but the system is still in a state of change, i.e. sediment accumulation happens at the same rate as it is removed.

> **REVISION TIP**
>
> You need to know about sediment cells in the context of one cell, such as the cell between Portland Bill and Selsey Bill.

## 3 Coastal systems and landscapes

### 3.1 Coastal systems

## Sources of energy in coastal environments

- A high-energy environment is characterised by powerful, destructive waves and high rates of erosion.
- A low-energy environment has a higher rate of deposition due to less powerful, constructive waves for most of the year and is more likely to be found on sandy and estuarine coasts.

## Wind and waves

Waves are formed when wind blows over the surface of the sea. The friction causes the particles to rotate. Wave energy is influenced by wind speed, duration, strength, and the length of fetch.

There are two main types of wave:

- destructive wave – decreases the gradient of the beach profile.
- constructive wave – increases the gradient of a beach profile.

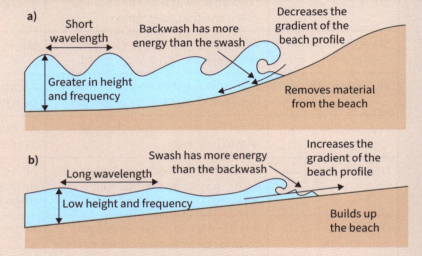

▲ **Figure 1** a) destructive waves b) constructive waves

## Currents and tides

| Currents | Tides |
|---|---|
| • Onshore rip currents can be created when waves break parallel to the shore, creating a cell circulation and a concentrated backwash.<br><br>• Tides create tidal currents which are influenced by the tidal range and can become concentrated in narrow channels and estuaries.<br><br>• Offshore currents move water and heat energy around the globe and are generated by the Earth's rotation. | Tides are caused by the gravitational pull of the Sun and the moon. This causes an ocean bulge where water is deeper, which travels with the Sun and moon as they pass overhead, causing two high tides and two low tides a day.<br><br>The tidal range influences the area that is affected by wave action:<br><br>• Some coastlines have small tidal ranges (more enclosed coasts like in the Baltic Sea).<br><br>• Some have much larger tidal ranges (such as those in the UK). |

**Key terms** Make sure you can write a definition for these key terms

sediment budget    sediment cell

## 3.2 Coastal processes

### Erosional processes

The impact of erosional processes is influenced by waves (destructive waves have more erosional power) and rock type (weaker rocks are more susceptible to erosion).

Hydraulic action:
Waves force air into the cracks in cliffs. The compressed air bubbles create a mini explosion **(cavitation)**. The repeated increase in pressure forces cracks to widen and break pieces off the cliff.

Wave quarrying:
The removal of loosened material by wave action.

Abrasion/corrasion:
Sediment that is carried by the waves is thrown at the cliffs and wears the cliff face away.

**Erosional processes**

Attrition:
Sediment that is carried in the waves knocks against each other, breaking parts of the rock down and creating smoother, smaller and rounder particles.

Solution:
A chemical process where limestone is dissolved by carbonic acid in seawater. It is the only erosional process which is not more effective under storm conditions.

### Transportation, longshore drift and deposition

Transportation transfers sediment from one store to another. Without it there would be no flows/transfers. Energy (velocity of the process), and particle size of the sediment (mass) are key factors in the type of transportation that operates.

| Transportation process | Explanation |
|---|---|
| Traction | Large sediment is rolled along the sea bed |
| Saltation | Medium-sized sediment 'bounces' along the sea bed |
| Suspension | Very small, fine sediment floats within the wave |
| Solution | Sediment is dissolved into the sea water |

Sediment transportation is influenced by the process of longshore drift, and the movement of tides and currents. Longshore drift (**littoral drift**) transports material along a coastline, in the direction of the prevailing wind.

Distinctive coastal landforms are created from the process of deposition (see 3.4).

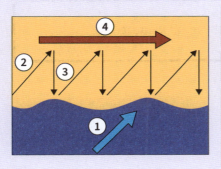

1. Waves approach the coastline at an oblique angle, influenced by the direction of the prevailing wind

2. Swash pushes material up the beach at this angle

3. Backwash moves sediment back down the beach at a right angle under the influence of gravity

4. The process repeats and sediment is transported along the beach

▲ **Figure 1** *The process of longshore/littoral drift deposition occurs when waves lose their energy and drop the sediment that they are carrying.*

## 3 Coastal systems and landscapes

### 3.2 Coastal processes

## Sub-aerial weathering

There are three types of **sub-aerial weathering** that provide a sediment source for the coastal system.

| | |
|---|---|
| **Chemical** | A process which attacks the minerals within rocks; e.g. carbonation occurs when the carbonic acid in rainwater reacts with the calcium carbonate in rocks such as limestone, dissolving the limestone and removing it by solution. |
| **Mechanical** | The change in composition of the rock by physical processes, where no chemical change occurs; e.g. freeze–thaw weathering occurs when water enters cracks in rocks, freezes, expands, exerting pressure on the rock, and then thaws. This repeated process will eventually lead to rocks splitting apart. |
| **Biological** | A process where the roots of plants grow into cracks in rocks, widening them and forcing them apart. Burrowing animals can also weather softer rocks and sand dunes. |

## Mass movement

Mass movement is the downward movement of material, under the influence of gravity. It is more likely to occur where cliffs have a weak rock type.

Rotational slumping:
More likely to occur in softer rocks, where sediment moves 'en masse' downslope, leaving a rotational scar (a curved rupture surface) and a terraced cliff profile.

Landslides:
Landslides leave behind a flat rupture surface as the movement of weathered sediment is planar (downslope).

**Mass movement**

Blockfall:
Occurs on slopes over 40°. Rock fragments fall from the cliff face and form talus scree slopes at the foot of the cliff.

Surface run-off:
In the hydrological cycle, surface run-off can transport sediment downslope and add it to the littoral zone.

 **Key terms** Make sure you can write a definition for these key terms

cavitation  littoral drift  sub-aerial weathering

# ⇄ RETRIEVAL

Learn the answers to the questions below, then cover the answers column with a piece of paper and write as many as you can. Check and repeat.

## Questions

## Answers

| | Questions | Answers |
|---|---|---|
| 1 | What are the three parts of an open coastal system? | Inputs, processes and outputs |
| 2 | Is a sediment cell a closed or an open system? | Open system |
| 3 | What is a sediment budget? | The difference between the losses and gains of sediment within a sediment cell |
| 4 | Give two processes that influence wave height. | Two from: wind speed / wind duration / wind strength / length of fetch |
| 5 | Give three characteristics of destructive waves. | Three from: greater in height and frequency / short wavelength / backwash has more energy than the swash / remove material from a beach / decrease the gradient of the beach profile |
| 6 | Name three erosional processes. | Three from: hydraulic action / attrition / wave quarrying / abrasion/corrosion / solution |
| 7 | What is solution? | A chemical process where limestone is dissolved by carbonic acid in seawater |
| 8 | What is longshore drift? | When sediment is transported along a coastline, in the direction of the prevailing wind |
| 9 | What is the movement of material up the beach by waves called? | Swash |
| 10 | What does sub-aerial mean? | Processes that occur on land, as opposed to in the sea |
| 11 | What causes the tides? | The gravitational pull of the Sun and the moon |
| 12 | Name one type of mechanical weathering. | Freeze–thaw |
| 13 | How does biological weathering affect rocks? | Weathering is a result of plants and animals, e.g. plant roots that grow into cracks in rocks forcing them apart, and burrowing animals |
| 14 | What is mass movement? | The downward movement of material, under the influence of gravity |
| 15 | Name two ways material can move by mass movement in coastal areas | Two from: blockfall / rotational slumping / landslides / sediment within surface runoff |

Put paper here

# 3 Coastal systems and landscapes

## 3.3 Landforms of coastal erosion

### Cliffs and wave-cut platforms

Erosion at the foot of a cliff by wave processes results in the recession of the cliff, leaving behind a wave-cut platform.

**3** The unsupported overhang will eventually collapse, and the cliff has retreated.

**2** Continued undercutting creates an overhang.

**4** Fallen debris is removed from the base of the cliff by wave action.

**1** Waves attack cliffs between the high and low tide marks, forming a wave-cut notch.

**5** A wide, gently sloping platform is left behind which is covered at high tide.

▲ **Figure 1** Formation of a wave-cut platform

### Cliff profile features: caves, arches and stacks

Continued erosion at a high-energy coastline can create several distinctive features.

**1** Waves attack geological weaknesses in a headland, such as joints and cracks.

**4** Continued erosion at the base of the arch and subaerial processes on the roof cause the arch to collapse.

**5** The resulting stack, an isolated pillar stands separate from the headland.

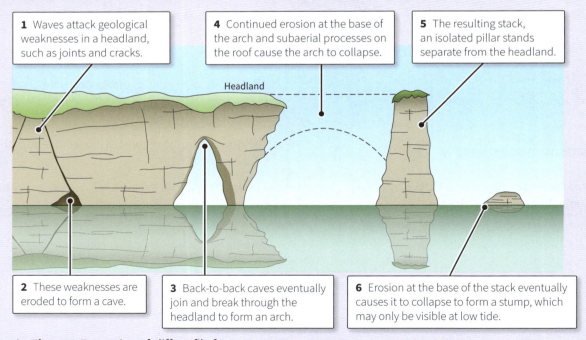

Headland

**2** These weaknesses are eroded to form a cave.

**3** Back-to-back caves eventually join and break through the headland to form an arch.

**6** Erosion at the base of the stack eventually causes it to collapse to form a stump, which may only be visible at low tide.

▲ **Figure 2** Formation of cliff profile features

# Factors affecting the development of landscapes of coastal erosion

There are several factors and processes that affect the development of coastal landforms and landscapes:

- geology
- dip of the rock strata
- wave energy
- and human activity.

## Geology

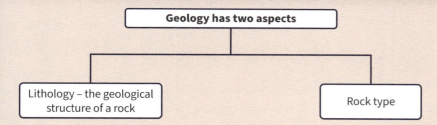

**Geology has two aspects**

Lithology – the geological structure of a rock

Rock type

- Soft, sedimentary rocks are more susceptible to erosion and higher rates of coastal recession.
- Harder rock types, such as igneous and metamorphic, are less easily eroded and weathered.
- Joints and faults are fractures in rocks that create points of weakness in rocks that are vulnerable to weathering and erosional processes. They can be exploited to form wave-cut notches and caves in the cliff.

## Dip

The dip of a rock describes the angle between the horizontal and the rock strata.

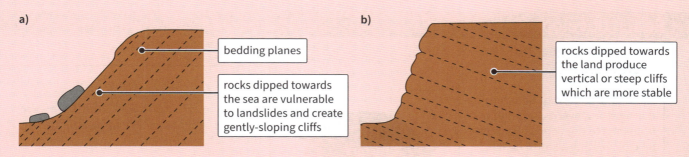

a)

bedding planes

rocks dipped towards the sea are vulnerable to landslides and create gently-sloping cliffs

b)

rocks dipped towards the land produce vertical or steep cliffs which are more stable

▲ **Figure 3** *a) Seaward and b) landward dip*

## Wave energy and human activity

- Destructive storm waves have more energy and lead to higher rates of erosion and cliff retreat. Wave energy is affected by a number of different factors (see 3.1).
- The influence of human activity at the coastline, such as through coastal management or development (see 3.6), can affect the natural processes that create the coastal landscape. For example, the use of groynes to protect one area of coastline by building up a beach, can lead to sediment starvation and increased rates of erosion and retreat in other areas.

# KNOWLEDGE

## 3 Coastal systems and landscapes

### 3.4 Landforms of coastal deposition

## Beaches

- Beaches are an accumulation of deposited sediment.
- They are found between the low, high and storm tide marks.
- An **offshore bar** is a narrow ridge of sediment which runs parallel to the coast, and is formed when backwash removes material from the beach and deposits it in the offshore zone.
- The different zones of a beach are all affected by the action of the waves and so are constantly changing.

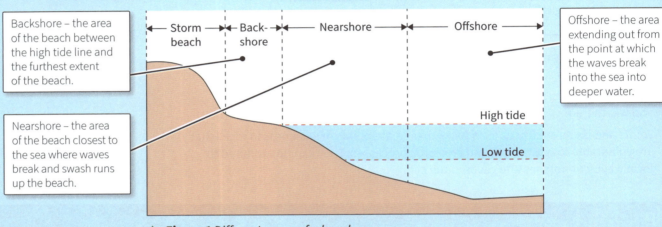

Backshore – the area of the beach between the high tide line and the furthest extent of the beach.

Nearshore – the area of the beach closest to the sea where waves break and swash runs up the beach.

Offshore – the area extending out from the point at which the waves break into the sea into deeper water.

▲ **Figure 1** *Different zones of a beach*

## Beach development

There are many factors that affect the development of beaches.

**Weather events:**
During the highest of tides and during storm events, sediment is pushed to the very back of the beach, creating a storm beach, where it may stay until another storm takes place.

**Factors affecting the development of beaches**

**Tides:**
Features found on beaches reflect the daily changes in swash and backwash. Berms are ridges of sediment which run parallel along the beach and mark the furthest extent of each high tide.

**Sediment type:**
Shingle beaches have a steeper gradient than sand beaches as percolation is more rapid, leaving less backwash.

**Wave energy:**
High energy waves are more common in the winter, reducing the gradient of the beach profile. Low energy waves are more common in the summer, building up the beach and increasing the beach gradient.

# Spits

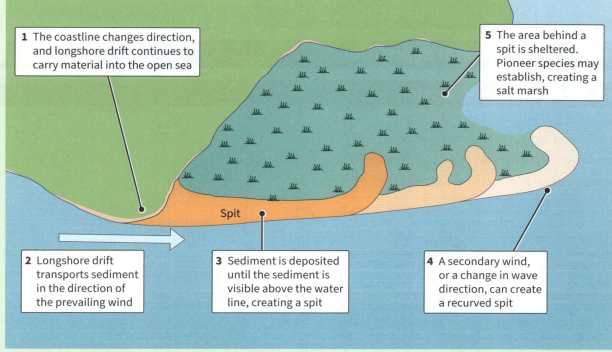

1 The coastline changes direction, and longshore drift continues to carry material into the open sea

5 The area behind a spit is sheltered. Pioneer species may establish, creating a salt marsh

Spit

2 Longshore drift transports sediment in the direction of the prevailing wind

3 Sediment is deposited until the sediment is visible above the water line, creating a spit

4 A secondary wind, or a change in wave direction, can create a recurved spit

▲ **Figure 2** *Spit formation*

- Spits are more common in areas with low tidal ranges.
- Spits which form across an estuary become cut off when the river current is too strong to allow deposition to continue. A **recurved spit** may form.
- Blakeney Point in Norfolk is an example of a compound spit with many recurves.

## Tombolos

- If a spit extends out to reach an island, it can join that island to the mainland, creating a **tombolo**, e.g. Tombolo di Orbetello in Italy, where three spits have linked the island of Monte Argentario to the mainland.

## Barrier beaches and islands

- If a spit extends out across a bay, it can connect to the headland on the other side and create a bar or **barrier beach**, with a lagoon of brackish water forming on the landward side.
- If the barrier beach becomes disconnected from the mainland, it forms a barrier island; e.g. the Friesian Islands in the Netherlands and Germany are Europe's biggest barrier island system.

## 3 Coastal systems and landscapes

### 3.4 Landforms of coastal deposition

## Sand dunes and dune succession

Sand dune formation needs:

- large quantities of sand from constructive waves
- a large tidal range enabling the sand to dry out at low tide
- strong onshore winds to blow the sand to the back of the beach.

> Sand dunes begin to form when sand accumulates around an obstruction on a beach, such as a strand line or fence.

⬇

> The first plants to grow on embryo dunes are pioneer colonising species, such as couch or lime grass, with long roots which stabilise the sand.

⬇

> The foredune is stabilised further by xerophytes (plants that can survive with little water) such as marram grass.

⬇

> The fixed and mature dunes have a higher humus content as plants die and break down into the soil and have a larger variety of plants.

## Mudflats and salt marshes

Mudflats and saltmarshes develop in areas with:

- low wave energy
- shelter, such as estuaries, bays, and behind spits; e.g. Keyhaven salt marshes have formed behind Hurst Castle spit on UK's Hampshire coastline.

> Fine, suspended sediment form into clumps or masses and are deposited between high and low tide levels.

⬇

> The first plants to grow on embryo dunes are pioneer colonising species, such as couch or lime grass, with long roots which stabilise the sand.

⬇

> Halophytic (salt-tolerant) plants establish themselves next and decompose to add organic matter to the soil.

⬇

> Over time, a wider variety of plants will be able to survive in the salt marsh.

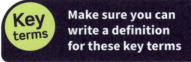 **Make sure you can write a definition for these key terms**

barrier beach  offshore bar  recurved spit  tombolo

# 3.5 Sea level change

## Long-term sea level change

Sea levels have fluctuated over the Earth's history. → Around 20,000 years ago, sea levels were 120 m lower than the present day. → About 10,000 years ago, sea levels began to rise at an average rate of 1.5 cm a year.

## Eustatic sea level change

- **Eustatic** change is a global change in the volume of water in the ocean.
- Temperatures increase, the ice stored in ice sheets and ice caps melts, and sea level rises.
- Warmer temperatures also lead to thermal expansion, where the water volume increases, and sea level rises.
- During glacial periods in the past when temperatures fell, more water was stored on the land in solid ice form and sea levels fell.

## Isostatic sea level change

- **Isostatic** change is changes in the land level, leading to a relative change in sea level.
- Land can subside due to the weight of the ice stored on it, leading to a relative sea level rise.
- Land that was once covered in ice during the last ice age is still rising and falling in a process called glacial isostatic adjustment.
- A high rate of deposition within a sediment cell may increase the level of the land, leading to a relative sea level fall.
- Tectonic processes, such as volcanic activity on the seafloor or subduction zones, can displace the ocean and lead to a sea level change.

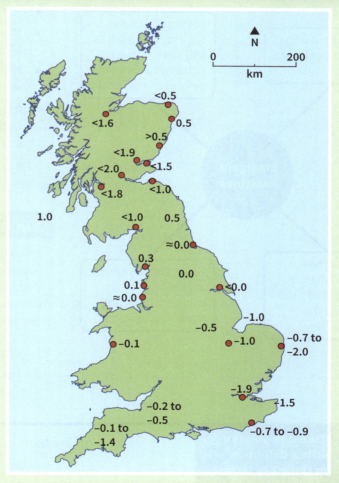

▲ **Figure 1** Isostatic sea level change in Great Britain. The numbers indicate the rise/fall in level of the land. All of Scotland is still rising, while Wales and the whole of southern England are sinking

## 3 Coastal systems and landscapes

### 3.5 Sea level change

## Tectonic sea level change

- Epeirogenic changes are the change in the shape of ocean basins and can occur due to sea floor spreading, leading to a relative sea level change.

- Orogenic movements involve the uplift of land, for example fold mountain creation at collision plate margins, and can affect the relative sea level.

- Following the 2011 earthquake, a large area of eastern Japan that was already below sea level became five times larger due to subsidence caused by the tectonic movement.

- Volcanic eruptions near to the coast can create volcanic islands, or underwater eruptions can initiate sea level change.

**REVISION TIP**

See the Geographical Skills section for help with aerial interpretation of landforms indicating sea level change.

## Emergent coastlines

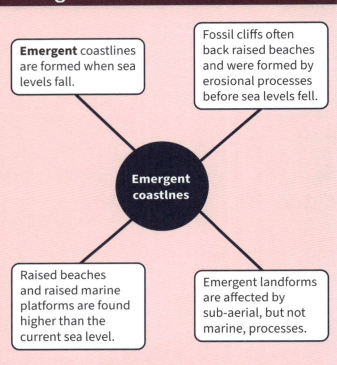

**Emergent** coastlines are formed when sea levels fall.

Fossil cliffs often back raised beaches and were formed by erosional processes before sea levels fell.

**Emergent coastlnes**

Raised beaches and raised marine platforms are found higher than the current sea level.

Emergent landforms are affected by sub-aerial, but not marine, processes.

## Submergent coastlines

- Are formed when sea levels rise.
- **Rias** are formed when river V-valleys are flooded and submerged.
- **Fjords** are formed when glacial U-shaped valleys are flooded and submerged.
- **Dalmatian** coastlines are formed where geological folds run parallel to the sea on a concordant coastline and sea levels rise, flooding the valleys of the folds and leaving narrow islands visible above the sea level.

## Recent climatic change

Contemporary sea level change from global warming is a risk to some coastlines. Most scientists believe that anthropogenic global warming is leading to an increase in global mean sea level. It is predicted that sea level could rise by between 0.43 m and 0.84 m by 2100. Low-lying coastal communities, and Small Island Developing States (SIDS), such as the Maldives, are becoming increasingly vulnerable to the effects of coastal flooding and erosion.

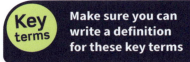
**Key terms** Make sure you can write a definition for these key terms

dalmatian   emergent   eustatic   fjord
isostatic   ria   submergent

# RETRIEVAL

Learn the answers to the questions below, then cover the answers column with a piece of paper and write down as many as you can. Check and repeat.

## Questions / Answers

| | Questions | Answers |
|---|---|---|
| 1 | Name two examples of landforms of erosion. | Two from: cliffs / wave-cut platforms / caves / arches and stacks |
| 2 | Name two landforms that result from cliff retreat. | Wave-cut notch, wave-cut platform |
| 3 | Place these landforms in the order in which they are formed, oldest first; stump, arch, cave, stack. | Cave, arch, stack, stump |
| 4 | Name two factors affecting landscapes of coastal erosion. | Two from: geology / dip of rock / wave energy / human activity |
| 5 | What factors affect the development of beaches? | Wave energy, sediment type, tides and weather |
| 6 | How does a tombolo form? | A spit extends out to reach an island; it joins that island to the mainland |
| 7 | What are xerophytes? | Plants that can survive with little water |
| 8 | What is the difference between eustatic and isostatic change? | Eustatic relates to a global change in the volume of water in the ocean; isostatic relates to a change in the level of the land, with a relative change in sea level |
| 9 | When are emergent coastlines formed? | When sea levels fall |
| 10 | What is a ria? | A drowned V-shaped river valley |

*Put paper here*

## Previous questions

Now go back and use these questions to check your knowledge of previous topics.

## Questions / Answers

| | Questions | Answers |
|---|---|---|
| 1 | What are the three parts of an open coastal system? | Inputs, processes and outputs |
| 2 | Give two processes that influence wave height. | Two from: wind speed / wind duration / wind strength / length of fetch |
| 3 | Name three erosional processes. | Three from: hydraulic action / attrition / wave quarrying / abrasion/corrosion / solution |
| 4 | What causes the tides? | The gravitational pull of the Sun and the moon |
| 5 | Name two ways material can move by mass movement in coastal areas. | Two from: blockfall / rotational slumping / landslides / sediment within surface runoff |

*Put paper here*

# 3 Coastal systems and landscapes

## 3.6 Coastal management

### Human intervention in coastal landscapes

The risks of coastal recession, rising sea levels and coastal flooding means that humans intervene in the coastline to disrupt the natural processes and offer protection to coastal communities around the world.

### Traditional approaches to coastal flood and erosion risk

#### Hard engineering

Hard engineering approaches directly affect the physical coastal processes. They are expensive, but effective at preventing coastal recession and protecting against flooding.

**Hard engineering approaches**

Groynes: structures built at right-angles to the coast that stop longshore drift, trap sediment and build up a beach, BUT can lead to sediment starvation in other areas

Offshore breakwaters: structures in the sea that are parallel to the shore that reduce the power of incoming waves, BUT can reduce the amount of sediment reaching the beach

Sea walls: can be recurved and repel wave energy to protect the coastline from destructive waves, BUT are large and unsightly

Revetments: often wooden frames placed at the back of a beach to absorb wave energy, BUT need frequent maintenance

Rip rap: large boulders, commonly granite or concrete, placed along coastlines to absorb wave energy, BUT are unsightly

#### Soft engineering

Soft engineering approaches involve working with the physical processes and are a more natural form of coastal management.

**Soft engineering approaches**

Beach: adding sediment to a beach to increase its ability to absorb the waves and protect the land behind, BUT sediment is continually transported away so needs constant replenishing

Cliff regrading and drainage: the angle of the cliff is reduced to lower the risk of mass movement and water is drained out, BUT this removes part of the cliff and overextraction of water can increase vulnerability of collapse

Dune stabilisation: planting vegetation on fenced-off areas of sand dunes to hold the dune structure together and enhance the natural barrier along the coastline, BUT this disrupts the natural succession of dunes

Managed retreat: allowing low-lying coastal areas to flood, eventually reverting to their original state, BUT the original land use (farmland) is lost and compensation may need to be paid

# Sustainable approaches to coastal flood and erosion risk

## Shoreline management plans (SMP)

Each of the UK's 11 sediment cells (see 3.1) is managed by a **shoreline management plan** (SMP).

is devised by local councils and the Environment Agency, with input from other organisations

identifies the opportunities to improve the coastal environment, the best approach to defend coastal assets and manage risks, and the consequences of putting the management in place

**Each SMP...**

outlines a sustainable approach to managing the threats to the coastline over 100 years

works with natural processes and allows natural coastal change.

Shoreline management policies include:

### No active intervention
- There are no coastal defences in place

### Strategic realignment (managed retreat)
- Moving the line of defence further towards the land
- Deliberately overtopping current defences

### Hold the line
- Maintaining the line of existing defences
- Improving or maintaining the protection currently in place

### Advance the line
- Building new coastal defences closer to the sea

## Integrated coastal zone management (ICZM)

Many countries use sustainable ICZM strategies to manage extended areas of the coastline within sediment cells.

involves all stakeholders, taking their views and needs into account

ensures that management is long-term, sustainable and allows for economic development

**ICZM...**

ensures that approaches in one area of the cell do not have negative impacts elsewhere

can change as the threats to coastal areas develop.

**Key terms** Make sure you can write a definition for these key terms

integrated coastal zone management (ICZM)
shoreline management plan (SMP)

# RETRIEVAL

Learn the answers to the questions below, then cover the answers column with a piece of paper and write down as many as you can. Check and repeat.

## Questions

## Answers

| | Questions | Answers |
|---|---|---|
| 1 | What are hard engineering approaches? | Constructions that directly affect the physical coastal processes, e.g. sea wall, groynes, rip rap |
| 2 | What is rip rap? | Large boulders placed along coastlines to absorb wave energy |
| 3 | What are revetments? | Frames placed at the back of a beach to absorb wave energy |
| 4 | What are soft engineering approaches? | A more natural form of coastal management, e.g. beach nourishment, cliff regrading, dune stabilisation |
| 5 | What is beach nourishment? | Adding sediment to a beach to increase its ability to absorb the power of the waves |
| 6 | What is dune stabilisation? | Planting vegetation on sand dunes to hold the dune structure together and enhance the natural barrier |
| 7 | What are the four shoreline management policy options? | No active intervention, strategic realignment, hold the line, advance the line |
| 8 | Name two advantages of an ICZM. | Two from: involves all stakeholders / ensures there are no negative impacts elsewhere / is a long-term sustainable strategy / can change as threats develop elsewhere |

*Put paper here*

## Previous questions

Now go back and use these questions to check your knowledge of previous topics.

## Questions

## Answers

| | Questions | Answers |
|---|---|---|
| 1 | Give three characteristics of destructive waves. | Three from: greater in height and frequency / short wavelength / backwash has more energy than the swash / remove material from a beach / decrease the gradient of the beach profile |
| 2 | What does sub-aerial mean? | Processes that occur on land, as opposed to in the sea |
| 3 | What is mass movement? | The downward movement of material, under the influence of gravity |
| 4 | What factors affect the development of beaches? | Wave energy, sediment type, tides and weather |
| 5 | What is the difference between eustatic and isostatic change? | Eustatic relates to a global change in the volume of water in the ocean; isostatic relates to a change in the level of the land, with a relative change in sea level |
| 6 | When are emergent coastlines formed? | When sea levels fall |

*Put paper here*

 **KNOWLEDGE**

## 3 Coastal systems and landscapes

### 3.7 Case study: A coastal environment at a local scale

 **Case study: The Holderness coast, UK**

The Holderness coast in East Yorkshire runs from Flamborough Head in the north to Spurn Head in the south and is the UK's sediment subcell 2a (see 3.1).

#### Coastal processes

The coastline experiences rapid coastal recession, around 2 m a year.

This is due to:

- waves that approach from the north-east – the direction of the longest fetch
- the high energy environment – waves have strong, destructive power
- the geology – predominant rock types are chalk (at Flamborough Head) and boulder clay (a mixture of rock material formed by the deposition of sediment carried by glacier), which are both soft and easily eroded
- currents – rip currents lead to higher rates of erosion
- transportation processes – sediment is carried southwards, towards Spurn Head, by longshore drift, which prevents sediment accumulation on beaches, allowing for high rates of cliff erosion
- the weather – the coastline experiences the stormy weather brought to the UK by mid-latitude low pressure systems.

> **REVISION TIP**
>
> You should revise the case study you have studied. This page provides the example of the Holderness coast.

#### Landscape outcomes

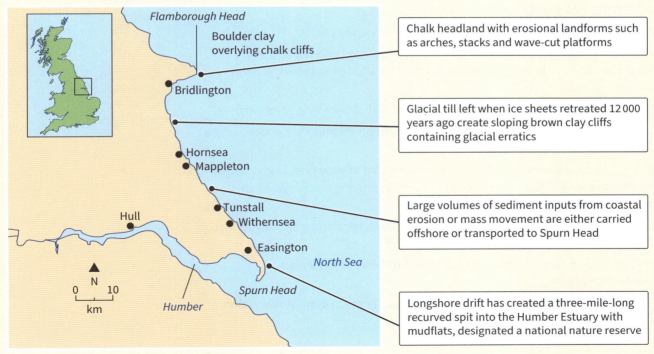

▲ **Figure 1** The landscape features of the Holderness coast

## 3 Coastal systems and landscapes

▲ *Figure 2* *An arch at Flamborough Head*

### Challenges for sustainable management

- Holderness has had a shoreline management plan since 1998.
- The main settlements, Bridlington, Hornsea, Mappleton and Withernsea, are protected by hard engineering, with a combination of sea walls, groynes and rock armour.
- The gas terminal at Easington is also protected with a 1 km-long revetment.
- Other areas of the coastline are not seen as worth protecting after undertaking a cost–benefit analysis.
- A cost–benefit analysis compares the economic cost of each plan (e.g. construction, maintenance), with the economic benefit (e.g. improving tourism, saving productive land).
- New developments near the coastline are prohibited and some existing tourist facilities are being relocated.
- There has been conflict between local **stakeholders** over which areas should be protected.
- Groynes used in some areas of the coastline have led to **sediment starvation** and increased erosion southward.

## 3.8 Case study: A contrasting coastal landscape

 **Case study: The Odisha coastline, north-east India**

The Odisha coastline is around 450 km long and found in north-east India.

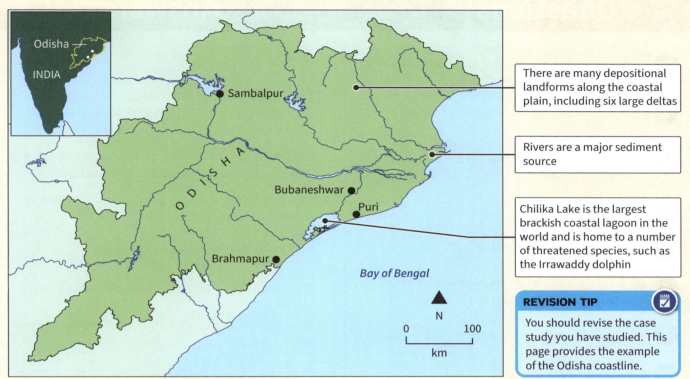

There are many depositional landforms along the coastal plain, including six large deltas

Rivers are a major sediment source

Chilika Lake is the largest brackish coastal lagoon in the world and is home to a number of threatened species, such as the Irrawaddy dolphin

**REVISION TIP**

You should revise the case study you have studied. This page provides the example of the Odisha coastline.

▲ *Figure 1 Odisha coastline on the Bay of Bengal*

### Risks and opportunities for human occupation and development

- The coastline is attractive for human settlement.
- There are large populations dependent on the coastline for resources such as shrimp farming.
- Mangrove ecosystems provide fuelwood, land for reclamation for cultivation and timber.

- Urbanisation, maritime transport, fishing, tourism, mining and offshore oil and gas production have led to resource exploitation.
- There are extreme tidal variations, frequent tropical cyclones and the area is at severe risk of sea level rise.
- Between 1990 and 2015, the shoreline receded 10–15 m/year.

### Responses of resilience, mitigation and adaptation

- The Government of India and the World Bank have created an ICZM to manage the coastline sustainably and balance the needs of all stakeholders.
- The ICZM promotes small-scale ecotourism activities that offer employment for locals and sustainable income.

- Communities are supported to plant and protect mangroves to improve coastal protection and reinstate habitats.
- Marine transport is being regulated and boats are being replaced with vessels that don't use diesel.
- Resettlement communities have been set up to rehome displaced people whose villages have been submerged. However, the success of these has been questioned.

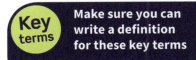 Make sure you can write a definition for these key terms

brackish   sediment starvation   stakeholder

# RETRIEVAL

Learn the answers to the questions below, then cover the answers column with a piece of paper and write down as many as you can. Check and repeat.

## Questions / Answers

| | Questions | Answers |
|---|---|---|
| 1 | What are the main rock types found in Holderness? | Boulder clay and chalk |
| 2 | What coastal protection is in place at Holderness? | The main settlements, are protected by hard engineering, with a combination of sea walls, groynes and rock armour |
| 3 | What has the negative impact of hard engineering been at Holderness? | Groynes used in some areas of the coastline have led to sediment starvation and increased erosion southward |
| 4 | What is a stakeholder? | A person or business with an interest in or who may be affected by something |
| 5 | Where is the region of Odisha? | North-east India on the Bay of Bengal |
| 6 | At what rate was the Odisha coastline found to be receding? | Around 10–15 metres a year between 1990 and 2015 |
| 7 | Which industries have exploited Odisha's resources? | Construction, transport, fishing, tourism, mining and offshore oil and gas production |
| 8 | Who has created the ICZM in Odisha? | The national government and the World Bank |
| 9 | What challenges are being addressed by the ICZM in Odisha? | Tourism, mangrove conservation, transport, resettlement |

*Put paper here*

## Previous questions

Now go back and use these questions to check your knowledge of previous topics.

## Questions / Answers

| | Questions | Answers |
|---|---|---|
| 1 | What is longshore drift? | When sediment is transported along a coastline, in the direction of the prevailing wind |
| 2 | How does a tombolo form? | A spit extends out to reach an island; it joins that island to the mainland |
| 3 | What are hard engineering approaches? | Constructions that directly affect the physical coastal processes, e.g. sea wall, groynes, rip rap |
| 4 | What is beach nourishment? | Adding sediment to a beach to increase its ability to absorb the power of the waves |
| 5 | Name two landforms that result from cliff reatreat. | Wave-cut notch, wave-cut platform |

*Put paper here*

# PRACTICE

## Exam-style questions

1   Outline sources of energy in coastal environments.                                    **[4 marks]**

2   Outline the processes of marine erosion operating at the coastline.   **[4 marks]**

3   Explain the development of wave-cut platforms.                                      **[4 marks]**

4   Outline processes that lead to the development of spits.                       **[4 marks]**

5   Outline factors leading to the development of mudflats
    and saltmarshes.                                                                                  **[4 marks]**

6   Explain the process of isostatic sea level change.                               **[4 marks]**

7   Explain the development of cliff profile features.                                 **[4 marks]**

8   **Figure 1** shows the predicted sea level rise in three different
    future scenarios. Analyse the data shown.                                         **[6 marks]**

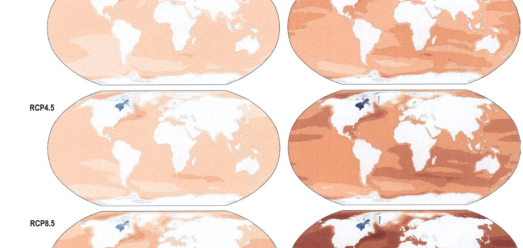

Regional mean sea level change (m)

◀ **Figure 1** Future mean sea level change for three RCPs, 2046–2065 and 2081–2100. RCPs are Representative Concentration Pathways, which are four different scenarios of greenhouse gas concentration that the IPCC use for modelling. RCP 2.6 is where carbon emissions are zero by 2100. RCP 8.5 is where emissions continue to rise and is a worst-case scenario.

9    Evaluate the usefulness of **Figure 2** in presenting data about cliff retreat
     in California.

[6 marks]

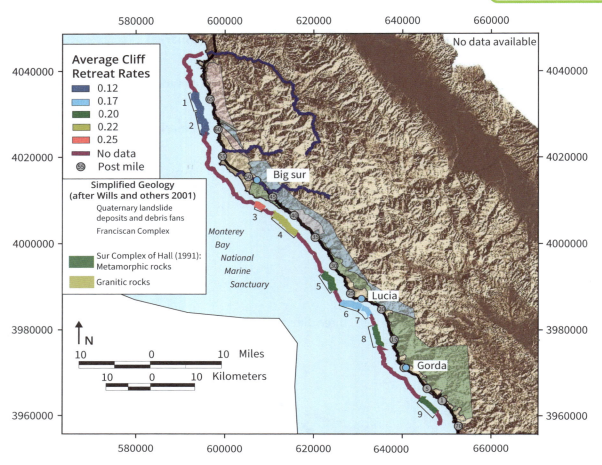

▲ *Figure 2* *The average rate of cliff retreat (in metres per year) at sections of the California coastline over a 52-year period.*

10  Using **Figure 3** and your own knowledge, assess the role of
transportation processes in the development of this landscape. **[6 marks]**

▲ **Figure 3** *Cape Talumphuk in Pak Phanang, Thailand*

11  Using **Figure 4** and your own knowledge, assess the view
that wave energy is the most important factor affecting
the development of this landscape. **[6 marks]**

▲ **Figure 4** *A landscape of coastal erosion, Dorset coastline, UK*

12    Using **Figure 5** and your own knowledge, assess the relative importance of vegetation in the development of this landscape.    **[6 marks]**

▲ **Figure 5** *An estuarine coastal landscape*

13    Using **Figure 6** and your own knowledge, assess the view that soft engineering could protect this coastline from erosion.    **[6 marks]**

▲ **Figure 6** *Cliffs near Kessingland, Suffolk, England, UK*

14   With reference to a coastal environment at a local scale, assess how the geomorphological processes present challenges for its sustainable management.

[20 marks]

15   With reference to a coastal landscape from beyond the UK, assess how integrated coastal zone management can manage the risks for human occupation and development.

[20 marks]

16   Assess the relative importance of different factors responsible for the development of landscapes of coastal deposition.

[20 marks]

## 4 Glacial systems and landscapes

### 4.1 Glacial systems

#### Systems concepts and glacial landscapes

- A glacial system is an open system. Open systems have links to other systems.
- Glacial landscapes are those which are being actively modified, or have been glaciated in the past in upland and lowland regions.
- They are formed by ice masses, either in sheet form across wide expanses, or within upland regions as mountain and valley glaciers.

| System concept | Definition | Examples for glacial systems |
|---|---|---|
| Inputs | The resources, substances or energy that enter a system. | Precipitation (snow, sleet, rain), avalanches, rockfalls, solar radiation |
| Outputs | The resources, substances or energy that leave a system. | Meltwater, calving, evaporation, sublimation |
| Energy | The capacity to do work or cause change within a system. | **Insolation**, potential energy (glacier's mass + gravity), kinetic energy (glacier movement, movement of meltwater), geothermal heat |
| Stores/components | Where energy or matter is kept for a relatively long time in a system. | Snow, ice |
| Flows/ transfers | The movement of energy or matter between stores within a system. | Evaporation, **sublimation**, meltwater flow, internal deformation, basal sliding |
| Positive feedback | Amplifies or reinforces changes within a system. | Ice and snow have a high albedo; warmer temperatures cause melting, exposed surfaces lower albedo; lower albedo means more absorption of Sun's radiation, causing more warming and more melting. |
| Negative feedback | Counteracts or reduces changes within a system, maintaining stability. | As ice-covered areas recede due to global warming, vegetation recolonises the land surface. More vegetation means increased photosynthesis, which takes carbon out of the atmosphere, reducing causes of global warming. |
| Dynamic equilibrium | A state in which a system stays in balance despite continuous change. | Glaciers have a line of equilibrium where the **accumulation zone** (inputs) meets the **ablation zone** (outputs). As inputs/outputs vary, this equilibrium line will move up or down the glacier. |

## The glacial budget

The glacial budget is the balance between inputs and outputs of a glacier.

In winter (and in periods of climate cooling), accumulation exceeds ablation, mass balance is positive and the glacier gains mass.

In summer (and in periods of climate warming), ablation exceeds accumulation, mass balance is negative and the glacier loses mass.

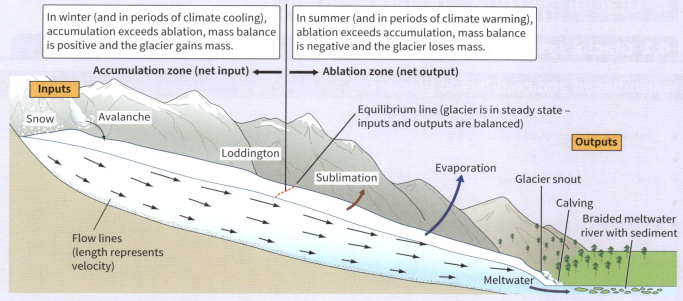

▲ **Figure 1** The glacial system

## Historical patterns of ice advance and retreat

Glaciers advance and retreat in response to long-term trends in mass balance.

- Since the 1980s, most glaciers have retreated and some have disappeared. This is due to global temperature increases.
- During the 1550–1850 period of cooling (the 'Little Ice Age') in Europe and North America, Alpine glaciers advanced.
- Terminal moraine is created as the snout advances. As a glacier retreats, the moraine is left behind, indicating the position of past advances.

### Pleistocene glaciations

The Pleistocene epoch lasted from 2.6 million years ago until 11,700 years ago.

- Glacial periods (ice ages) were interspersed with interglacials (warmer periods).
- Huge ice sheets spread over large parts of Europe, North America and Asia.
- Glaciers were often much larger (e.g. in the Andes) or part of a continous ice sheet (such as the Scandinavian ice sheet or the Alpine ice sheet).

▶ **Figure 2** The position of the Franz Josef glacier in New Zealand in 2014, compared to 1600 and 1800. The glacier has retreated by 3 km since the 1800s.

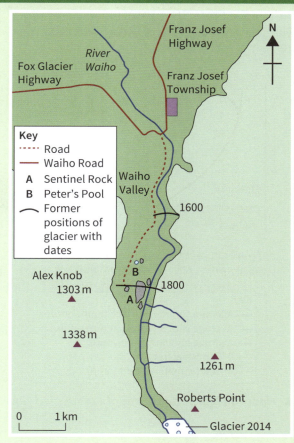

## 4 Glacial systems and landscapes

### 4.1 Glacial systems

## Warm-based and cold-based glaciers

The temperature profiles of glaciers are affected by different factors:

- the temperature of their environment
- heat generated by friction as they move
- geothermal heat from below
- the release of latent heat as meltwater refreezes
- the pressure melting point (PMP): ice melts at temperatures lower than 0°C at the base of the glacier due to the weight of ice above.

Some large glaciers are a mix, with sections where the ice is thickest (ice melts at a lower temperature) or where it is warmed by geothermal heat.

### Cold-based glaciers

In polar climates, surface temperatures are too cold throughout the year to produce meltwater, and the temperature at the base of the glacier remains well below its PMP. The glacier remains frozen to the bedrock.

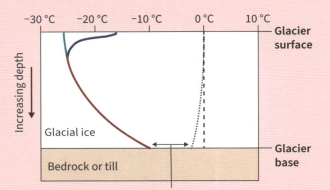

Temperature at the base is **well below** pmp so no melting at base and no basal sliding.

**Key**
— Average summer surface temperatures
— Average winter surface temperatures
— Average annual temperature
- - - 0°C (normal freezing/melting point)
······ Pressure melting point (pmp)

▲ *Figure 3 Temperature profile of a cold-based glacier*

### Warm-based glaciers

Warm-based glaciers are at lower latitudes but higher altitudes where the average temperature at the base is at or near the PMP. Ice melts at the base, allowing the glacier to 'slide'. Above zero surface summer temperatures produce meltwater, which travels to the base through cracks in the ice.

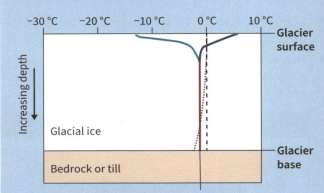

Temperature at the base is **higher** than pmp so melting will occur and this allows basal sliding.

**Key**
— Average summer surface temperatures
— Average winter surface temperatures
— Average annual temperature
- - - 0°C (normal freezing/melting point)
······ Pressure melting point (pmp)

▲ *Figure 4 Temperature profile of a warm-based glacier*

 **Key terms** — Make sure you can write a definition for these key terms

ablation zone    accumulation zone    cold-based glacier
insolation    sublimation    warm-based glacier

## 4.2 Cold environments

### The global distribution of cold environments

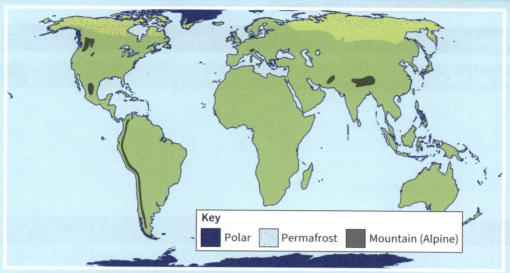

**Key**
■ Polar    ▨ Permafrost    ■ Mountain (Alpine)

◀ **Figure 1** The present-day global distribution of cold environments

- Most cold environments are in high latitudes (close to the poles) where sunlight is at a more oblique angle. There are also long periods of darkness.

- The northern hemisphere has more land mass at high latitudes than the southern hemisphere. Land cools down faster than water, so cold temperatures persist for longer.

- Altitude affects temperature, so mountains tend to have cold environments with increased precipitation (snow).

### Pleistocene glaciations

The Pleistocene epoch lasted from 2.6 million years ago until 11,700 years ago. Glacial periods (ice ages) were interspersed with interglacials (warmer periods).

- Huge ice sheets spread over large parts of Europe, North America and Asia.

- Extensive glaciers and ice caps in South America and in mountainous regions. Glaciers were often much larger or were part of a continuous ice sheet, such as the Scandinavian ice sheet or the Alpine ice sheet.

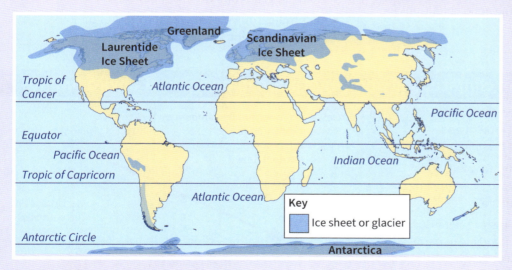

**Key**
■ Ice sheet or glacier

▲ **Figure 2** The approximate distribution of ice sheets 20,000 years ago

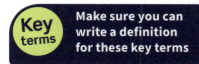

**Make sure you can write a definition for these key terms**

active layer    periglacial

 **KNOWLEDGE**

## 4 Glacial systems and landscapes

### 4.2 Cold environments

## Physical characteristics of cold environments

### Climate

- Long winters with temperatures significantly below freezing. Short, cool summers.
- Precipitation (mainly snow) varies significantly: some cold areas have high rates of snowfall in winter, others (e.g. Antarctica) have very little.
- Strong winds adding wind chill effect to temperatures.

### Soils

- Cold environments typically have ice or bare rock, no soils.
- The lack of water restricts soil development. Weathering rates are low, decomposition happens slowly and vegetation is sparse. Soils are thin, acidic and nutrient-poor.

- Permafrost is a layer of soil or sediment that is frozen for two or more years. It can be several metres thick. There is no nutrient cycling within permanently frozen ground.
- Plants can grow in the **active layer**, but the impermeable permafrost below means the active layer is often saturated with water in summer.

### Vegetation

#### Polar regions and glacial environments

- Little vegetation, though bare rock surfaces may be colonised by lichens

#### Alpine environments

- Warm, wet summers, creating good conditions for vegetation growth

#### Periglacial (tundra) environments

- Sparse, low-growing mosses, lichens, grasses and small shrubs
- Plants have small, waxy leaves that reduce water loss, and can cope with the poor soils
- Reproduce quickly to make the most of the short, warmer growing season

## Interactions between climate, soil and vegetation in periglacial environments

- The cold temperatures, short days and lack of liquid water inhibit plant growth and decomposition, so soils are low in nutrients.
- Plants that survive are highly adapted, and vulnerable to change. They are slow-growing, so slow to recover from damage.
- Waterlogged and frozen soils store significant amounts of carbon because organic material has not fully decomposed. On thawing or drying out, there is potential for large releases of greenhouse gases as the organic material decomposes.

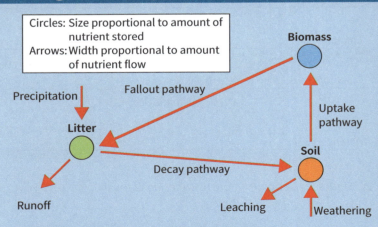

▲ **Figure 3** *Nutrient cycle in a periglacial (tundra) environment*

 **RETRIEVAL**

Learn the answers to the questions below, then cover the answers column with a piece of paper and write down as many as you can. Check and repeat.

## Questions | Answers

| # | Question | Answer |
|---|----------|--------|
| 1 | Describe the global distribution of cold environments. | At high latitudes/near the poles, primarily located in the extensive land masses of the northern hemisphere, and in mountain areas |
| 2 | Why are cold environments more extensive in the northern hemisphere than in the southern hemisphere? | Land cools down faster than water so cold temperatures persist for longer in the northern hemisphere where there are extensive land masses |
| 3 | The Pleistocene epoch began approximately 2.6 million years ago until how many years ago? | 11,700 years ago |
| 4 | The Pleistocene epoch saw alternating colder periods (ice ages) and warmer periods. What is the geographical term for these periods? | Glacials and interglacials |
| 5 | Name the ice sheet that covered the UK at the last glacial maximum, approximately 20,000 years ago. | The Scandinavian ice sheet |
| 6 | What is the active layer in a periglacial environment? | A layer of soil above permafrost that thaws in summer and refreezes in winter |
| 7 | Name a periglacial environment that, unusually for a cold environment, has high biodiversity in summer due to warm, wet conditions? | Alpine |
| 8 | Tundra vegetation typically flowers and sets seed rapidly as temperatures increase in spring. Why is this? | To maximise the time available in the very short growing season |
| 9 | Why are tundra soils low in nutrients? | Cold temperatures, short days and lack of liquid water inhibit plant growth and decomposition |
| 10 | Explain why tundra soils often store significant amounts of carbon. | They contain large amounts of undecomposed or semi-decomposed organic material because decomposition is slow or absent in waterlogged (no oxygen) or permanently frozen soils |
| 11 | Is a glacial system an open system or a closed system? | Open system |
| 12 | What is insolation? | Direct impact of heat from the sun |
| 13 | What is the area of net output in a glacial system called? | The ablation zone |
| 14 | What is the equilibrium line of a glacial system? | The point on a glacier where inputs and outputs are balanced |
| 15 | Which part of a glacier is used to measure advance and retreat? | The position of the snout |

Put paper here

# 4 Glacial systems and landscapes

## 4.3 Glacial processes

### Freeze–thaw weathering (frost action)

Freeze–thaw weathering is the dominant weathering process. In periglacial environments, **ice wedges** form where water in cracks in soil or sediment are exposed to the freeze–thaw cycle. Over many years the frozen ice wedge may extend 10 m into the soil and be 3 m wide at the surface.

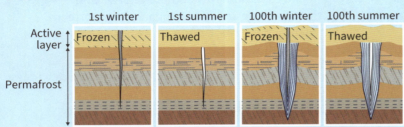

◀ **Figure 1** The process and formation of ice wedges

### Carbonation

- Carbon dioxide ($CO_2$) is more soluble at colder temperatures.
- $CO_2$ dissolves in water to form a weak carbonic acid.
- This reacts with calcium carbonate in rocks such as limestone to form calcium bicarbonate.
- Calcium bicarbonate is soluble in water and so is easily eroded and transported away by water.

### Nivation

**Nivation** is a weathering and erosional process associated with snow patches.

- Persistent snow patches accumulate in depressions or sheltered areas of the landscape and are compacted.
- Weathering and erosional processes lead to the formation of nivation hollows, which can become corries if climate cooling allows ice accumulation.

### Mass movement

In addition to rock falls, landslides and slumps, in cold environments material moves by **solifluction** and **frost creep**.

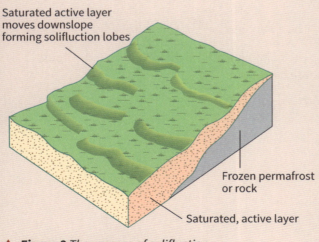

▲ **Figure 2** The process of solifluction

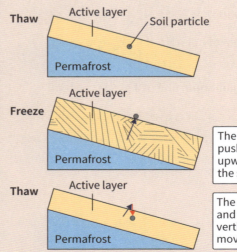

▲ **Figure 3** The process of frost creep

# Internal deformation and basal sliding

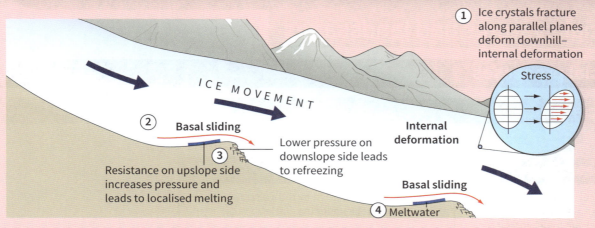

▲ **Figure 4** *The process of internal deformation and basal sliding*

| Internal deformation | Basal sliding |
|---|---|
| • ① Under great pressure at the base of a glacier, the ice crystals align themselves to the direction of movement of the glacier and move (1 or 2 cm per year). <br><br>• Movement is most likely to take place near the bed where the pressures are highest. <br><br>• Occurs in both warm- and cold-based glaciers. <br><br>• It is the only movement process in cold-based glaciers. | • ② In warm-based glaciers, meltwater at the base acts as a lubricant, allowing movement of typically 1–2 m per day, up to several kilometres a year. <br><br>• ③ Where the base of the glacier meets an obstacle, the increased pressure may cause melting on the upslope side, enabling basal sliding. The ice may then refreeze on the downslope where pressure is reduced. <br><br>• Friction as a glacier moves downslope can also increase melting at the base. <br><br>• ④ Meltwater channels and saturated sediment at the base of the glacier (till) act as a lubricant. <br><br>• Seasonal changes in temperature, for example creating more meltwater in summer, can lead to glacial surges. |

# Extensional and compressional flow

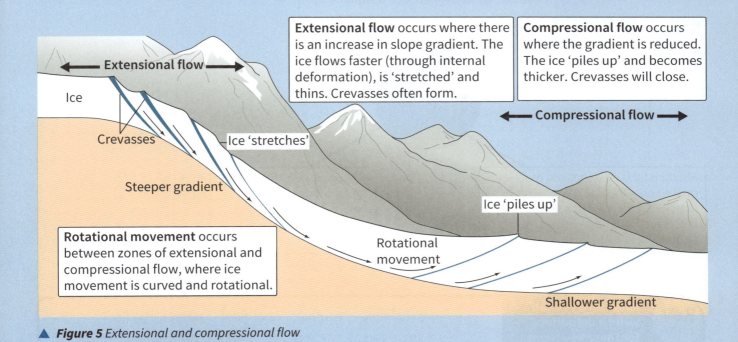

▲ **Figure 5** *Extensional and compressional flow*

## 4 Glacial systems and landscapes

### 4.3 Glacial processes

## Fluvioglacial processes

### Erosion

| Plucking |
|---|
| • Plucking occurs when subglacial water freezes against rock and pieces of rock are 'plucked' out of the ground when the glacier moves on. |
| • Freeze–thaw weakens rock, making it more easily eroded by plucking. |

| Abrasion |
|---|
| • Rock debris carried by the ice scrapes scours the bedrock under the glacier. |
| • Course debris can scratch bedrock, leaving striations. Finer debris can leave smooth and polished surfaces. |
| • The debris itself can also be worn down to fine particles (rock flour). |

### Transportation

There are three main ways in which glaciers transport debris:

- Supraglacial – lateral and medial moraine which can be washed into the glacier by meltwater.

- Englacial – debris carried in the glacier; moraine that has been washed into the glacier or buried by snowfall.

- Subglacial – debris under the glacier carried by the ice or by subglacial meltwater.

### Deposition

Deposition takes place in the ablation zone. Glacial till is unsorted, but stones and pebbles may be aligned with the direction of ice flow.

**Deposition**

Ablation till is till deposited as ice melts. Lodgement till is deposited material at the base of a glacier and is pressed into/plastered onto the underlying bed.

Meltwater may then transport debris further before depositing it as sorted fluvioglacial material.

 **Key terms** Make sure you can write a definition for these key terms

basal sliding   compressional flow   extensional flow
frost creep   ice wedges   internal deformation
nivation   solifluction   rotational movement

# RETRIEVAL

Learn the answers to the questions below, then cover the answers column with a piece of paper and write down as many as you can. Check and repeat.

| | Questions | Answers |
|---|---|---|
| 1 | Carbonation is much more important as a weathering process in cold environments than in hot ones. Why is this? | Carbon dioxide is much more soluble in water at cold temperatures than at hot temperatures |
| 2 | What is the name for the weathering and erosion processes associated with persistent snow patches? | Nivation |
| 3 | What type of ice movement accounts for movement in cold-based glaciers? | Internal deformation |
| 4 | What is the speed of movement in cold-based glaciers? | 1–2 cm per year |
| 5 | Basal sliding can result at the base of a glacier when the glacier meets an obstacle. Describe this process. | The increased pressure causes melting (due to lower PMP) on the upslope side of the obstacle, enabling basal sliding |
| 6 | When and why does extensional flow occur in a glacier? | Extensional flow occurs where there is a sudden increase in slope gradient; the ice flows faster and is 'stretched' and thins |
| 7 | In between zones of extensional and compressional flow, the ice moves in a shallow curve as though around a pivot. What is this called? | Rotational movement |
| 8 | Outline the process of plucking. | Subglacial water freezes against rock; as the glacial ice moves under the force of gravity, pieces of (weathered) rock are 'plucked' out of the ground |
| 9 | How are striations produced? | These are scratches on bedrock made by the abrasion of rocks being dragged over the bedrock at the glacier base |
| 10 | What is the term for debris transported inside the glacier? | Englacial debris |

Put paper here

## Previous questions

Now go back and use these questions to check your knowledge of previous topics.

| | Questions | Answers |
|---|---|---|
| 1 | The Pleistocene epoch began approximately 2.6 million years ago until how many years ago? | 11,700 years ago |
| 2 | What is the area of net output in a glacial system called? | The ablation zone |
| 3 | Why are cold environments more extensive in the northern hemisphere than in the southern hemisphere? | Land cools down faster than water so cold temperatures persist for longer in the northern hemisphere where there are extensive land masses |
| 4 | The Pleistocene epoch saw alternating colder periods (ice ages) and warmer periods. What is the geographical term for these periods? | Glacials and interglacials |
| 5 | What is insolation? | Direct impact of heat from the sun |

Put paper here

## 4 Glacial systems and landscapes

### 4.4 Glacial landforms

## Origin and development of glaciated landscapes

The development of glacial and periglacial landforms involves the interaction of several factors.

Sources of energy: insolation, potential and kinetic energy, geothermal heat

Time: historic patterns of advance and retreat

Glacial budget: ablation and accumulation

**Factors contributing to the formation of glacial and periglacial landforms**

Ice movement and glacier type: cold-based, warm-based

Geology: influence of rock type, tectonics

Weathering processes: frost action and nivation

Periglacial processes: permafrost, active layer and mass movement

Fluvioglacial processes: meltwater, erosion, transportation and deposition

Glacial processes: erosion, transportation and deposition

**REVISION TIP**

When writing about the development of a specific landform or landscape (i.e. shown in a photo on the exam paper), consider the relative importance of factors or their inter-relationship.

## Landscapes of glacial erosion

Red Tarn – a corrie lake

Corrie – a scooped out hollow in the landscape

Striding Edge – a classic example of an arête

Glacial troughs or U-shaped valleys

▲ *Figure 1 The Lake District; a landscape of glacial erosion*

# Landforms of glacial erosion

## Corries

A **corrie** is formed in nivation hollows, with rotational movement, basal sliding, freeze-thaw and compressional flow being some of the processes involved in its development.

## Arêtes

An **arête** is a narrow ridge between two corries. More resistant rocks may also be a factor, resisting the weathering processes that would otherwise reduce the height and steepness of the arête.

## Glacial troughs

A **glacial trough** is a steep-sided, largely flat-bottomed U-shaped valley. They are the remnants of glacial modification of a v-shaped river valley. A ribbon lake may develop.

## Hanging valleys

A **hanging valley** is a tributary U-shaped valley that is higher than the main valley (glacial trough) it feeds into. During glaciation the tributary glacier had less power than the main glacier so erosion was less deep, leaving it 'hanging' above the main valley, often with a waterfall.

## Truncated spurs

**Truncated spurs** are created when a glacier cuts through a valley with interlocking spurs, creating a straight glacial trough often with hanging valleys.

▲ **Figure 2** *Truncated spurs: The Three Sisters, Glencoe*

## Roches moutonnées

**Roches moutonnées** have a smooth stoss (upstream) side where pressure is increased when sliding over an obstruction, allowing basal sliding and abrasion to take place. The craggy (downstream) side is where pressure was released, ice froze and plucking took place.

## 4 Glacial systems and landscapes

### 4.4 Glacial landforms

## Landforms of glacial deposition

Glacial deposition is linked to fluvioglacial deposition, but it is distinct from it.

**Moraine** is unsorted rock debris comprising materials of different shapes and sizes. Moraine as a term refers to both the debris transported by a glacier (supraglacial lateral and medial moraine, englacial and subglacial moraine), but also to the landforms created by the direct deposition of debris by ice.

| | Description | Development factors |
|---|---|---|
| **Drumlins** | **Drumlins** are egg-shaped (oval) hills. They have a streamlined appearance.<br><br>Drumlins vary in height from between 5 m and 50 m, and in length from 10 m to 300 m.<br><br>Drumlins have a stoss end and a lee end: the stoss end is steep and blunt, the lee end is tapered with a gentle slope.<br><br>Some drumlins appear more elongated than others. This indicates speed of ice movement: the more elongated the drumlin is, the faster the ice was moving.<br><br>Drumlins commonly occur in clusters on the flat floors of glaciated valleys. These clusters are known as swarms. | Drumlins are formed beneath glaciers and are aligned in the direction of ice flow.<br><br>The stoss end points in the upstream direction of flow, and the lee end points downstream.<br><br>Some drumlins have rocky cores, suggesting that till lodged against an obstruction and was then shaped into a drumlin as the ice moved over it.<br><br>Another theory is that meltwater may create hollows in the base of the glacier, which then mould subglacial till into drumlins.<br><br>Or glaciers overloaded with debris deposit this when they exit an upland area, and then shape the debris as they move over it. |
| **Erratics** | A large boulder or rock fragment that has been transported and deposited by a glacier, often over great distances, and then left in a location where it is clearly different from the underlying local bedrock/lithology.<br><br>**Erratics** can provide valuable clues about the direction and extent of past glacial movement. | As a glacier moves over the landscape, it can uplift and entrain rocks and boulders from the bedrock below.<br><br>These rocks are then transported within the glacier, either englacially or supraglacially.<br><br>As the glacier melts or retreats, it can no longer support the rock's weight and it is deposited. |
| **Till plains** | **Till plains** are extensive flat plains, though some have a gently rolling topography. They are composed of a mixture of unsorted clay, silt, sand, gravel, and boulders.<br><br>Clasts in the till orientate to point in the direction of ice movement. | Till plains are formed when a large sheet of ice becomes detached from a glacier.<br><br>As it melts, it deposits its till to level out the topography.<br><br>Hummocky till plains may be more associated with lodgement till than ablation till. |

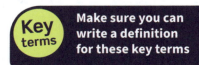

**Key terms** Make sure you can write a definition for these key terms

arête   corrie   drumlin   erratic   glacial trough   hanging valley
moraine   roches moutonnée   till plain   truncated spur

# Types of moraine

| | Description | Development factors |
|---|---|---|
| **Ground moraine** | Irregular, hummocky ground made of till which can be several metres thick. | This is sediment transported underneath a glacier which has been smeared across the bedrock. |
| **Terminal moraine** | A curved ridge of moraine, perpendicular to the ice flow, that marks the furthest extent reached by the glacier.<br>The upstream side is usually steeper than the downstream. | The debris is deposited by ablation of ice at the snout.<br>The ice supports the debris on the upstream side, explaining why it is steeper. |
| **Recessional moraine** | A ridge or series of ridges located behind terminal moraine, perpendicular to the ice flow, but which do not mark the glacier's furthest extent. | Recessional moraines are deposited by ablation during stationary periods in the glacier's recession when debris can accumulate at the snout. |
| **Lateral moraine** | Parallel ridges of glacial till and rock debris along valley sides.<br>The ridges appear almost symmetrical and can be tens of metres high. | Lateral moraine is largely formed by the build-up of screen slopes as a result of frost-shattering.<br>Lateral moraines are evidence of a glacier's limited ability to move debris inwards from its sides. |
| **Medial moraine** | A low ridge of unsorted till that is parallel to the direction in which the ice flowed.<br>Where a glacier had more than one medial moraine, a series of ridges can be found, sometimes affecting drainage in the valley.<br>Medial moraines are often disrupted by glacial advances and retreats, being reformed into terminal or recessional moraines. | A medial moraine is formed when two glaciers with lateral moraines converge. The inner lateral moraines of each glacier (those facing each other) combine to form a single medial moraine.<br>If more tributary glaciers join the main glacier, multiple medial moraines can be created. |

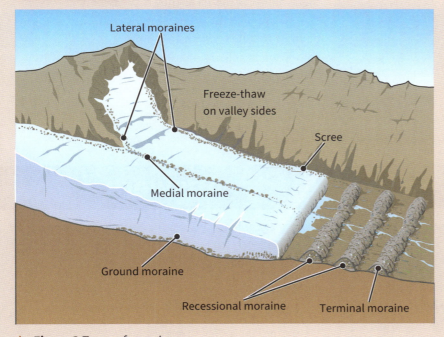

▲ *Figure 3* Types of moraine

# ⇄ RETRIEVAL

Learn the answers to the questions below, then cover the answers column with a piece of paper and write down as many as you can. Check and repeat.

## Questions | Answers

| | Questions | Answers |
|---|---|---|
| 1 | What type of ice movement contributes to the deepening of a corrie? | Rotational |
| 2 | Where are corries formed? | In nivation hollows |
| 3 | Name this landform of glacial erosion: a narrow, knife-edge ridge between two corries. | Arête |
| 4 | Name this landform: a tributary valley that is higher than the main valley it feeds into. | Hanging valley |
| 5 | Outline the formation of a truncated spur by contrasting the erosive power of rivers and glaciers. | A river has less erosive power than a glacier, so while a river winds between interlocking spurs, which may be more resistant to erosion, a glacier erodes through the spurs, truncating them |
| 6 | In what direction does the blunt, steeper stoss end of a roche moutonnée point: upstream or downstream? | Upstream (the stoss side is the side that faced the advancing glacier) |
| 7 | Which process of erosion produces the craggy lee end of a roche moutonnée? | Plucking |
| 8 | One theory of drumlin formation involves till lodging against an obstacle and being reshaped. Describe another theory. | One from: meltwater may create hollows in the base of the glacier that then mould subglacial till into drumlins / glaciers overloaded with debris deposit this when they exit an upland area, and then shape the debris as they move over it |
| 9 | What identifies a boulder in a glaciated landscape as being an erratic? | The rock type of the erratic is different from the bedrock that it is now resting on/above |
| 10 | Name four types of moraine landforms. | Four from: terminal / recessional / lateral / medial / ground |

*Put paper here* (repeated in divider column)

## Previous questions

Now go back and use these questions to check your knowledge of previous topics.

## Questions | Answers

| | Questions | Answers |
|---|---|---|
| 1 | What is the speed of movement in cold-based glaciers? | 1–2 cm per year |
| 2 | Name a periglacial environment that, unusually for a cold environment, has high biodiversity in summer due to warm, wet conditions? | Alpine |
| 3 | Name the ice sheet that covered the UK at the last glacial maximum, approximately 20,000 years ago. | The Scandinavian ice sheet |
| 4 | Carbonation is much more important as a weathering process in cold environments than in hot ones. Why is that? | Carbon dioxide is much more soluble in water at cold temperatures than at hot temperatures |
| 5 | When and why does extensional flow occur in a glacier? | Extensional flow occurs where there is a sudden increase in slope gradient; the ice flows faster and is 'stretched' and thins |

*Put paper here* (repeated in divider column)

# 4 Glacial systems and landscapes

## 4.5 Fluvioglacial and periglacial landforms

## Fluvioglacial landforms

Fluvioglacial deposits are distinct from glacial deposits because fluvioglacial deposits are generally well sorted according to size and weight.

### Meltwater channels

**Meltwater channels** are steep-sided channels carved into the bedrock or sediments created by subglacial streams beneath a glacier. These could even flow uphill because of the hydrostatic pressure of the glacier, which along with abundant debris, gave them significant erosive power. They may or may not be occupied by a river.

### Kames

**Kames** are formed by deposits in depressions or crevasses in the ice which leave steep-sided, conical-looking hills or ridges (typically between 10 m and 50 m high) of stratified sand and gravel when the ice melts. A kame delta is created from deposits of material in a marginal lake.

Kame terraces are the remnants of sediment infill from lakes alongside a glacier.

### Eskers

**Eskers** are long sinuous ridges stretching for up to 400 km, created by sediment from subglacial rivers. They can sometimes go uphill because of hydrostatic pressure.

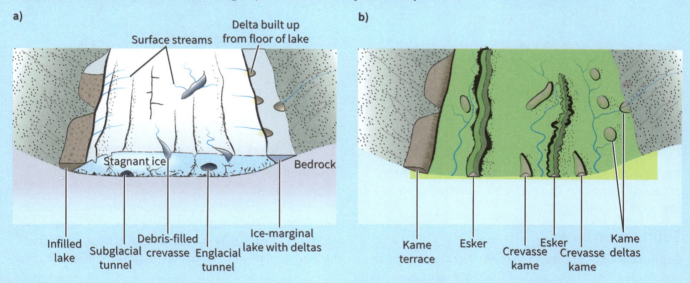

▲ *Figure 1 Formation of eskers and kames in* (**a**) *glacial landscapes and* (**b**) *post-glacial landscapes*

## Outwash plains

An **outwash plain** is an extensive, gently-sloping expanse of stratified sand and gravel that forms in front of the glacier snout (proglacial). Coarse material is close to the snout and finer sediment further away.

# 4 Glacial systems and landscapes

## 4.5 Fluvioglacial and periglacial landforms

## Periglacial landforms

Periglacial landscapes can be active or relict, and consist of landforms formed around the margins of glaciers and ice sheets. The combination of freezing and thawing creates the distinctive features and landforms.

### Ice wedges and patterned ground

Ice wedges are vertical wedges up to 10 m deep and 1 m across (see 4.3). Repeated freezing and thawing widen and deepen cracks and also raise the surface, moving larger stones to the edge and leaving fine sediment in the centre.

Ice wedges often develop into **patterned ground** – polygons, circles, steps or nets. On steeper ground the shapes are elongated and could become stripes.

### Pingos

**Pingos** are domed hills with an ice core. Closed-system pingos are thought to form in areas of continuous permafrost where a lake has dried up, but there is underlying saturated sediment. Open-system pingos occur where water can filter into the ground, and the top layer of the groundwater, near the surface, freezes in winter.

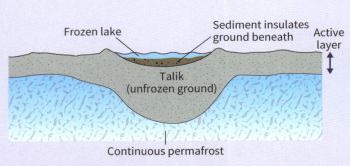

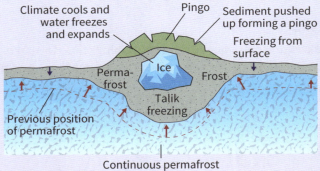

▲ **Figure 2** *Formation of a closed pingo*

### Blockfields

**Blockfields** are extensive areas of angular fragments of rock and boulders caused by frost shattering and chemical weathering.

### Solifluction lobes

**Solifluction lobes** often have a semi-circular or tongue-like shape. They are formed by the movement downhill of saturated soil, which cannot drain because of the frozen permafrost beneath (see 4.3).

### Terracettes

**Terracettes** are parallel ridges on vegetated slopes. Thought to be formed by soil moving downhill by solifluction, and then getting trapped by obstacles, when frost heave forms steps. Alternatively, water may be forced to move laterally if it meets an impermeable layer, forming steps. Animal trampling may also contribute.

### Thermokast

**Thermokarst** landscapes are characterised by depressions, mounds, and sinkholes, irregular-shaped ponds or lakes and also gullies, especially near riverbanks or coastlines. Melting permafrost or ice wedges create voids that are filled by water that retains heat, leading to further permafrost degradation – a positive feedback loop.

## 4.6 Human impacts on fragile cold environments

### Impacts of climate change in cold environments

Climate change in cold environments is having and will continue to have significant impacts, creating both threats and opportunities. Each impact can be a threat to some people, but an opportunity to others.

| Environmental impacts | Economic impacts | Social impacts | Political impacts |
|---|---|---|---|
| • Glacial retreat: glacial meltwater contributes to river regimes; when mountain glaciers disappear, freshwater availability will reduce in dry seasons.<br>• Sea level rise: melting glaciers and ice sheets contribute to sea level rise: the global average increase is 98 mm since 1993 and over 200 mm since 1900.<br>• Loss of biodiversity: the Arctic is expected to be ice-free in summer by 2030 and the high Arctic biome may disappear completely. Extinction of highly-adapted species that have nowhere else to migrate to. Amount and depth of permafrost is reducing in periglacial environments. | • Tourism decline: less predicable snowfall makes Alpine tourism a less reliable source of income.<br>• Resource accessibility: melting ice opens new areas for exploitation, e.g. fossil fuel extraction. Resource exploitation can also have environmental impacts.<br>• Infrastructure costs: thawing permafrost can damage buildings, roads, pipelines. | • Indigenous peoples: climate change threatens traditional ways of life.<br>• Migration: sea level rises and habitat changes will displace populations around the world.<br>• Water insecurity: more than one billion people depend for fresh water on rivers fed by Himalayan glaciers. | • Territorial disputes: Arctic countries are launching conflicting claims over Arctic territories as new resources and routes become available.<br>• Policy changes: international efforts to combat climate change involve major challenges that have promoted opposition with consequences for national politics. |

### Positive feedback loops

Global warming can create positive feedback loops which will intensify the rate of warming and the impacts of climate change.

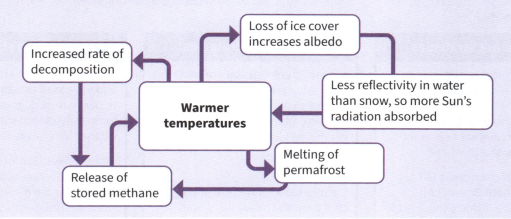

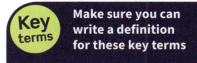

**Key terms**

Make sure you can write a definition for these key terms

blockfield   esker   kame   meltwater channel   outwash plain
patterned ground   pingo   solifluction lobe
terracette   thermokarst

# KNOWLEDGE

## 4 Glacial systems and landscapes

### 4.6 Human impacts on fragile cold environments

## Environmental fragility

Cold environments are fragile and take a long time to recover from damage. If damage sets off positive feedback loops it may be irreversible. Plants and animals that have adapted are highly specialised, making them very sensitive to change.

## Human impacts on cold environments

| Micro-scale | Meso-scale | Macro-scale |
|---|---|---|
| Trampling: tourists walking on fragile vegetation and soil structures<br><br>Local pollution: e.g. rubbish and waste from tourist facilities; fuel spillage from snowmobiles used in tourism and leisure<br><br>Infrastructure damage: buildings, pipelines, roads cause localised disruption to permafrost; localised changes to albedo | Resource extraction: deforestation and pollution from mining or drilling; habitat destruction, e.g. oil spills<br><br>Road building: habitat fragmentation and disruption of migration routes; increases accessibility for further exploitation<br><br>Overfishing: leading to changes in marine food chains | Greenhouse gas emissions: leads to global warming, which disproportionately affects polar regions and alpine areas<br><br>Ozone depletion: release of CFCs leads to ozone layer thinning, and increased UV radiation levels; harmful to organisms<br><br>Global transportation: shipping routes through previously inaccessible areas affecting marine ecosystems and increasing the threat of pollution and oil spills |

## Management of cold environments

Responses to climate change in cold environments include resilience, mitigation and adaptation.

- Resilience strategies aim to strengthen the capacity of ecosystems and communities to withstand climate change and recover from it.
- Mitigation strategies focus on reducing the underlying causes of climate change and the severity of its impacts.
- Adaptation strategies try to help communities adapt to the new conditions brought about by climate change.

| Resilience strategies | Mitigation strategies | Adaptation strategies |
|---|---|---|
| Greenland's ice monitoring: a network of monitoring stations and satellites track changes in the ice sheet's mass and movement. This helps scientists to predict and plan for sea-level rise.<br><br>Marine protection areas (MPAs): in the Southern Ocean, krill populations are declining due to overfishing and warming ocean temperatures. Marine reserves like the Ross Sea MPA are being established and fishing quotas lowered to protect the ocean food web. | Global agreements on emissions: The Paris Agreement (2015) commits countries to reduce carbon emissions with the aim of limiting global temperature increases to 1.5°C by the end of the century.<br><br>Cleaner fuels for Arctic shipping: soot from the heavy fuels used in shipping has coated Arctic sea ice, increasing melting rates. A voluntary agreement by the International Maritime Organization encourages ships to switch to cleaner fuels. | Village relocation: sea level rises and increased coastal erosion due to permafrost degradation have forced several coastal Alaskan villages to relocate inland.<br><br>Alpine diversification: Ski resorts facing shorter seasons and thinner snow cover are diversifying into warmer-climate tourist activities.<br><br>Water management: as glaciers retreat in the Andes, villages that rely on glacial meltwater are constructing reservoirs and developing water conservation measures. |

# 4.7 Case study: A glaciated environment

## Case study: Helvellyn — a landscape of glacial erosion

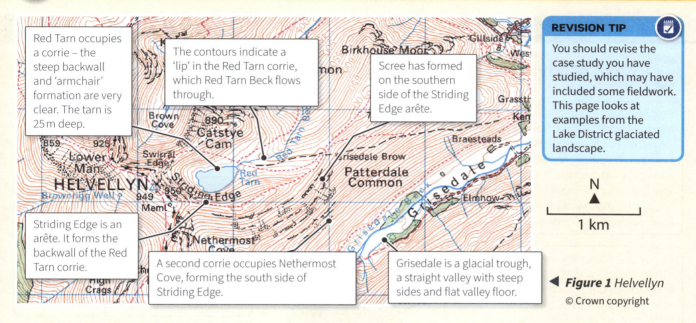

Red Tarn occupies a corrie – the steep backwall and 'armchair' formation are very clear. The tarn is 25 m deep.

The contours indicate a 'lip' in the Red Tarn corrie, which Red Tarn Beck flows through.

Scree has formed on the southern side of the Striding Edge arête.

Striding Edge is an arête. It forms the backwall of the Red Tarn corrie.

A second corrie occupies Nethermost Cove, forming the south side of Striding Edge.

Grisedale is a glacial trough, a straight valley with steep sides and flat valley floor.

**◄ Figure 1** Helvellyn
© Crown copyright

## Fieldwork methods

Fieldwork was carried out in Grisedale using till fabric analysis on exposed areas of till to determine the direction of the glacier.

This method is based on the assumption that, as the glacier moves, it will align larger *clasts* (rock fragments) in the direction of ice movement, so that their long axis (*a*-axis) is parallel to the direction of flow. Students used quadrats to randomly select sample sites from areas where till is exposed, and then measured clast orientation using a compass. The long side of the compass was placed along the *a*-axis, and the compass bearing read off. Clasts were left in situ. A rose diagram was then used to present the data, with 0° as compass direction north. The concentric rings represent the number of clasts in each segment of 15°.

The results indicated a clear direction of flow either to the north-east or to the south-west. Map analysis showed that north-east was the more likely.

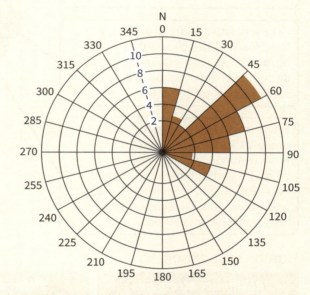

**▲ Figure 3** Till fabric analysis results

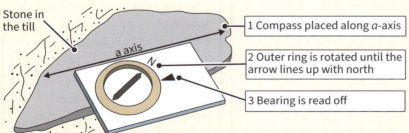

1 Compass placed along *a*-axis

2 Outer ring is rotated until the arrow lines up with north

3 Bearing is read off

**▲ Figure 2** Measuring clast orientation with a compass

## 4 Glacial systems and landscapes

### 4.8 Case study: A contrasting glaciated landscape

 **Case study: Challenges for people and the environment in the Alps**

Precipitation is increasing, but more is falling as rain. Snow cover is expected to reduce by 25% over the next 20–30 years. The resulting albedo changes contribute to more warming. The reduction in snow fall affects water supply. Increased rainfall contributes to landslides and avalanches.

Alpine plant species are highly adapted. These species are migrating upwards at up to 4 m per decade. Up to 60% of species may become extinct when their habitat niches disappear completely.

For every 1°C increase in temperature, the snowline rises by about 150 m. So, less snow will accumulate at lower altitudes, where over half of all Alpine ski resorts have most of their runs.

Extreme weather events, especially droughts and floods, are becoming more common as winter rainfall increases and summer precipitation decreases. Changes in river regimes (e.g. Rhine, Danube, Rhône) will have a wide impact across Europe.

**The Alps**

**REVISION TIP**

You should revise the case study you studied. This page has examples from the Alps.

In 2003, heatwaves reduced the mass of Alpine glaciers by 10%. Heatwaves are becoming more common.

As temperatures increase, pathogens from the south are spreading to plant and animal species, which have no resistance to these diseases.

Permafrost holds rocky terrain together and it is disappearing on south-facing rock faces on lower mountain ranges, significantly increasing landslide and avalanche risks.

The Alps are warming at around twice the global rate. Since 1850, Alpine glaciers have lost 30% of their surface area and half their volume. Within 25 years, half of Switzerland's small glaciers will be gone. Risks of hazards increase as glaciers retreat, HEP generation is affected and habitat loss occurs.

## Human responses

| Resilience strategies | Mitigation strategies | Adaptation strategies |
| --- | --- | --- |
| Infrastructure strengthening: reinforcing key infrastructure (roads, bridges, buildings) to withstand increased floods, rockfalls, landslides and avalanches. | Alpine Convention (1995): Eight countries and the EU have agreed to follow protocols on areas such as clean energy, sustainable transportation, protection of landscapes and conservation. | Diversification of farming: farmers have introduced vineyards and used fields as campsites for tourists. |
| Reservoir construction: because of the changing patterns of precipitation, and earlier snow melt, many Alpine regions are increasing reservoir storage of glacial meltwater for use through the summer. | Afforestation: planting forests on mountain slopes absorbs more $CO_2$, stabilises slopes and reduces runoff following storm events. | Diversification of tourism: new resorts have been developed to provide warm weather activities and encourage more people to visit the Alps in summer. |
| | Renewable energy: investment in the Alps' HEP potential means emissions from energy production are reducing. | Snow production: winter sports resorts are investing in snow-making machines to compensate for declining snowfall. |

# RETRIEVAL

Learn the answers to the questions below, then cover the answers column with a piece of paper and write down as many as you can. Check and repeat.

## Questions | Answers

| | Questions | | Answers |
|---|---|---|---|
| 1 | Describe the formation of an esker. | Put paper here | Eskers form when a blockage occurs in an englacial tunnel so the meltwater deposits its sediment load. When the ice melts in situ, the esker is revealed |
| 2 | Which two processes are likely to be responsible for blockfields? | | Freeze–thaw weathering and chemical weathering (possibly in a warmer past climate) |
| 3 | A steep-sided channel carved into the bedrock of a glaciated landscape is most likely to be what fluvioglacial landform? | | A meltwater channel |
| 4 | An outwash plain is a proglacial feature, which means what? | Put paper here | It forms in front of the ice, away from the glacier |
| 5 | What is the term for kames formed along the margins of a glacier? | | Kame terrace |
| 6 | In periglacial environments, what term is used to describe the polygons, circles, stripes, steps and nets seen on flat or gently-sloping soil and sediment? | Put paper here | Patterned ground |
| 7 | Name the circular, domed hills with an ice core that are characteristic of some periglacial landscapes. | | Pingos |
| 8 | Why are cold environments considered to be fragile environments? | | Once damaged, they take a long time to recover and in some cases the damage may be irreversible, especially if damage sets off positive feedback loops |
| 9 | What type of response is afforestation of Alpine landscapes – resilience, mitigation or adaptation? | Put paper here | Migration |
| 10 | What type of response to climate warning is snow production – resilience, mitigation or adaptation? | | Adaptation |

## Previous questions

Now go back and use these questions to check your knowledge of previous topics.

## Questions | Answers

| | Questions | | Answers |
|---|---|---|---|
| 1 | Name this landform: a tributary valley that is higher than the main valley it feeds into. | Put paper here | Hanging valley |
| 2 | What is the active layer in a periglacial environment? | | A layer of soil above permafrost that thaws in summer and refreezes in winter |
| 3 | Why are tundra soils low in nutrients? | | Cold temperatures, short days and lack of liquid water inhibit plant growth and decomposition |
| 4 | Where are corries formed? | Put paper here | In nivation hollows |
| 5 | Name four types of moraine landforms. | | Four from: terminal / recessional / lateral / medial / ground |

## Exam-style questions

1  Outline the role of water in the process of erosion in
   cold environments.                                                    **[4 marks]**

2  Outline the impact of temperature variation on weathering
   processes in cold environments.                                      **[4 marks]**

3  Outline factors leading to the formation of glacial troughs.         **[4 marks]**

4  Explain the concept of glacial budget.                               **[4 marks]**

5  Outline factors leading to the formation of periglacial landforms.   **[4 marks]**

6  Explain the concept of positive feedback in relation to impacts
   of climate change on cold environments.                              **[4 marks]**

> **EXAM TIP**
>
> 4-mark questions are point-marked. That means each valid point or development of a valid point receives one mark, up to a maximum of 4 marks.

7  Analyse the data shown in **Figure 1**.                              **[6 marks]**

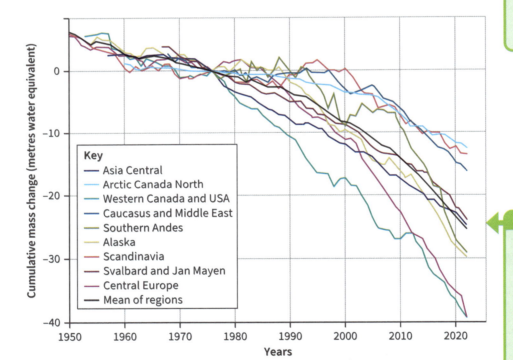

> **EXAM TIP**
>
> Make sure you use evidence from the figure in your answer to 'Analyse the data' 6-mark questions, and manipulate the data in your analysis by carrying out calculations. Look for links between data sets – or comment on where they diverge. This is all analysis.

▲ *Figure 1 Cumulative mass change relative to 1976 for reference glaciers in global cold environment regions (1950–2022). Reference glaciers have records from over 30 years of continual observation.*

8    Analyse the data shown in **Figure 2**.                                      **[6 marks]**

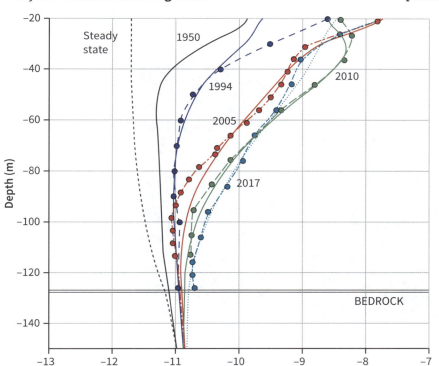

◀ **Figure 2** *Englacial temperatures in the Dôme du Goûter glacier (French Alps).*

Note: The measurements were taken at the same locations on the glacier in each year. The dots and dashed lines are actual measurements and the continuous lines are based on modelling using surface temperature data.

9    Assess the usefulness of **Figure 3** in depicting active layer depth along a transect in Alaska, from 2013–2021.                      **[6 marks]**

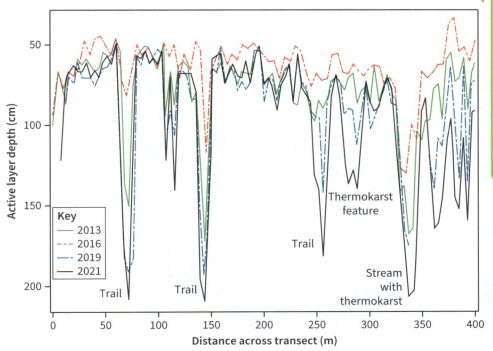

▲ **Figure 3** *Active layer depths along a 400m transect in Alaska*

Note: Active layer depths were measured by pushing a measuring rod into the ground until it reached permafrost. The transect is crossed by trails where vegetation cover has been removed by people walking over it. Thermokarst features in this case are pits and subsidence features where the ground has collapsed.

10    Using **Figure 4** and your own knowledge, assess the role of weathering in the development of this landscape.    **[6 marks]**

**EXAM TIP**

Notice how we have at least two different landforms to discuss the role of weathering in relation to, in the landscape pictured.

▲ **Figure 4** *shows a mountain landscape in Svalbard, a Norwegian archipelago in the Arctic Ocean that is north of the Arctic Circle*

11    Using **Figure 5** and your own knowledge, assess the view that glacial depositional processes dominate in the development of this landscape.    **[6 marks]**

▲ **Figure 5** *A glaciated landscape in New York State, USA*

12 Using **Figure 6** and your own knowledge, assess the role of ice in the development of this landscape. [6 marks]

▲ *Figure 6 A landscape in the Tuktoyaktuk Peninsula in Canada, on the coast of the Beaufort Sea*

Note: The landform in the centre of the image is estimated to be 1000 years old. It is currently around 70 m tall and 300 m wide at its base, and grows at a rate of 2 cm a year.

13 Using **Figure 7** and your own knowledge, assess the relative importance of factors leading to the development of this landscape. [6 marks]

▲ *Figure 7 A glaciated landscape of the Allalin Glacier in Switzerland*

# Exam-style questions

**14** 'Opportunities for human occupation and development may outweigh challenges in cold environments in coming decades.'

With reference to a glaciated landscape beyond the UK, to what extent do you agree with this view? **[20 marks]**

**15** Assess the importance of fluvioglacial processes in the formation of a glaciated landscape you have studied at a local scale. **[20 marks]**

**16** 'Adaptation is the best response to the impacts of climate change in cold environments.'

How far do you agree with this view? **[20 marks]**

**17** 'Cold environments are too fragile for human occupation and development.'

With reference to a glaciated landscape beyond the UK, how far do you agree with this view? **[20 marks]**

**18** 'Glaciated landscapes are products of past processes more so than periglacial landscapes.'

To what extent do you agree with this view? **[20 marks]**

**19** With reference to a glaciated environment at a local scale, assess the impact of human activity upon the natural systems and physical landscape. **[20 marks]**

**20** 'No amount of sustainable management can halt the processes threatening periglacial environments.'

To what extent do you agree with this view? **[20 marks]**

# ⚙ KNOWLEDGE

## 5 Hazards

### 5.1 The concept of hazard in a geographical context

## Natural hazards and their impacts

- A natural hazard is a natural phenomenon, to which human activity may contribute.
- This might have a negative effect on humans, animals or the environment.
- Natural disasters caused by hazards can cause high numbers of fatalities, impact health, disrupt social activity, and damage property, the local and national economy, and the environment.
- Impacts of hazards are linked to other factors, such as the density of the human population in the area where the hazard's impacts occur.
- Impacts can also be divided into primary impacts, the immediate effects of a disaster, and secondary impacts, which happen after the disaster has occurred.
- Two ways of measuring impacts of natural hazards are number of (human) deaths and cost of damage.

**REVISION TIP**

Try to write a summary of natural hazards and their impacts without looking at this book. Check your answer and correct any mistakes. Try the same activity again a few days later to help you remember the details.

## Hazard perception and characteristic responses

People do not always have an accurate perception of the risk of hazards.
Risk perception is influenced by experience, and:

- hazards occur infrequently (people may not have experienced a hazard in their lifetime)
- hazards vary in magnitude (a hazard may be different from previous experience)
- hazards vary in distribution (people may only be aware of hazards happening to others)
- impacts vary with the type of settlement (impacts in a city may differ from those in a village)
- educating people about hazard risk can improve the accuracy of risk perception
- education can also improve responses to a hazard (e.g. carrying out earthquake drills).

**Fatalism:**
Acceptance of disasters as something humans can do nothing to avoid or prevent, e.g. seeing earthquakes as the will of a god or destiny.

**Risk sharing:**
Distributing potential hazard costs among a larger group or community, e.g. everyone in an area taking out insurance against wildfire.

**Hazard management:**
Organising responses to minimise a disaster's impact after it has occurred, e.g. developing tropical storm evacuation plans.

**Characteristic human responses to hazards**

**Prediction:**
Forecasting when and where a hazard will occur using science, e.g. using satellite data to predict the path of approaching hurricanes.

**Adjustment/adaptation:**
Modifying behaviour and/or lifestyles to live with hazards, e.g. in Bangladesh stilt houses are an adaptation to flood hazards.

**Mitigation:**
Taking action now to reduce the impact of future hazards, e.g. constructing earthquake-resistant buildings in Japan.

**Key terms** Make sure you can write a definition for these key terms

adjustment/adaptation   fatalism   hazard management
mitigation   preparedness   recovery   response   risk sharing

# KNOWLEDGE

## 5 Hazards

### 5.1 The concept of hazard in a geographical context

## The hazard management cycle

**Preparedness:** using data from previous events to plan for a hazard is key to reducing impacts.

**Response:** deploying emergency services and resources. The speed and effectiveness will depend on preparedness.

**Recovery:** post-disaster reconstruction and restoration including both the built environment and the natural environment. It will be informed by assessments made during response.

**Mitigation:** the steps taken to reduce the negative impacts of the hazard. These steps will then influence the next round of preparedness.

## The Park model of human response to hazards

This model plots four stages of human responses to hazards over time. The curve usually indicates quality of life, and the model illustrates the speed at which the area or country returns to normality. The steepness of the curve illustrates a) the scale of the disaster and b) the speed of response. A fifth stage is sometimes included, in which disaster planners and community leaders reflect on the recovery and improve strategies to reduce vulnerabilities and improve preparedness.

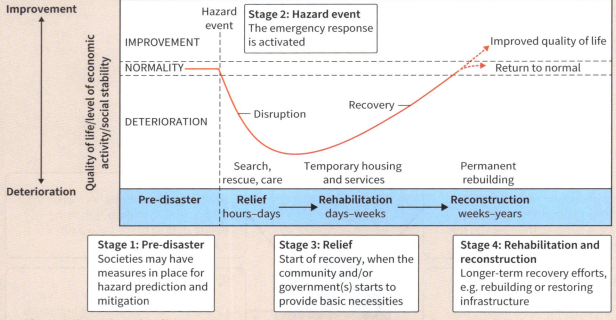

▲ *Figure 1* *The Park model of human response to hazards*

## Evaluating the Park model

☑ It makes it easier to compare responses to similar hazards.

☑ It is flexible and can be used in the same way for different types of hazards (including human-made ones like industrial accidents).

☒ It assumes that everywhere in that area responds in the same way.

☒ It overgeneralises. Real-world responses can vary greatly due to a number of factors and be much less predictable than the model's assumptions.

# RETRIEVAL

Learn the answers to the questions below, then cover the answers column with a piece of paper and write down as many as you can. Check and repeat.

| | Questions | | Answers |
|---|---|---|---|
| 1 | What is meant by the term 'natural hazard'? | Put paper here | A natural phenomenon that might have a negative effect on humans, animals or the environment |
| 2 | There are three main types of natural hazard: geophysical, atmospheric and… | | Hydrological |
| 3 | Of those three types, what type does a volcanic eruption fit into? | | Geophysical |
| 4 | What are primary impacts of hazards and how are they different from secondary impacts? | Put paper here | Primary impacts are immediate effects; secondary occur after the disaster |
| 5 | In what two ways are the impacts of hazards commonly measured? | | Number of deaths and cost of damage in US$ |
| 6 | How might variations in hazard incidence (occurrence) in an area reduce the perception of risk for people living there? | Put paper here | If incidence is low, then people living in the area might never have experienced the hazard, which can lower the perception of risk |
| 7 | How can education influence hazard perception and response? | Put paper here | Education can improve the accuracy of risk perception and understanding of response strategies (what to do in a disaster) |
| 8 | Adjustment/adaptation is one of the six characteristic human responses to hazards. List the other five. | Put paper here | Fatalism, prediction, mitigation, management, risk sharing |
| 9 | Describe the adjustment/adaptation response, using an example | Put paper here | Altering behaviour and/or lifestyles in response to hazard risk; e.g. building stilt houses in flood-prone areas |
| 10 | What is the aim of the 'preparedness' phase of the hazard management cycle? | Put paper here | To use evidence and data from previous events to plan for a future hazard event |
| 11 | How does the 'mitigation' phase of the hazard management cycle contribute to 'preparedness' for future events? | | Mitigation measures may show how previous plans had weaknesses, which will be addressed in the next round of 'preparedness' planning |
| 12 | What does the Park model provide a model of? | Put paper here | Human responses to a hazard or disaster event |
| 13 | What happens during the 'relief' stage of the Park model? | Put paper here | The community and/or support from government(s) starts to provide basic necessities to victims of the disaster |
| 14 | What advantage does the Park model offer when investigating responses to similar hazards over time? | Put paper here | It provides a visual representation of responses (a curve) which makes it easy to compare events over time |
| 15 | Why might the Park model be criticised for its assumption of returning to normality after a disaster? | | Because not all areas or societies are able to return to what was normal before the disaster happened |

## 5 Hazards

### 5.2 Plate tectonics

## The Earth's structure and internal energy sources

The **core** is at the centre of the Earth.

- The inner core is solid; mainly made of iron; around 6000°C.
- The outer core is semi-molten; mainly made of iron and nickel; 4500–6000°C.
- The core is the source of most of the Earth's heat, most of which is the result of the natural decay of radioactive elements.

The **mantle** surrounds the core.

- It is semi-molten nearer the core (the **asthenosphere**) and solid nearer the crust.

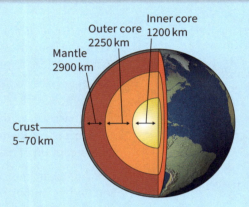

Inner core 1200 km
Outer core 2250 km
Mantle 2900 km
Crust 5–70 km

The **crust** is the outer shell of the Earth.

- Oceanic crust: less than 200 million years old, 6–10 km thick, denser than continental crust.
- Continental crust: over 1500 million years old, 45–50 km thick, less dense than oceanic crust.

▲ **Figure 1** *The Earth has three main layers – core, mantle and crust*

## Plate tectonic theory of crustal evolution

- Alfred Wegener's theory of continental drift suggested that the Earth's continents were once joined together in a single supercontinent (Pangaea) and have since drifted apart to their current positions.
- His evidence for this was based on the continents' 'jigsaw fit'; fossil and geological evidence linking continents now separated by oceans; evidence of glaciation in present-day tropical areas.
- The discovery of 'seafloor spreading' in the mid-twentieth century backed up the theory; studies of ocean floor rocks showed that they were youngest in the middle and oldest near the continents. Palaeomagnetism also supported the theory.

## Lithosphere and asthenosphere

The **lithosphere** comprises the crust and the solid part of the mantle. It is broken up into tectonic plates that move over the semi-fluid asthenosphere beneath. Different inter-related mechanisms are believed to be involved in the movement of these plates.

# Plate movement

Convection current:
Rising heat causes magma to rise, and as it cools, it sinks, creating a continuous current. These currents drag tectonic plates, causing them to move.

**Causes of plate movement**

**Gravitational sliding:**
- **Ridge push:** Mid-ocean ridges are high; magma rises up through them, cools and becomes more dense, sliding down the ridge and moving away from it.
- **Slab pull:** As oceanic plates cool, they become denser and start to subduct into the mantle at convergent boundaries. The weight of the sinking plate edge pulls the rest of the plate behind it.

Seafloor spreading:
Occurs at mid-ocean ridges where new oceanic crust is formed through volcanic activity and then gradually moves away from the ridge.

# Constructive plate margins

Two tectonic plates move away from each other, creating new lithosphere.

- Landforms:
  - most constructive margins are located in the oceans. As plates pull apart, magma rises to fill the gap, creating **mid-ocean ridges**. Submarine volcanoes may rise above sea level to form islands.
  - on continents, the plates pull apart, and fracture in a series of parallel faults. The land between these faults drops, forming a **rift valley**.
- Vulcanicity: basaltic (basic) lava eruptions are associated with constructive margins.
- Seismicity: shallow-focus earthquakes.

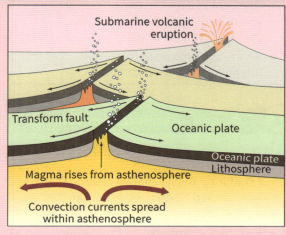

▲ **Figure 2** Constructive plate margin

# Conservative plate margins

Two plates slide past each other.

- Vulcanicity: no melting of rock takes place so there is no volcanic activity.
- Seismicity: the plates can stick due to friction; powerful, shallow-focus earthquakes result when this 'sticking' is released.

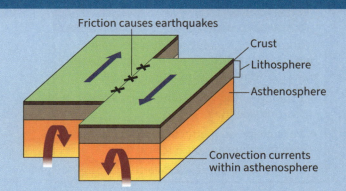

▶ **Figure 3** Conservative plate margin

**Key terms**

**Make sure you can write a definition for these key terms**

asthenosphere   core   crust   fold mountain   gravitational sliding
hot spot   lithosphere   magma plume   mantle
mid-ocean ridge   ridge push   rift valley   slab pull

## 5 Hazards

### 5.2 Plate tectonics

## Destructive plate margins

Two tectonic plates move towards each other, often involving the subduction (and destruction) of one under the other.

**1  Oceanic–continental convergence**

Denser oceanic plate is subducted beneath the less dense continental plate. Slab-pull may be involved in this process.

- Landforms: deep-sea trenches and **fold mountains** along the continental margin.
- Vulcanicity: at the zone of melting (the Benioff Zone), magma is less dense and rises through cracks and faults. Explosive, acid volcanoes.
- Seismicity: both deep-focus and intermediate-focus earthquakes.

**2  Oceanic–oceanic convergence**

Two oceanic plates converge. One is subducted beneath the other.

- Landforms: deep-sea trenches and island arcs (caused by rising magma from the Benioff Zone).
- Vulcanicity: explosive acid volcanoes.
- Seismicity: shallow, intermediate and deep-focus earthquakes.

**3  Continental–continental convergence**

Two continental plates collide with similar densities. No subduction. The crust is pushed upwards, forming fold mountain ranges.

- Landforms: high fold mountains, often featuring faulting.
- Vulcanicity: no subduction, so no magma and therefore no volcanic eruptions.
- Seismicity: shallow-focus earthquakes.

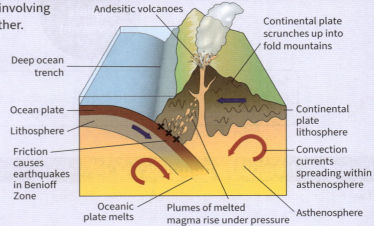

▲ *Figure 4 Oceanic plate meets continental plate*

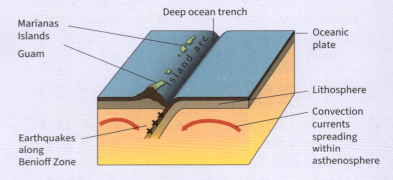

▲ *Figure 5 Oceanic plate meets oceanic plate*

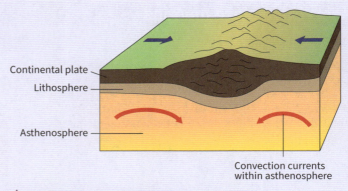

▲ *Figure 6 Continental plate meets continental plate*

## Magma plumes

Most volcanic activity is along plate margins, but there are exceptions, e.g. the volcanoes of Hawaii.

- Concentrated areas of radioactive decay produce hot spots in the lower mantle.
- **Magma plumes** rise through the mantle and the lithosphere to create surface volcanic activity.

- The **hot spot** doesn't move, creating a chain of volcanoes as the plate moves away.

#  RETRIEVAL

Learn the answers to the questions below, then cover the answers column with a piece of paper and write down as many as you can. Check and repeat.

## Questions

## Answers

| # | Question | Answer |
|---|----------|--------|
| 1 | Name the two types of crust. | Oceanic crust and continental crust |
| 2 | Explain the main source of the Earth's heat. | The natural decay of radioactive elements |
| 3 | Name the theory associated with Alfred Wegener. | Continental drift |
| 4 | Name two processes involved in gravitational sliding. | Ridge push and slab pull |
| 5 | Which theory about plate movement is directly supported by palaeomagnetism? | Seafloor spreading |
| 6 | Mid-ocean ridges are most likely to be found at which type of plate margin? | Constructive plate margin |
| 7 | Name two plate margins where no vulcanicity occurs. | Two from: continental–continental convergence / destructive margins / conservative plate margins |
| 8 | Name a plate margin characterised by basic (basaltic) eruptions and shallow-focus earthquakes. | Constructive plate margins |
| 9 | Which plate margin is most associated with island arcs? | Oceanic–oceanic convergence destructive plate margins |
| 10 | What name is given to a focused area of heating in the lower mantle responsible for magma plumes? | Hot spot |

*Put paper here*

## Previous questions

Now go back and use these questions to check your knowledge of previous topics.

## Questions

## Answers

| # | Question | Answer |
|---|----------|--------|
| 1 | What is meant by the term 'natural hazard'? | A natural phenomenon that might have a negative effect on humans, animals or the environment |
| 2 | In what two ways are the impacts of hazards commonly measured? | Number of deaths and cost of damage in US$ |
| 3 | Adjustment/adaptation is one of the six characteristic human responses to hazards. List the other five. | Fatalism, prediction, mitigation, management, risk sharing |
| 4 | What is the aim of the 'preparedness' phase of the Hazard Management Cycle? | To use evidence and data from previous events to plan for a future hazard event |
| 5 | What advantage does the Park model offer when investigating responses to similar hazards over time? | It provides a visual representation of responses (a curve) which makes it easy to compare events over time |

*Put paper here*

## 5 Hazards

### 5.3 Volcanic hazards

## Spatial distribution of volcanic hazards

- Volcanic activity is common at constructive and some destructive margins. Volcanic hazards are not found at conservative margins or continental-continental convergence destructive margins.

- Some volcanic activity is found away from plate margins, due to hot spots, e.g. the Hawaiian hot spot.

- Location also affects the type and magnitude of eruption.

  — Constructive plate margins feature basic lava which is less viscous and flows gently from fissures or a vent.

  — Destructive margins feature more viscous acidic magma, which is explosive.

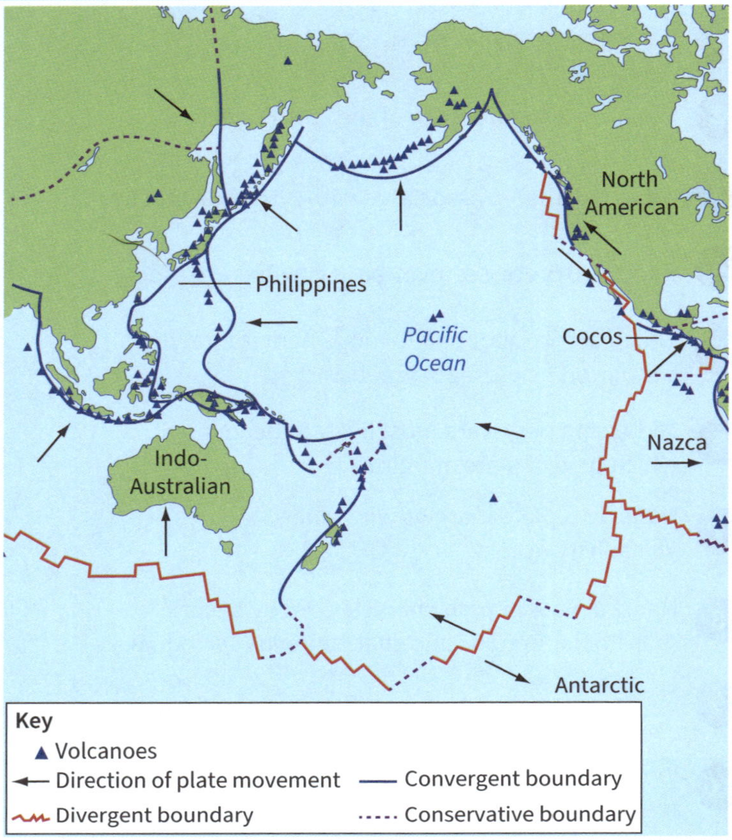

▶ **Figure 1** *Spatial distribution of volcanoes and plate margins around the Pacific Ocean*

**Key**

▲ Volcanoes

← Direction of plate movement — Convergent boundary

〜 Divergent boundary ····· Conservative boundary

## Magnitude of volcanic hazards

Magnitude is measured using the Volcano Explosivity Index (VEI) logarithmic scale. Each level has 100 times the energy of the level below it.

### Factors affecting volcano explosivity (magnitude)

| Silica content | Gas content | Magma temperature |
|---|---|---|
| High-viscosity magma (high silica content) tends to trap gases more effectively, leading to explosive eruptions when the pressure is released. | High gas content can lead to more explosive eruptions. Magma that rises rapidly often has a higher gas content as the gas has had less time to dissipate. | Cooler magmas tend to be more viscous than hotter ones, and therefore more likely to trap gases and be explosive. |

| Presence of water | Eruption history |
|---|---|
| An eruption that mixes magma and water (phreatomagmatic) tends to be more explosive due to the rapid vaporisation of water into steam. Water can come from groundwater, surface water or subduction. | Volcanoes with a history of explosive eruptions may have conditions (like established conduits or dome structures) that favour further explosive eruptions. |

## Frequency, regularity and predictability of volcanic hazards

Evidence on frequency and regularity can help with prediction, but volcanoes often erupt infrequently, irregularly and therefore unpredictably.

Volcano type:
Shield volcanoes erupt more frequently than stratovolcanoes

Plate margin type:
Constructive margins have more frequent eruptions, while destructive margins are less frequent

**Factors affecting frequency and regularity**

Volcano history:
Volcanoes that have erupted in the past may have internal structures that make another eruption more likely than 'expected' for their type

Silica content:
Low-silica, basic magma-type volcanoes erupt frequently (less explosive), while high-silica, acid magma-type volcanoes are less frequent (more explosive)

## Risk management

Risk management is designed to reduce the impacts of a hazard through preparedness, mitigation, prevention and adaptation (see 5.1). Each of these affects the effectiveness of the response to a hazard event.

### Monitoring

- Monitoring techniques: Scientists use seismographs (to detect earthquakes), tiltmeters and GPS (to measure ground deformation), and gas detectors to gather data that can hint at an impending eruption.
- Historical data: past patterns of eruptions can offer clues about future eruptions.
- Warning signs: increased seismic activity, changes in gas emissions and ground swelling can all indicate a potential eruption.

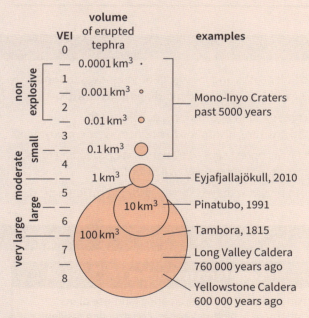

▲ **Figure 2** *Past eruptions can offer clues about future eruptions*

## 5 Hazards

### 5.3 Volcanic hazards

## Forms of volcanic hazard

### Nuées ardentes

**Nuées ardentes** (or **pyroclastic flows**) are fast-moving, superheated clouds of gas, ash and volcanic rocks. They can reach speeds of up to 100 km/h and temperatures of 700°C.  Example: the 1991 Mount Pinatubo eruption in the Philippines produced pyroclastic flows that travelled 12–16 km from the vent and were up to 200 m thick.

### Lava flows

Lava flows can reach 1170°C and take years to cool completely. Temperature and viscosity determine the speed and shape of the flow. Slower moving than other volcanic hazards, but can be hugely destructive by burning or burying. Example: Lava flows from the 2018 Kīlauea volcano in Hawaii destroyed over 700 homes, displacing 2000 residents. Lava added 875 acres to the island.

### Mudflows

**Mudflows** (or lahars) are produced from the rapid melting of high-altitude ice and snow, or triggered by tropical storms following an eruption. They can be hundreds of metres wide, flowing at tens of metres per second. Example: the Nevado Del Ruiz eruption of 1985 which caused a lahar that killed over 20,000 people in Armero, Columbia.

### Tephra and ash falls

The largest **tephra** falls near the crater. The smallest ash particles can travel for thousands of kilometres, burying structures, affecting air travel and harming human health. Example: in 2010, the eruption of Eyjafjallajökull in Iceland led to significant disruptions in European air travel for weeks due to the ash cloud.

### Gases/acid rain

Eruptions release carbon dioxide, sulphur dioxide and hydrogen sulphide, which can mix with water droplets in the atmosphere to produce acid rain, harming plants, animals and infrastructure. The gases can be deadly close to the eruption. Example: Since 1980, at least 71 people have been hospitalised and seven have died due to inhalation of volcanic gases while visiting Mount Aso, Japan.

## Categorising hazard impacts

- Primary:  occur as the direct result of the hazard, for example lava flows
- Secondary: arise as a result of primary hazards, for example heavy rainfall destabilising tephra causing a mudflow (lahar)
- Environmental: river/lake/air pollution; damage to wildlife/vegetation/ecosystems

- Social: deaths and injuries; homelessness; trauma; disease; destruction to buildings and infrastructure
- Economic: damage to crops, buildings and infrastructure; disruption to businesses; unemployment
- Political: unrest caused by personal loss leading to protests and potentially civil disorder
- All of these can have short- and long-term impacts.

## Short- and long-term responses

### Short-term responses

Evacuation of threatened areas; provision of emergency shelters and relief supplies; medical assistance for the injured; deployment of emergency personnel for rescue operations; immediate assessment of infrastructure damage and hazards, such as ashfall or toxic gases.

### Long-term responses

Establishing early warning systems; creating land-use plans that restrict settlement in high-risk zones; constructing resilient infrastructure; initiating public education and awareness campaigns about volcanic risks and preparedness measures; developing contingency plans for potential future eruptions; continuous monitoring of volcanic activity by specialised agencies.

# Impacts and human responses to a recent volcanic event

## Eyjafjallajökull, Iceland (2010)

**REVISION TIP**

You should revise the event you have studied. This page provides the example of the eruption of Eyjafjallajökull in Iceland in 2010.

### Background

- Before April 2010 Eyjafjallajökull had been dormant for nearly 200 years.
- The volcano is one of several in Iceland that are covered by thick ice.
- The eruption was preceded by a series of earthquakes and small fissure eruptions.

### Impacts

Despite only being a 3 on the VEI, the long duration of the eruption (39 days), the fine nature of the ash, and winds blowing towards Europe meant this eruption had major impacts.

- Environmental: ashfall caused the loss of vegetation, and contamination of water sources. Flooding damaged habitats and infrastructure (washing away part of the perimeter road).
- Economic: the threat of ash damaging engines led to the closure of a significant portion of European airspace and 10 000 flights were cancelled, resulting in a US$5 billion loss to the European economy.
- Social: 800 people evacuated, and the ashfall affected the respiratory health of some local residents, no deaths.

| Short-term responses | Long-term responses |
| --- | --- |
| <ul><li>Evacuation on three occasions of about 800 people from areas directly threatened by the eruption and by flooding.</li><li>Financial assistance to farmers affected by ashfall, with the European Red Cross providing food for local people and counselling support.</li><li>Immediate monitoring to forewarn of any further activity.</li><li>Ash forecasts monitored by European air traffic organisations, leading to flight cancellations.</li></ul> | <ul><li>Improved monitoring systems for volcanic activity in Iceland.</li><li>Aviation authorities reassessed protocols to prevent future widespread disruption. Europe's airspace is now organised into nine blocks, which can be closed individually.</li><li>Research into the impacts of volcanic ash on jet engines to determine safe flight thresholds.</li><li>Promotion of geothermal tourism, following the international publicity the eruption received.</li></ul> |

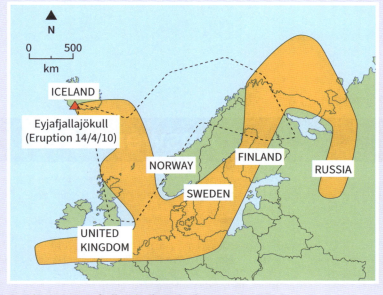

▲ **Figure 3** *Eyjafjallajökull's ash plume across Europe*

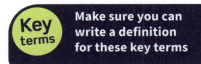

**Key terms** Make sure you can write a definition for these key terms

mudflow    nuée ardente    pyroclastic flow    tephra

Learn the answers to the questions below, then cover the answers column with a piece of paper and write down as many as you can. Check and repeat.

## Questions | Answers

| # | Questions | | Answers |
|---|-----------|---|---------|
| 1 | Which margin type tends to have less viscous, gently-flowing lava? | | Constructive plate margins |
| 2 | Name three factors that affect volcano explosivity. | | Three from: silica content / gas content / magma temperature / presence of water / eruption history |
| 3 | Which volcano type erupts more frequently: shield volcano or stratovolcano? | | Shield volcano |
| 4 | Name the volcanic hazard: flows at a speed of up to 100 km/h at a temperature of up to 700 °C. | | Nuées ardentes (or pyroclastic flows) |
| 5 | Name the volcanic hazard: a general term for all sizes of volcanic fragments ejected from the air during an eruption. | | Tephra |
| 6 | Name the volcanic hazard: this hazard was responsible for destroying over 700 homes in Hawaii in 2018. | | Lava flow |
| 7 | Name the hazard: results from a mixture of sulphur dioxide from volcanic eruptions and rainwater. | | Acid rain |
| 8 | Is a lava flow a primary or a secondary hazard? | | Primary hazard |
| 9 | Is constructing infrastructure that is resilient to volcanic impacts a short-term or a long-term response? | | Long-term |
| 10 | In what year did the Eyjafjallajökull ash cloud spread across European air space? | | 2010 |

*Put paper here*

## Previous questions

Now go back and use these questions to check your knowledge of previous topics.

## Questions | Answers

| # | Questions | | Answers |
|---|-----------|---|---------|
| 1 | What are primary impacts of hazards and how are they different from secondary impacts? | | Primary impacts are immediate effects; secondary occur after the disaster |
| 2 | What does the Park model provide a model of? | | Human responses to a hazard or disaster event |
| 3 | Name two processes involved in gravitational sliding. | | Ridge push and slab pull |
| 4 | Name two plate margins where no vulcanicity occurs. | | Two from: continental–continental convergence / destructive margins / conservative plate margins |
| 5 | What name is given to a focused area of heating in the lower mantle responsible for magma plumes? | | Hot spot |

*Put paper here*

# ⚙ KNOWLEDGE

## 5 Hazards

### 5.4 Seismic hazards

## Spatial distribution of seismic hazards

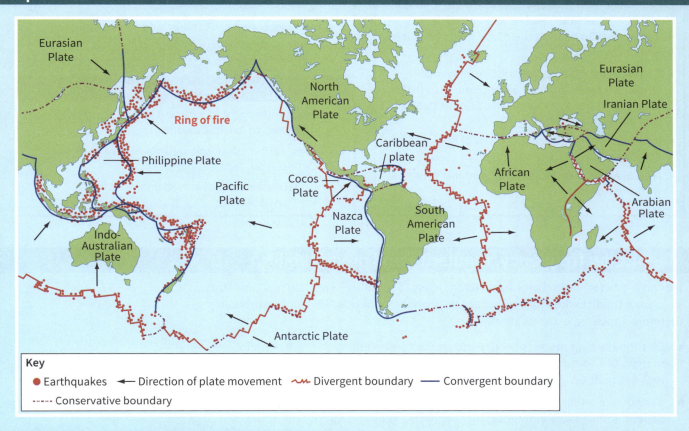

▲ **Figure 1** *Spatial distribution of earthquakes and plate margins*

- At constructive plate margins, where plates are moving apart, earthquakes are shallow and tend to be low magnitude
- At destructive plate margins, where plates are moving together, earthquakes can be deep and high magnitude
- At conservative plate margins, earthquakes are shallow but can be high magnitude

## Forms of seismic hazard

### Earthquakes

- Sudden shaking or trembling of the ground. Shaking is most intense at the epicentre of an earthquake (the point on the surface directly above the focus).
- Causes: tectonic plate movements, volcanic activity, human activities like reservoir construction or fracking.
- Measurement: Richter Scale and Moment Magnitude Scale.
- Impacts: structural damage, loss of life, economic costs.

### Shockwaves

Waves of energy emitted from the earthquake epicentre.

- Types: Primary (P-waves), Secondary (S-waves), and Surface waves. P-waves are the fastest and reach the surface first. S-waves shake the ground. Surface waves are the slowest and cause most of the damage.
- Measurement: seismographs.
- Damage potential: dependent on amplitude, frequency and ground conditions.

## 5 Hazards

### Tsunamis

Large ocean waves generated by underwater seismic events. The waves travel very fast (640–960 km per hour) and have a low wave height at sea. On nearing the coast, they can reach heights of over 25 m.

- Causes: earthquakes, volcanic eruptions or landslides under or near water bodies. Up to 90% of tsunamis are associated with seismicity along the 'Pacific Ring of Fire'.
- Impacts: coastal destruction, loss of life, economic loss.
- Measurement: seismographs, tsunami early warning systems.

### Liquefaction

Soil temporarily losing strength and acting as a fluid during seismic events.

- Causes: saturated soils subjected to sudden shaking.
- Impacts: building collapse, ground deformation.

### Landslides

Downward movement of rock and soil due to gravity, often triggered by seismic events.

- Causes: earthquakes, soil saturation (steep slopes at highest risk).
- Impacts: destruction of infrastructure, loss of life, disruption of services.

## Magnitude, frequency, regularity and predictability

We can predict where **seismic hazards** are likely to occur – along plate margins – but we cannot predict when they will occur or at what magnitude. Seismic hazards can therefore appear to occur randomly.

- Earthquake magnitude (size) depends on factors including the type of plate margin, the speed and direction of the plate movements, and the amount of stress that has built up.
- Earthquake intensity is a measure of the amount of shaking at a location. It is dependent on many different factors including geology (e.g. soft mud shakes more than solid bedrock). Depth of focus also affects intensity, with more shaking typically if the focus is closer to the surface.

## Categorising hazard impacts

- Primary: occur as the direct result of the hazard; destruction to buildings and infrastructure caused by ground shaking or collapse of buildings due to liquefaction.
- Secondary: arise as a result of primary hazards; fires caused by gas leaking from pipes fractured by an earthquake.
- Environmental: landslides/avalanches; flooding; damage to wildlife/vegetation/ecosystems

- Social: destruction to buildings and infrastructure (communications, transport, power, sewerage); deaths and injuries; homelessness; trauma; disease
- Economic: high cost of rebuilding buildings and infrastructure; disruption to businesses; unemployment
- Political: unrest caused by personal loss leading to protests and potentially civil disorder
- All of these can have short- and long-term impacts.

# Short- and long-term responses

## Short-term responses

Emergency rescue operations to help and find survivors (buried under collapsed buildings); providing medical help; setting up temporary shelters; reconnecting/ensuring availability of water, power, telecommunications; providing public information about how to access help; prioritising repairs to key infrastructure.

## Long-term responses

Focus on rebuilding and mitigation, for example reconstruction of infrastructure and buildings; restoration and improvement of local economies; healthcare for those suffering longer-term physical and mental health problems; revision of seismic safety standards and emergency planning; investment in or updating of seismic hazard warning systems.

# Risk management

- Although there are indicators that an earthquake may be imminent (e.g. tremors, elevated levels of radon gas, changes in ground surface and groundwater levels), they are not reliable.
- Adaptation, preparedness and particularly mitigation are the best defence against the impacts of these hazards.

Urban planning: Avoiding construction near fault lines and hazardous areas

Disaster response plans: Establishing and practising response strategies

Building codes: Implementing and enforcing seismic-resistant construction standards

Education and training: Promoting awareness and preparedness (earthquake drills, evacuation procedure)

**Mitigation**

Early warning systems: Particularly useful for warning of tsunamis and aftershocks

Panels of marble and glass flexibly anchored to steel superstructure

Rolling weights on roof to counteract shock waves

Reinforced lift shafts with tensioned cables

'Birdcage' interlocking steel frame

Reinforced latticework foundations deep in bedrock

Rubber shock absorbers between foundations and superstructure

▲ **Figure 2** *A modern earthquake-proof building*

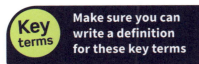

**Key terms** Make sure you can write a definition for these key terms

liquefaction    seismic hazard    shockwave

## 5 Hazards

### 5.4 Seismic hazards

## Impacts and human responses to a recent seismic event

### Morocco earthquake (2023)

#### Background

- Morocco is a country in north Africa, located near the African/Eurasian continental plate boundary, which are moving towards each other at a rate of 4 mm a year.

- Despite its location, high-magnitude earthquakes are rare. The 2023 magnitude 6.8 earthquake was the strongest to hit the country for at least 120 years.

- The earthquake struck on 8 September at 11:00 pm local time. It had a shallow focus, 18.5 km below the surface, with an epicentre in the Atlas Mountains, around 70 km south of Marrakesh (Morocco's fourth-largest city).

- As a result of the lower risk of earthquakes, preparation for seismic hazards is at a lower level in Morocco than in other countries like Türkiye or Greece.

▲ **Figure 3** *Location of the 2023 Morocco earthquake*

**REVISION TIP**

You should revise the example you have studied. This page provides the example of the Morocco earthquake of 2023.

#### Impacts

- Social: Over 2900 people were killed and 5500 people injured by collapsing buildings (many were asleep); landslides buried villages in the mountains. Over 60,000 houses were damaged, with the most damage close to the epicentre in the mountains, where nearly 3000 villages were affected. Housing is mainly made from traditional red clay bricks, which crumbled.

- Economic: The estimated costs of rebuilding are US$12 billion. Insurance industries were not liable for major payouts as most people affected by the disaster did not have private insurance.

- Environmental: Many historic buildings in Marrakesh were damaged by the earthquake, including parts of the city walls.

| Short-term responses | Long-term responses |
|---|---|
| • People used their hands to dig through rubble while waiting for specialised equipment.<br>• Strong aftershocks followed the earthquake, and many spent three nights sleeping outdoors, as they could not return to their homes.<br>• The Health Ministry mobilised over 2000 doctors and nurses to provide emergency medical aid. A field hospital was set up close to the Atlas Mountains.<br>• Businesses reopened two days later in areas declared safe by the government.<br>• The armed forces delivered drinking water, food and tents to remote mountain settlements.<br>• The army also cleared rubble from key roads leading to the worst-affected mountain regions.<br>• Offers of aid came from many countries, but Morocco only accepted aid from Spain, UK, Qatar and UAE. | • King of Morocco, Mohammed VI, promised shelter and an emergency payment of 30,000 dirhams (£2300) to earthquake survivors.<br>• A US$12 billion five-year rebuilding plan was announced: 140,000 dirhams (£10,700) for the rebuilding of each collapsed house and 80,000 dirhams (£6100) for each damaged house.<br>• The UN has committed funds for the rebuilding and restoration of damaged historic buildings, which are a major tourist attraction and important to Morocco's economy. |

# RETRIEVAL

Learn the answers to the questions below, then cover the answers column with a piece of paper and write down as many as you can. Check and repeat.

## Questions | Answers

| | Questions | | Answers |
|---|---|---|---|
| 1 | Name the plate margin at which earthquakes are most likely to have a deep focus. | Put paper here | Destructive plate margins |
| 2 | Name the plate margin at which earthquakes are likely to have both a shallow focus and a lower magnitude. | | Constructive plate margins |
| 3 | As well as tectonic plate movements, name another possible cause of earthquakes. | | One from: volcanic activity / reservoir construction / fracking |
| 4 | Which of the three types of shockwave are high frequency and reach the surface first? | Put paper here | Primary waves (P-waves) |
| 5 | In which global region do up to 90% of tsunamis occur? | | Pacific Ring of Fire |
| 6 | What is the name of the type of seismic hazard in which soils temporarily act as a fluid? | | Liquefaction |
| 7 | Is liquefaction a primary impact or a secondary impact? | Put paper here | Primary impact |
| 8 | Is constructing infrastructure that is resilient to seismic impacts a short-term or a long-term response? | | Long-term |
| 9 | Which of the following is seen as the most effective response to earthquake hazards: mitigation, prevention or adaptation? | Put paper here | Mitigation |
| 10 | Name the main reason why earthquake preparedness is less well developed in Morocco than in Türkiye or Greece. | | Because high magnitude earthquakes are relatively rare in Morocco |

## Previous questions

Now go back and use these questions to check your knowledge of previous topics.

## Questions | Answers

| | Questions | | Answers |
|---|---|---|---|
| 1 | How can education influence hazard perception and response? | Put paper here | Education can improve the accuracy of risk perception and understanding of response strategies (what to do in a disaster) |
| 2 | Explain the main source of the Earth's heat. | | The natural decay of radioactive elements |
| 3 | Which margin type tends to have less viscous, gently-flowing lava? | | Constructive plate margins |
| 4 | Name the volcanic hazard: a general term for all sizes of volcanic fragments ejected from the air during an eruption. | Put paper here | Tephra |
| 5 | Is a lava flow a primary or a secondary hazard? | | Primary hazard |

## 5 Hazards

### 5.5 Storm hazards

## The nature of tropical storms

A tropical storm is a type of storm system that forms in tropical regions, characterised by a low-pressure centre, thunderstorms and wind speeds in excess of 120 km/h (75 mph). A tropical storm can cover an area of up to 500 km in diameter.

## Underlying causes of tropical storms and their spatial distribution

Tropical storms derive their energy from the evaporation of ocean water, which rises and recondenses into clouds and rain. Their spatial distribution is limited to tropical areas because of the factors required for their formation.

- Warm ocean waters: above 26°C to fuel the storm with energy
- Atmospheric instability: favourable conditions for thunderstorm activity: rapidly rising air
- Pre-existing weather disturbance: to initiate the organisation of thunderstorms into a cyclonic pattern
- High humidity: oceans supply the moisture and energy (though latent heat) – they lose energy and power over land

- Low wind shear: high wind shear (different wind speeds at different altitudes) can tear a tropical storm formation apart or stop it from forming
- Distance from the equator: at least 5° latitude from the equator to provide the necessary 'spin' from the Earth's rotation (Coriolis effect)

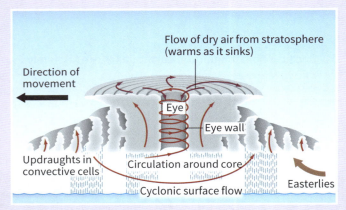

▲ **Figure 1** *Cross-section through a tropical storm*

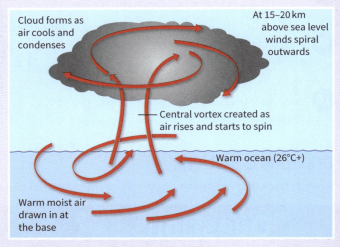

▲ **Figure 2** *Formation of a tropical storm*

## Forms of tropical storm hazard

High winds:
- Wind speeds of above 120 km/h cause damage to infrastructure, vegetation and can lead to flying debris

Storm surges:
- Caused by a combination of low atmospheric pressure and powerful winds driving ocean water on-shore, storm surges are typically 3 m in height above normal sea level; higher if in combination with high tides

Coastal flooding:
- Inundation of coastal areas from storm surges

River flooding:
- Overflow of rivers due to the heavy rain, which can be more than 200 mm of rain in just a few hours

Landslides:
- Saturated ground from heavy rainfall can trigger landslides, causing destruction and blocking transportation routes

## Magnitude, frequency, regularity, predictability

- Magnitude: measured by wind speeds and central pressure. The Saffir–Simpson scale has five different categories, with category five storms having wind speeds of over 250 km/h and **storm surges** of over 5.5 m.
- Regularity: tropical regions experience these storms annually due to prevailing conditions.

- Frequency: most common in late summer into autumn, when ocean temperatures have built to above 26°C. Global warming is likely to increase the frequency of tropical storms as ocean temperatures rise.
- Predictability: once formed, tropical storms often follow similar tracks, but this is dependent on atmospheric and oceanographic conditions, and they can be unpredictable.

## Categorising hazard impacts

- Primary: occur as the direct result of the hazard; destruction caused by strong winds and storm surges
- Secondary: arise as a result of primary hazards; river flooding as a result of heavy rain, or landslides as a result of soil saturation
- Environmental: storm surge; flooding; landslides/ avalanches; damage to wildlife/vegetation/ ecosystems (saltwater inundation of freshwater)

- Social: aquifer contamination; destruction to buildings and infrastructure (communications, transport, power, sewerage); deaths and injuries; homelessness; trauma; disease
- Economic: financial cost to individuals, business and insurance companies; disruption to businesses; unemployment
- Political: unrest caused by personal loss leading to protests and potentially civil disorder
- All of these can have short- and long-term impacts.

## Short- and long-term responses

### Short-term responses

Emergency rescue operations to find survivors from severe flooding; providing temporary shelters; treating injuries; preventing the spread of disease as a result of contamination of flood water with, for example, sewage or chemicals.

### Long-term responses

Often focus on rebuilding and mitigation; for example, reconstruction of infrastructure and buildings; restoration and improvement of local economies; healthcare for those suffering longer term physical and mental health problems; improved disaster preparedness planning (early warning systems and disaster response plans).

## Risk management

### Monitoring

- Early warning systems allow tropical storms to be tracked, and computer modelling based on historic storm patterns allow accurate prediction of landfall and storm surge heights.
- This data can reduce the risk of a disaster in densely populated regions. Insurance companies also depend on such data.

### Mitigation

- Storm surges can be reduced by hard engineering barriers or soft engineering buffers such as replanted mangrove forests or restored coral reefs.
- Insurance policies are also a form of mitigation, so that individuals can afford to rebuild, and also by encouraging resilience (e.g. storm shutters) through reduced premiums.

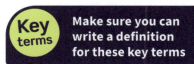

**Key terms** Make sure you can write a definition for these key terms

storm surge

## 5 Hazards

### 5.5 Storm hazards

## Impacts and human responses to two recent tropical storms

> **REVISION TIP**
>
> You should revise the storms you have studied. This page compares Typhoon Haiyan and Hurricane Katrina.

|  | Typhoon Haiyan ('Yolanda') | Hurricane Katrina |
|---|---|---|
| **Location** | Philippines: hit by an average of 20 tropical storms per year | USA, especially city of New Orleans, Louisiana |
| **Population of country** | 116 million | 339 million |
| **GDP per capita** | US$8100 | US$63,700 |
| **Life expectancy** | 70 years | 80.7 years |
| **Date and time** | 8 November 2013, 4:00 am | 29 August 2005, 10:00 am |
| **Category of storm** | Category 5 | Category 5 |
| **Top wind speed** | 314 km/h | 200 km/h |
| **Storm surge** | 5 m | 6 m |
| **Rainfall** | 400 mm | 250 mm |
| **Impacts** | 6190 people died (most drowned in storm surge in Tacloban); 29,000 people injured; Over one million homes damaged or destroyed; Salt contamination by storm surge ruined one million tonnes of crops; Outbreaks of disease because of sewage contamination; Economic losses: US$2.9 billion, mostly from losses in agriculture and fishing. | 1200 people died (most drowned in storm surge in New Orleans); Levees protecting the city had been built to deal with a Category 3 storm and failed, flooding 80% of the city; Over 275,000 homes damaged or destroyed; Lack of assistance and fear of starvation may have led to looting following the disaster; Damage to oil facilities in the Gulf of Mexico meant increased oil prices; Economic losses: US$125 billion. |
| **Short-term responses** | Warnings given 48 hours ahead of landfall and 800,000 people evacuated, but not all shelters were far enough inland; 400,000 people were housed in 1200 emergency camps; The government was criticised for a slow response, but damage to roads was extensive and in Tacloban many government officials had died; Six months later, many still lived in temporary shelters and had limited access to clean water. | Warnings given about 56 hours ahead of landfall meant one million people had evacuated; 60,000 mostly poorer residents did not evacuate, some because they had no transport, others to protect property; The US government was criticised for its initially slow response; Evacuation centres were seen as inappropriate: 30,000 people arrived at the Louisiana Superdome, which had capacity for 800; The National Guard was deployed to rescue people and to counter looting. |
| **Long-term responses** | Government launched a recovery plan for infrastructure, social services, resettlement and livelihood; 200,000 homes built to withstand winds of up to 250 km/h. | 400 km of levees are now higher and much stronger and the Lake Borgne surge barrier built; Upgrading defences cost US$14 billion and US government also spent US$16 billion repairing and strengthening housing damaged. |

#  RETRIEVAL

Learn the answers to the questions below, then cover the answers column with a piece of paper and write down as many as you can. Check and repeat.

## Questions | Answers

| # | Question | Answer |
|---|----------|--------|
| 1 | A tropical storm is characterised by a wind speed in excess of what? | 120 km/h (75 mph) |
| 2 | What is the scientific name for the effect on the atmosphere produced by the Earth's rotation? | Coriolis effect |
| 3 | Tropical storms do not form within which distance (in degrees of latitude) from the equator? | 5° |
| 4 | Name three of the hazards associated with tropical storms. | Three from: high wind / storm surge / coastal flooding / river flooding / landslides |
| 5 | Which two aspects of tropical storms are measured by the Saffir–Simpson scale? | Wind speed and storm surge height |
| 6 | Would river flooding resulting from tropical storm rainfall be classed as a primary or secondary impact? | Secondary impact |
| 7 | Give an example of an environmental impact of a tropical storm. | Inundation of salt water from storm surges into freshwater habitats |
| 8 | Which two cities were most associated with Haiyan and Katrina? | Tacloban (Haiyan) and New Orleans (Katrina) |
| 9 | Name an environmental impact of Typhoon Haiyan. | Salt contamination by storm surge ruined 1 million tonnes of crops |
| 10 | In the cases of Typhoon Haiyan and Hurricane Katrina, did the level of economic development affect accuracy of prediction? | No, in the Philippines there was 48 hours of warning before Haiyan made landfall, similar to the warning period for Katrina |

*Put paper here*

## Previous questions

Now go back and use these questions to check your knowledge of previous topics.

## Questions | Answers

| # | Question | Answer |
|---|----------|--------|
| 1 | There are three main types of natural hazard: geophysical, atmospheric and… | Hydrological |
| 2 | Which volcano type erupts more frequently: shield volcano or stratovolcano? | Shield volcano |
| 3 | Name the plate margin at which earthquakes are most likely to have a deep focus. | Destructive plate margins |
| 4 | Which of the three types of shockwave are high frequency and reach the surface first? | Primary waves (P-waves) |
| 5 | Which of the following is seen as the best response to earthquake hazards: mitigation, prevention or adaptation? | Mitigation |

*Put paper here*

## 5 Hazards

### 5.6 Wildfires

## Nature and causes of wildfires

**Wildfires** can be caused by humans, either accidentally through cigarettes, sparks, disposable barbecues, etc., or on purpose (arson), or as a natural process (lightning, volcanic activity). Woodland areas used for recreation are most prone to wildfires.

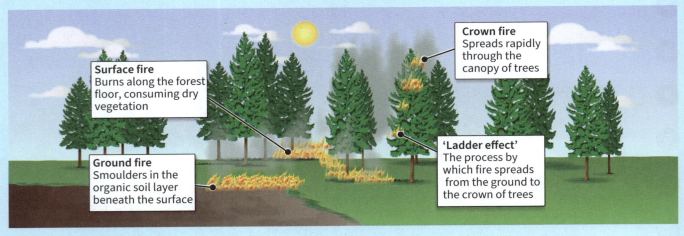

**Surface fire**
Burns along the forest floor, consuming dry vegetation

**Ground fire**
Smoulders in the organic soil layer beneath the surface

**Crown fire**
Spreads rapidly through the canopy of trees

**'Ladder effect'**
The process by which fire spreads from the ground to the crown of trees

▲ *Figure 1* Types of wildfires

## Conditions favouring wildfires

Most wildfires occur during or after prolonged dry periods when vegetation (the fuel for wildfires) has become dry and combustible.

**Vegetation type:**
Dry, flammable vegetation fuels fires (the greener the vegetation, the slower the fire spread)
Some species are fire-promoting (e.g. eucalyptus from Australia)
Invasive species may also be more flammable than indigenous species

**Fuel supply:**
Forests provide a greater amount of fuel than grassland
Forests where undergrowth has not been cleared have more fuel
Forests with large quantities of dry, dead vegetation have high potential for wildfire

**Topography:**
Steep slopes enhance fire spread as heat rises, meaning flames reach vegetation above on the slope more quickly

**Conditions favouring wildfires**

**Climate and recent weather:**
Droughts dry out vegetation (lower moisture content)
High temperatures promote combustion
Winds carry flames and embers, helping fires spread quickly

**Human activity:**
Logging creates favourable conditions for wildfires because of the quantity of dead vegetation left by logging operations

## Categorising hazard impacts

- Primary: occur as the direct result of the hazard, for example buildings being burned down, crops being burned in fields.

- Secondary: arise as a result of primary hazards; for example, the risk of flooding can increase in areas where vegetation has been removed by burning.

- Environmental: increased carbon emissions and air and water pollution (ash deposits); damage to wildlife/vegetation/ecosystems; reduction in biomass due to loss of habitats.

- Social: deaths and injuries; destruction to buildings and infrastructure (communications, transport, power); homelessness; trauma.

- Economic: high cost of rebuilding buildings and infrastructure; disruption to businesses; unemployment; loss of personal belongings; loss of crops/livestock.

- Political: pressure on emergency services; review of mitigation/preparedness strategies.

- All of these can have short- and long-term impacts.

## Short- and long-term responses

| Short-term responses | Long-term responses |
| --- | --- |
| Firefighting efforts to control and extinguish wildfires; evacuating from danger zones, which can change quickly if wind direction changes; emergency shelters to provide shelter and basic necessities to those affected by the disaster. | In some ecosystems, wildfires play an important role in the nutrient cycle and provide opportunities for regrowth. However, if wildfires are frequent, replanting programmes are required to replace burned vegetation. Infrastructure and buildings are repaired and replaced, and support given to local business to help their recovery. Fire management plans are updated to prevent future wildfires from spreading. Community education to reiterate the importance of fire safety. |

## Preparedness, mitigation, prevention and adaptation

### Preparedness

Being prepared for a hazard event enables authorities and communities to be able to take precautionary measures before a wildfire occurs (e.g. evacuation), reducing potential impacts.

Prediction involves:

- weather monitoring: observing temperature, humidity and wind, which influence wildfire likelihood

- fuel monitoring: assessing vegetation dryness and abundance, and the extent of green vegetation

- satellite and aerial (drone) surveillance: identifying hot spots or areas with high wildfire potential.

### Mitigation

Clearing dead vegetation and controlled burning, either before or during an event, reduces fuel supply. Homeowners in forest areas are encouraged to clear vegetation from around their homes for the same reason. During wildfires, clearing vegetation ahead of the fire creates a firebreak. These are often created alongside natural firebreaks such as rivers.

### Prevention

When conditions dictate, the common causes of wildfires, such as campfires or other recreational outdoor fires, are banned and strictly enforced.

 **Key terms** Make sure you can write a definition for these key terms

crown fire    ground fire    ladder effect    surface fire

# ⚙ KNOWLEDGE

## 5 Hazards

### 5.7 Case study: A place in a hazardous setting

 **Case Study: Lahaina, Hawaii (August 2023)**

## Background and causes

- Lahaina is a settlement on the Hawaiian island of Maui.
- In August 2023, the island was in its dry season, experiencing drought and strong winds gusting at over 100 km/h.
- The risk of wildfires has increased as there are fewer farmers. Farming removes dead vegetation (fuel) so with less cultivated land, and the spread of non-indigenous guinea grass which is prone to burning, the wildfire risk has increased.
- Strong winds blew down electricity pylons. It is thought that sparks from this caused the ignition of the wildfire.

## Impacts of the wildfire

Firefighters contained the initial brushfire having soaked it with 23,000 gallons of water and cleared a containment line. However, it reignited and was soon out of control and moving towards Lahaina at 80 km/h.

- 115 people were killed – the worst wildfire disaster in the USA for 100 years.
- 2000 homes and 800 businesses employing 7000 people were damaged or destroyed.
- 7000 people were made homeless.
- Historic buildings, some 200 years old, were destroyed.
- Economic losses were estimated at US$5billion.
- Toxic ash from the fires was expected to cause environmental damage to marine ecosystems.

## Responses to the wildfire

- Maui has emergency sirens, but they were not sounded as they were principally to warn of tsunamis and may have caused confusion.
- Instead, an alert was sent to mobile phones, but these arrived too late.
- The US government sent specialist search and rescue teams accompanied by human-remains detection dogs to recover bodies of those who perished.
- Payments of US$700 were made to each survivor to help with immediate living costs.
- The Hawaiian government requested US$24 million to improve responses to wildfires by reducing fuel (e.g. dead vegetation and grasses) and improving fire-fighting equipment, including water storage tanks across the islands.

## Recovery

- The process of rebuilding was delayed by Hawaii's strict building permit system. After seven months, a new permit office was opened to help speed up the process.
- In the meantime, those made homeless by the fire were accommodated in hotels and in temporary housing constructed by FEMA, the USA's emergency management authority.

▶ **Figure 2** *Impact of the Lahaina wildfire*

# 5.8 Case study: A multi-hazardous environment

## Case Study: Japan as a multi-hazardous environment

### Background

Japan experiences volcanic, seismic and storm hazards, which it manages through risk mitigation, disaster preparedness, prediction and monitoring, and reconstruction post-hazard.

**REVISION TIP**

You should revise the case study you have studied. This page provides the example of the 2011 Japan earthquake and tsunami.

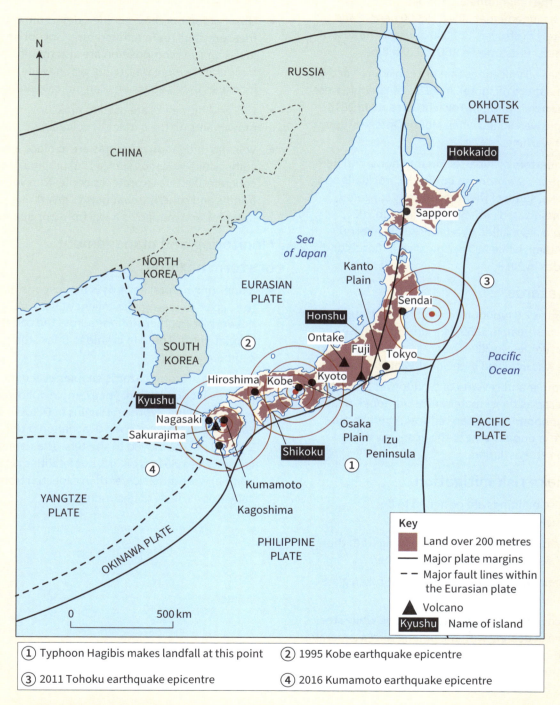

① Typhoon Hagibis makes landfall at this point  ② 1995 Kobe earthquake epicentre

③ 2011 Tohoku earthquake epicentre  ④ 2016 Kumamoto earthquake epicentre

▲ *Figure 1* *Japan is a multi-hazardous environment*

## 5 Hazards

### Seismic and volcanic hazards

- Japan is located on the Pacific Ring of Fire. The northern part of the country is on the Okhotsk Plate, while the southern part is on the Eurasian Plate.

- In 2011, the magnitude 9.1 Tohoku earthquake caused a tsunami that killed 18,000 people and completely destroyed 123,000 houses. The cost of the disaster is estimated at US$220 billion.

- Japan has 55 volcanoes that are considered active (with one more offshore). In 1926, 144 people were killed by an eruption of Tokachi volcano. In 2014, 63 people were killed when Mount Ontake erupted without warning.

- Three-quarters of Japan is mountainous, and here steep slopes are prone to landslides, rockfalls and mudflows, some of them seismically triggered, others associated with heavy rainfall. In 2021, 27 people were killed by a landslide in a resort called Atami, while 200 people died in floods and landslides in 2018.

### Storm hazards

- On average 2.6 typhoons reach the main islands of Japan each year. In 1959, Isewan Typhoon caused 5000 casualties; in 2004, heavy rain from Typhoon Tokage was involved in the deaths of 98 people.

- The number and severity of typhoons is predicted to increase as the climate warms. Although deaths from typhoons have been reduced, there is still a severe economic cost to the economy when a typhoon strikes Japan.

### Earthquake risk mitigation

- Japanese buildings are designed to be earthquake-resistant.

- Seismic isolators separate the building from the ground with a layer of springs or rubber.

- Automatic window shutters prevent broken glass from falling onto the streets below.

- Wood construction is mixed with steel: while steel is effective at dealing with the pulling forces generated by earthquakes, wood performs better with compressive forces.

### Volcano monitoring

- Scientists monitor all Japan's active volcanoes.

- A network of seismometers monitor volcanoes for earthquakes. An increased frequency of earthquakes can indicate an eruption is imminent.

- Observation boreholes in the sides of the volcano measure changes in temperature. Hot spring water levels and gas composition are also monitored: temperature increases, rising groundwater and increases in gas emissions can indicate eruptions.

- GPS and tiltmeters are used to measure any movements on the surface of volcanoes.

- Volcano mitigation measures are in place around some active volcanos such as Sakurajima, including concrete shelters to protect people from volcanic bombs and concrete channels to divert nuées ardentes and mudflows away from populated areas.

### Monitoring and management of storm hazards

- Monitoring natural hazards is the responsibility of the Japan Meteorological Agency (JMA), which has a network of weather stations across the country, and two satellites. The JMA is able to track storms and estimate intensity.

- Warnings of storm hazards enables evacuation plans to come into operation. For example, following warnings about Typhoon Hagibis in October 2019, 800,000 people were given evacuation orders and 230,000 people moved to shelters. The damage caused was US$17.3 billion, but fatalities (118 people) were much lower than expected from this catastrophic level of destruction.

# RETRIEVAL

Learn the answers to the questions below, then cover the answers column with a piece of paper and write down as many as you can. Check and repeat.

## Questions | Answers

| | Questions | | Answers |
|---|---|---|---|
| 1 | What name is given to wildfires that smoulder underground in the organic layer beneath the surface? | Put paper here | Ground fires |
| 2 | Name one way in which wildfires may have natural causes. | | One from: lightning strike / volcanic eruption |
| 3 | What is the name given to the process by which wildfires climb from the ground to the top of trees? | | The ladder effect |
| 4 | Complete this saying relating to wildfire spread: The greener the vegetation… | Put paper here | …the slower the fire spread |
| 5 | How can topography influence the spread of wildfires? | | Steep slopes enhance fire spread as heat rises, meaning flames reach vegetation above on the slope more quickly |
| 6 | As well as droughts and dry periods, what other weather conditions favour intense wildfires? | Put paper here | Hot temperatures and strong winds |
| 7 | How may invasive vegetation species contribute to conditions favouring intense wildfires? | | They may be more flammable than indigenous species |
| 8 | Name one way in which wildfires are predicted. | | One from: weather monitoring / fuel monitoring / satellite and aerial surveillance |
| 9 | Why is Japan considered a multi-hazardous environment? | Put paper here | It regularly experiences more than one hazard type: earthquakes, volcanic eruptions, landslides and storms (typhoons) |
| 10 | In what year was the Tohoku earthquake and tsunami and how many people died? | | 2011 and 18,000 people |

## Previous questions

Now go back and use these questions to check your knowledge of previous topics.

## Questions | Answers

| | Questions | | Answers |
|---|---|---|---|
| 1 | Name three factors that affect volcano explosivity. | Put paper here | Three from: silica content / gas content / magma temperature / presence of water / eruption history |
| 2 | What is the name of the type of seismic hazard in which soils temporarily act as a fluid? | | Liquefaction |
| 3 | A tropical storm is characterised by a wind speed in excess of what? | | 120 km/h (75 mph) |
| 4 | Name three of the hazards associated with tropical storms. | Put paper here | Three from: high wind / storm surge / coastal flooding / river flooding / landslides |
| 5 | Name an environmental impact of Typhoon Haiyan. | | Salt contamination by storm surge ruined 1 million tonnes of crops |

## Exam-style questions

1   Outline the nature of tephra, a volcanic hazard.   **[4 marks]**

2   Outline the features of the Park model of human response to hazards.   **[4 marks]**

3   Outline the concept of mitigation.   **[4 marks]**

4   Outline the concept of hazard perception.   **[4 marks]**

5   Outline processes leading to tectonic plate movement.   **[4 marks]**

6   Outline the relationship between volcanoes and plate margins.   **[4 marks]**

7   **Figure 1** shows an assessment of wildfire risk in Hawaii.

Assess the usefulness of **Figure 1** in presenting information on wildfire risk.   **[6 marks]**

> **EXAM TIP**
>
> This is Section C of Paper 1. You will have a choice of either Question 5 (Hazards) **or** Question 6 (Ecosystems under stress). Answer questions on the topic you have studied only!

> **EXAM TIP**
>
> 4-mark questions are point-marked. That means each valid point or development of a valid point receives 1 mark, up to a maximum of 4 marks.

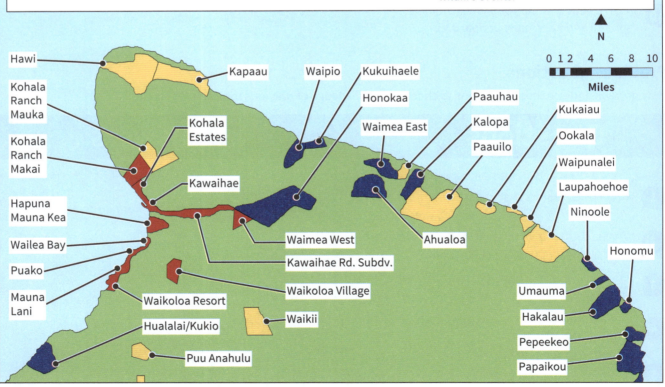

**Fire Environment Rating - Ignition Risk**

**Key**                              **IGNITION RISK**

☐ Low Hazard     ☐ Moderate Hazard     ☐ High Hazard

Little to no natural (lightning or lava) ignition risk. No history of arson Wildland areas absent or distant from public and/or vehicular access.

Some history of wildfire, but not particularly fire prone area due to prevailing lack of fire prone conditions, weather, and vegetation type.

Most historic wildfire events were anthropogenic with easy access to wildland areas via roads or proximity to development OR natural ignition sources such as lightning or lava are prevalent Fire prone area. High rate of ignitions or history of large scale fires and/or severe wildfire events.

▲ *Figure 1* *Hawaii fire hazard assessment*

**8** **Figure 2** shows the number of deaths from earthquakes worldwide between 1818 and 2017.

Analyse the data shown in **Figure 2**. **[6 marks]**

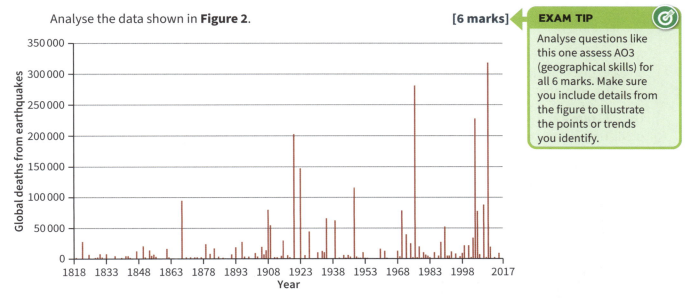

▲ **Figure 2** Global deaths from earthquakes, 1818–2017

**9** **Figure 3** shows the number of recorded deaths and reported economic losses from natural disasters by decade from 1970 to 2019.

Analyse the data shown in **Figure 3**. **[6 marks]**

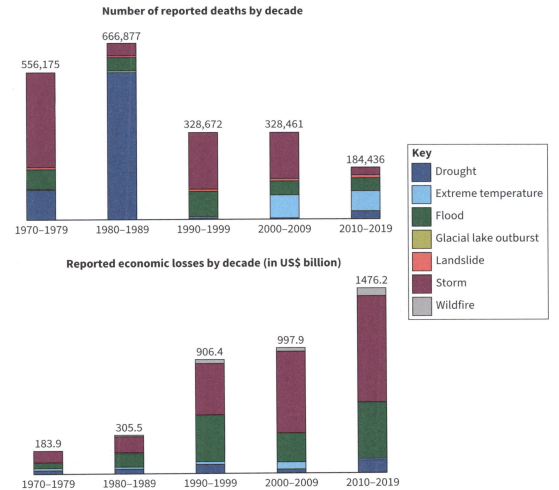

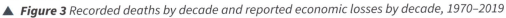

▲ **Figure 3** Recorded deaths by decade and reported economic losses by decade, 1970–2019

10 **Figure 4** shows a radiative power log for Guatemala's Volcán de Fuego. On 3 June 2018, this volcano erupted. Three thousand people were evacuated from surrounding areas, but 200 people were killed by pyroclastic flows and lahars.

Radiative power is calculated from infrared satellite images of the volcano that make it possible to measure the thermal energy radiated by active volcanoes.

Using **Figure 4** and your own knowledge, assess the challenges of predicting volcanic eruptions. **[9 marks]**

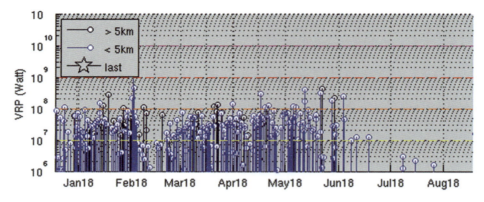

▲ **Figure 4** *Radiative power measurements for the Fuego volcano, Guatemala, Jan–Aug 2018*

11  **Figure 5** shows a tsunami hazard map for Okitsu, a village in Japan with a population of under 1000, where a third of residents are over the age of 75. Predictions are made on the basis of a tsunami with maximum inundation (flooding) height of 15m.

Using **Figures 5a**, **5b** and your own knowledge, assess the challenges of managing tsunami risk in Okitsu village.                                    **[9 marks]**

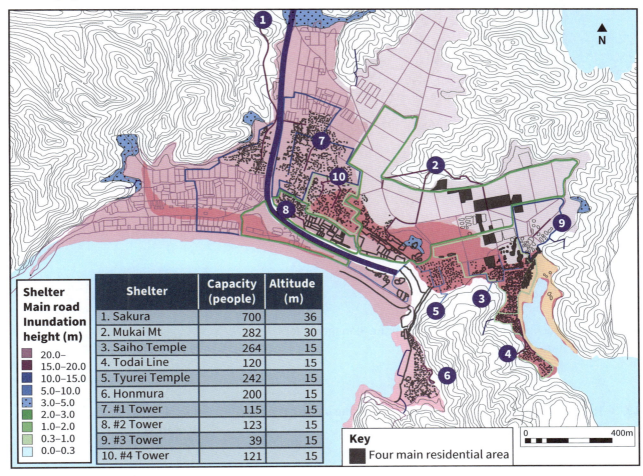

The map includes the following shelter table:

| Shelter | Capacity (people) | Altitude (m) |
|---|---|---|
| 1. Sakura | 700 | 36 |
| 2. Mukai Mt | 282 | 30 |
| 3. Saiho Temple | 264 | 15 |
| 4. Todai Line | 120 | 15 |
| 5. Tyurei Temple | 242 | 15 |
| 6. Honmura | 200 | 15 |
| 7. #1 Tower | 115 | 15 |
| 8. #2 Tower | 123 | 15 |
| 9. #3 Tower | 39 | 15 |
| 10. #4 Tower | 121 | 15 |

Legend:
- Shelter
- Main road
- Inundation height (m)
  - 20.0–
  - 15.0–20.0
  - 10.0–15.0
  - 5.0–10.0
  - 3.0–5.0
  - 2.0–3.0
  - 1.0–2.0
  - 0.3–1.0
  - 0.0–0.3

Key: ▪ Four main residential area

Scale: 0 — 400m

▲ *Figure 5a* Okitsu tsunami hazard map

# Exam-style questions

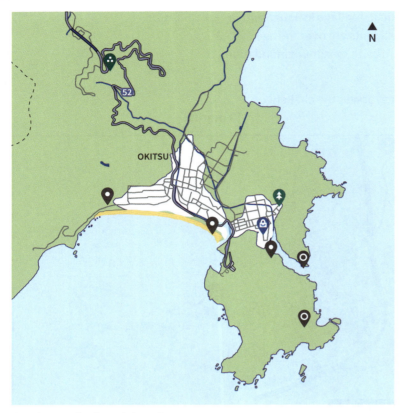

▲ *Figure 5b* *Okitsu's location*

12 Assess the usefulness of prediction in the management of
storm hazards. **[9 marks]**

13 How far do you agree that climate change is the biggest threat
to the successful management of wildfires? **[9 marks]**

14 'Prediction is less useful than managing the response to hazards when
they occur.'

With reference to a multi-hazardous environment you have studied,
to what extent to you agree with this view? **[20 marks]**

15 To what extent has plate tectonic theory influenced human responses
to seismic hazards? **[20 marks]**

16 For a local scale place in a hazardous setting that you have studied, assess
the community's responses to the risks in relation to the physical nature of
the hazard. **[20 marks]**

> **EXAM TIP**
>
> 20-mark questions have
> 10 marks available for
> AO1 and 10 marks for
> AO2 (your argument).
> Some degree of balance
> in your answer is needed.
> Could you present a
> counterargument in your
> penultimate paragraph,
> ahead of your conclusion?

 **KNOWLEDGE**

# 6 Ecosystems under stress

## 6.1 Ecosystems and sustainability

### Global trends in biodiversity

**Biodiversity** is the variability among living organisms, the variety of life on Earth. This includes diversity within species, between species and diversity of **ecosystems**.

Global trends in biodiversity show an unprecedented rate of biodiversity decline, creating a nature crisis that is undermining natural services on which human life depends.

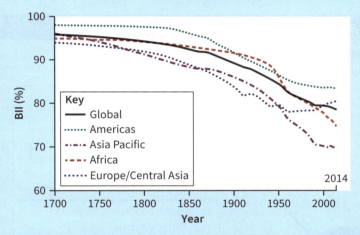

▲ **Figure 1** *The Biodiversity Intactness Index (BII) estimates how much originally present biodiversity remains on average across the terrestrial biomes within a region. 90% is the proposed safe limit*

Over the last 50 years there has been a 68% decrease in population sizes of mammals, birds, amphibians, reptiles and fish.

One million animal and plant species are threatened with global extinction, many within the next few decades.

Demand for resources is unsustainable, encouraged by governments that prioritise economic growth.

The WWF (World Wildlife Fund) estimates that the global human population is now using the resources of 1.6 planets.

Diversity is reducing and all terrestrial ecosystems, and many marine ecosystems, are damaged.

### UK trends in biodiversity

Compared with other countries in the G7, the UK performs the worst in terms of how much biodiversity survives.

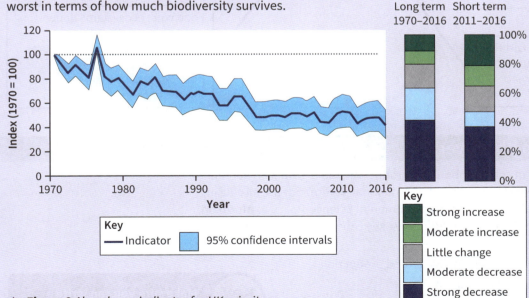

▲ **Figure 2** *Abundance indicator for UK priority species, 1970–2016, a decline of 60%. Priority species are those of primary importance for conserving biodiversity: there are 940 such species in the UK*

According to the RSPB (Royal Society for the Protection of Birds), 15% of species in the UK are threatened with extinction, with only 5% of the UK's land protected and managed effectively for nature.

It is estimated that, since the 1930s, 97% of the wildflower meadows in Britain have gone, 70% of floodplains are now intensively farmed and, since 1990, 1.9 million acres of grassland has been used for housing developments.

The Woodland Trust estimates that up to 70% of the UK's ancient woodland has been lost.

# KNOWLEDGE

## 6 Ecosystems under stress

### 6.1 Ecosystems and sustainability

## Ecosystems and their importance

Declining biodiversity undermines the ability of natural systems, including ecosystems, to provide crucial services, such as: habitats, food, medicines, pollination and dispersion of seeds, regulation of air quality, ocean acidity, the water cycle and climates, formation of soils and buffers against natural hazards (e.g. mangroves, coral reefs).

## Humans and ecosystem development and sustainability

Ecosystems are central to human development and well-being. Biodiversity decline has serious consequences for human and ecosystem functioning.

**Biodiversity decline**

Ecosystem resilience: ecosystems less resilient to environmental changes

Food security: reduced crop yields (e.g. declines in pollinators, reduction in soil nutrients) and fishery harvests, increased vulnerability to pest outbreaks

Ecosystem services: affects many ecosystem processes (e.g. nutrient cycling, uptake of carbon); loss of wetlands/forests can reduce capacity to purify water and regulate climate

Vulnerability to diseases: certain species may serve as natural buffers by limiting the spread of pathogens

Human health: reduction in availability of medicinal plants, increased exposure to diseases transmitted by wildlife, disruption in balance of ecosystems affecting air/water quality

Economic costs: e.g. overfishing or habitat destruction can affect the livelihoods of fishing communities and decrease the availability of seafood

Cultural and spiritual losses: can erode traditional knowledge and practices and disrupt cultural connections with nature

Food web: ecosystems depend greatly upon interactions among species, e.g. if carnivores are removed, prey species populations increase, leading to further uncontrolled changes

## Causes of declining biodiversity

Changing patterns of human consumption and production, rapid growth in human populations and the actions of national and international governance are all causing declining biodiversity. The IPBES (the Intergovernmental Science-Policy Platform on Biodiversity and Ecosystem Services) identifies five key causes of declining biodiversity, in order of their importance:

- changes in land and sea use, principally for food production
- direct exploitation of animals and plants
- climate change
- pollution
- invasive species.

 **Key terms** Make sure you can write a definition for these key terms

biodiversity   ecosystem

# RETRIEVAL

Learn the answers to the questions below, then cover the answers column with a piece of paper and write down as many as you can. Check and repeat.

## Questions | Answers

| # | Questions | Answers |
|---|-----------|---------|
| 1 | Define biodiversity. | The variability among living organisms, the variety of life on Earth |
| 2 | Over the last 50 years, what percentage decrease has there been in the population sizes of mammals, birds, amphibians, reptiles and fish? | 68% |
| 3 | How many animal and plant species are threatened with extinction globally? | One million |
| 4 | How does the UK rank in terms of biodiversity in the G7 countries? | It performs the worst in terms of how much biodiversity survives |
| 5 | What is meant by the term 'priority species'? | Species that are of primary importance in conserving biodiversity |
| 6 | What percentage of species in the UK are threatened with extinction? | 15% |
| 7 | According the Woodland Trust, how much of the UK's ancient woodland has been lost? | Up to 70% |
| 8 | According to IPBES, what is the top cause of declining biodiversity worldwide? | Changes in land and sea use |
| 9 | Name two more of the IPBES' top five causes of declining biodiversity. | Two from: direct exploitation of animals and plants / climate change / pollution / invasive species |
| 10 | Explain one way in which a decline in biodiversity could reduce food security. | One from: decreased crop yields (e.g. because of declines in pollinators, reduction in soil nutrients) / reduced fishery harvests / increased vulnerability to pest outbreaks |
| 11 | Identify two ecosystem services that are threatened by loss of wetlands and forests. | The capacity of wetlands and forests to purify water and regulate climate |
| 12 | How may biodiversity decline increase vulnerability to disease? | Certain species may serve as natural buffers by limiting the spread of pathogens |
| 13 | Give an example of an economic cost of biodiversity decline. | Overfishing affects livelihoods of fishing communities |

Put paper here

# 6 Ecosystems under stress

## 6.2 Ecosystems and processes

## Structure of ecosystems

### Biotic and abiotic factors

In an ecosystem, living organisms (biotic factors) interact with their physical environment (abiotic factors).

- Biotic elements are the living part of an ecosystem such as micro-organisms, plants and animals.
- Life is supported by the abiotic elements such as weathered rock, climate and various gases.

Ecosystems are structured hierarchically, with organisms organised into populations, which in turn make up communities. Communities interact within ecosystems.

### Energy flows

Ecosystems require a constant input of energy. This energy enters ecosystems as sunlight (in almost all ecosystems) during photosynthesis, where plants and organisms convert it into chemical energy.

Energy flows through ecosystems in a one-way direction, passing from producers to consumers to decomposers.

### Trophic levels

**Trophic levels** represent the different positions within a food chain or web, indicating an organism's role in energy transfer. Energy is lost as heat at each trophic level. There are typically four main trophic levels.

**Trophic level 4**
Tertiary consumers (carnivores): animals that eat secondary consumers

**Trophic level 3**
Secondary consumers (carnivores or omnivores): animals that eat primary consumers

**Trophic level 2**
Primary consumers (herbivores): animals that eat producers

**Trophic level 1**
Producers (autotrophs): organisms such as plants, algae and some bacteria that use sunlight to create their food through photosynthesis

▲ **Figure 1** Trophic levels

### Food chains and food webs

A **food chain** is a simplified linear representation of energy transfer in an ecosystem. For example:

Sun ▶ Grass (producer) ▶ Rabbit (primary consumer) ▶ Fox (secondary consumer)

A **food web** is a more realistic and complex representation of relationships between species and their diets, showing multiple interconnected food chains. Organisms have multiple potential food sources, so a food web is a more accurate depiction of the energy flow in an ecosystem.

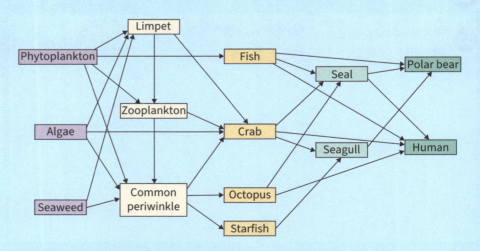

▲ **Figure 2** A marine ecosystem food web

## Ecosystems as systems

| System concept | Definition | Examples |
|---|---|---|
| Inputs | Inputs are the resources and energy that enter an ecosystem from its external environment | Solar energy, precipitation, nutrients (introduced by weathering of rocks) |
| Outputs | Outputs are resources and energy that leave the ecosystem and flow into other ecosystems or into the atmosphere | Heat, runoff, migration of species |
| Stores | Stores are reservoirs within the ecosystem where energy and materials are temporarily stored or accumulated | Biomass (the total mass of living organisms within an ecosystem), soil, water bodies |
| Transfers | Transfers represent the movement of energy and materials within the ecosystem | For example, photosynthesis, predation (transfer of energy from one trophic level to another), decomposition (carbon) |

### Biomass and net primary production

Biomass is the total mass of all biotic organisms per unit area.

**Net primary production** (NPP) is the biomass (minus respiration) produced in an ecosystem – the net difference between photosynthesis and respiration. It is a measure of the rate at which an ecosystem can capture and store carbon.

Tropical rainforests have high NPP throughout the year. Temperate forests have seasonal variations because the rate at which plants take $CO_2$ out of the air reduces in winter. Savanna grasslands rival tropical rainforests for NPP during and just after the wet season.

## Succession

**Succession** is the gradual, predictable process of changes in the composition and structure of plant and animal communities within an ecosystem over time.

- It is driven by changes in environmental conditions and the interactions between species.
- It leads to increased complexity, diversity and stability in ecosystems over time.

| Concept | Definition |
|---|---|
| Primary succession | Succession that starts where no life previously existed, such as on newly formed volcanic islands or rock surfaces left after glacial retreat. |
| Secondary succession | Occurs in areas where an ecosystem has been disturbed or disrupted by events like wildfires, logging or farming. |
| Lithosere | Primary succession that occurs on exposed rock surfaces. |
| Pioneer species | A pioneer species is a plant species which requires very few nutrients to survive. |
| Seral stage | A stage in the development of the ecosystem as it progresses towards climax. New species that are better adapted to the conditions outcompete a declining species. |
| **Climatic climax** | The final stage in the succession of a distinct vegetation community. The flora and fauna in the community have attained a state of dynamic equilibrium. |
| **Sub-climax** | Sub-climax vegetation has not been able to mature to the natural climatic climax vegetation for the region. This is because of an arresting factor. |
| **Plagioclimax** | A sub-climax ecosystem area where the arresting factor is human interference, e.g. heather moorland. |

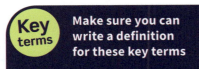
**Key terms** Make sure you can write a definition for these key terms

climatic climax    food chain    food web
net primary production (NPP)    nutrient cycle
plagioclimax    sub-climax    succession    trophic level

# 6 Ecosystems under stress

## 6.2 Ecosystems and processes

### Succession in a lithosere

| Pioneer species such as moss and lichen colonise bare rock | → | Decomposition of dead plants alongside weathering of rock start basic soil formation | → | Species such as grasses outcompete the mosses and lichens; decomposers and other fauna start to colonise the area |

| Climatic climax occurs when the largest species the area can accommodate, takes hold and colonises | ← | Seeds brought in by birds and wind; improved soil structure mean larger species can colonise, and outcompete existing species |

In the UK, oak woodland is an example of a climatic climax community – an optimum state for the given conditions. No further changes will occur if all elements remain stable.

### Mineral nutrient cycling

The **nutrient cycle** constantly recycles nutrients (such as carbon, nitrogen, phosphorous, sulphur) between the soil and plants. The Gersmehl Nutrient Cycle provides a systems-based view of the mineral nutrient cycle. The size of the circles indicates the size of the stores and the thickness of the arrows are proportional to the size of the transfers, allowing easy comparison between different ecosystems.

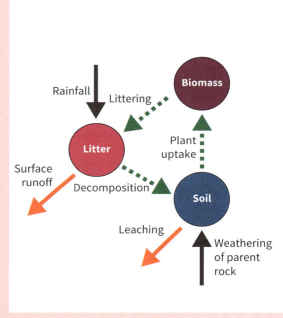

**Nutrient inputs** →
• Nutrients dissolved in raindrops
• Nutrients from weathered rock

**Nutrient transfers** ┄┄►
• Littering – fallout from the plants, mostly leaf fall transferring nutrients to the litter
• Decomposition – decay of organic material in the litter by fungi and bacteria transferring nutrients to the soil
• Plant uptake – the uptake of nutrients from the soil through the plant roots

**Nutrient outputs** →
• Nutrients lost through surface runoff
• Nutrients lost through leaching

**Nurtrient stores** ○
• **Biomass** – contains all living plant and animal matter in the ecosystem
• Soil – contains minerals from the parent rock in addition to humus from decomposed plant and animal remains
• Litter – sits on top of the soil and contains both dead and decaying plant and animal material

▲ *Figure 2 The Gersmehl nutrient cycle*

The nutrient cycles for different biomes vary considerably. For example, tropical rainforests have a small 'litter' nutrient store as leaves are rapidly decomposed, while coniferous forests have large litter stores due to cold conditions and waxy leaves, meaning slower decomposition.

# 6.3 Terrestrial ecosystems

## Factors affecting terrestrial ecosystems

Climate:
- Temperature
- Precipitation
- Seasonal change
- Climate change impacts

Vegetation:
- Plant types (adaptation to climate)
- Climatic climax
- Biodiversity

Soils:
- Composition (e.g. loam, clay, sand)
- Nutrient content (nutrient cycle)
- pH levels
- Soil erosion

**Factors affecting terrestrial ecosystems**

Topography:
- Elevation (altitude)
- Slope
- Aspect (orientation to the sun)
- Drainage patterns
- Microclimates

Time:
- Succession
- Weathering / soil formation
- Natural climate change (glaciation)
- Human impacts over time (historical land use)
- Natural hazards (e.g. flooding)

## Responses to change

Ecosystems respond slowly to natural changes in one or more of their components or environmental controls. Response to anthropogenic (human-influenced or -caused) changes are more rapid.

| Long-term influences | Short-term influences |
|---|---|
| • Plate tectonics can lead to changes in ocean circulation, affecting climates over millions of years.<br>• Changes in the Earth's orbit and axial tilt over tens of thousands of years (Milankovitch cycles) cause warmer and colder phases (ice ages).<br>• Changes in solar output and sunspot cycles can influence climate changes over centuries.<br>• Volcanic eruptions can cause temporary cooling for several years due to the release of aerosols into the atmosphere. | • Short-term changes in climate, or other factors, do not usually produce substantial responses in ecosystems.<br>• Gradual change in climate will lead to profound, but gradual, changes in ecosystems.<br>• The damage caused to biodiversity from anthropogenic climate change comes from the rapidity of the change – decades or centuries – which is much faster than the capacity of most plants and many animals to adapt to it. |

## Factors influencing ecosystem change

There are many factors influencing ecosystem change. Three are:

- Changing climate conditions can affect the distribution of species, timing of growing and flowering seasons, and the functioning of ecosystems.
- Land use changes, habitat destruction and the release of pollutants can disrupt ecosystems and lead to biodiversity loss.
- The introduction of invasive species to new environments can outcompete indigenous species for resources, disrupt food chains, and change ecosystem dynamics.

Learn the answers to the questions below, then cover the answers column with a piece of paper and write down as many as you can. Check and repeat.

## Questions | Answers

| | Questions | | Answers |
|---|---|---|---|
| 1 | What are biotic elements of an ecosystem? | | The living parts of an ecosystem |
| 2 | What are trophic levels? | | The different positions within a food chain or web, indicating an organism's role in energy transfer |
| 3 | Define biomass. | | The total mass of all biotic organisms per unit area |
| 4 | What is a pioneer species? | | A plant species which requires very few nutrients to survive |
| 5 | What is a seral stage? | | A stage in the development of the ecosystem as it progresses towards climax |
| 6 | What is a sub-climax in a vegetation succession? | | Vegetation that has not been able to mature to the natural climatic climax vegetation for the region |
| 7 | What is a plagioclimax? | | A sub-climax ecosystem area where the arresting factor is human interference |
| 8 | Name a plagioclimax vegetation community found in upland areas of Britain. | | Heather moorland |
| 9 | Name the diagram used to represent nutrient cycles in ecosystems. | | Gersmehl Nutrient Cycle |
| 10 | What are the five main factors affecting the type and characteristics of terrestrial ecosystems? | | Climate, vegetation, soils, topography, time |

*Put paper here*

## Previous questions

Now go back and use these questions to check your knowledge of previous topics.

## Questions | Answers

| | Questions | | Answers |
|---|---|---|---|
| 1 | Define biodiversity. | | The variability among living organisms, the variety of life on Earth |
| 2 | What is meant by the term 'priority species'? | | Species that are of primary importance in conserving biodiversity |
| 3 | Explain one way in which a decline in biodiversity could reduce food security. | | One from: decreased crop yields / reduced fishery harvests / increased vulnerability to pest outbreaks |
| 4 | Identify two ecosystem services that are threatened by loss of wetlands and forests. | | The capacity of wetlands and forests to purify water and regulate climate |
| 5 | How may biodiversity decline increase vulnerability to disease? | | Certain species may serve as natural buffers by limiting the spread of pathogens |

*Put paper here*

#  KNOWLEDGE

## 6 Ecosystems under stress

### 6.4 Biomes

## The global distribution of major terrestrial biomes

A **biome** is a large-scale ecological area with plants and animals that are well adapted to their environmental conditions. The distribution of terrestrial biomes is largely a match to the pattern of climate zones, with variation due to relief, continentality and, at a local scale, geology.

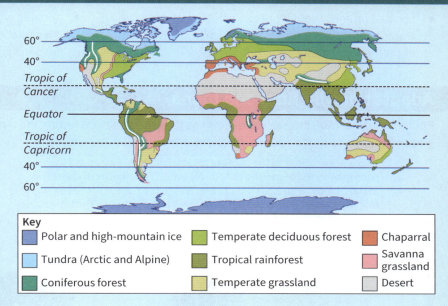

▶ **Figure 1** Distribution of terrestrial biomes

**Key**

- Polar and high-mountain ice
- Tundra (Arctic and Alpine)
- Coniferous forest
- Temperate deciduous forest
- Tropical rainforest
- Temperate grassland
- Chaparral
- Savanna grassland
- Desert

## Tropical rainforest biome

| Climate | |
|---|---|
| Temperature | Average daytime temperature range of 28°C, and rarely falls below 22°C at night |
| Precipitation | 2000mm annually evenly distributed through the year, due to the 'Inter Tropical Convergence Zone' (ITCZ) |
| Humidity | Often exceeds 80% due to the continuous evapotranspiration |
| Seasonality | Year-round growing season with warm temperatures, high precipitation and 12 hours of daylight |
| **Soil** | |
| Nutrient cycle | Poor in nutrients with leaching from the heavy precipitation; most nutrients held in vegetation |
| Soil type | Intensive and prolonged weathering typically produce soils with low fertility and with a hard, iron-rich layer near the surface (laterite) |
| **Soil moisture budget** | |
| Water surplus | Little variation in soil moisture due to minimal seasonal variation in rainfall, though some rainforest areas are flooded in wet seasons |
| Season variation | Tropical rainforests receive more precipitation than they lose through evapotranspiration and runoff |
| **Biodiversity** | |
| Biomass | Very high: typically 400 to 700 tonnes/ha |
| NPP | NPP is very high: an average of 2200 g/m$^2$/year |
| Species richness | Ideal growing conditions, with competition for sunlight and nutrients resulting in high levels of biodiversity |

# 6 Ecosystems under stress

## 6.4 Biomes

## Distribution of tropical rainforests

- Located largely, though not exclusively, between the tropics: latitude is a key factor determining the location of the tropical rainforest biome.

- Located where the ITCZ brings high rates of precipitation on a daily basis. At the same latitude, savanna grassland has a long dry season.

## Adaptations by flora and fauna

Epidemiological transition refers to how the types of disease affecting the population of a country changes as the country develops economically and socially.

Fauna are adapted to the rainforest's stratification. Eagles use the emergent trees to scan the canopy for prey. The majority of fauna live in the canopy, with birds either camouflaged to escape predation or brightly-coloured to stand out to potential mates among the leaves.

Leaves have 'drip tips' to prevent water accumulating, reducing the risk of fungal growth blocking ability to absorb sunlight.

Emergent trees break through the canopy. Their seeds are adapted to be blown on the wind.

Thin soils mean trees have relatively shallow roots. Competition for sunlight means leaves have large surface areas to maximise photosynthesis. Buttress roots provide stability and support.

The canopy, understory and shrub layers have vegetation adapted to low light conditions, with large, broad leaves. No wind so plants rely on fauna to spread their seeds.

Emergent layer

Canopy

Under-storey

Shrub layer and forest floor

Height above ground (m): 50 45 40 35 30 25 20 15 10 5

The canopy layer traps much of the rainfall and blocks most sunlight. Only around 1% of sunlight reaches the forest floor.

▲ **Figure 2** *Stratification and vegetation adaptations in a tropical rainforest*

## Human activity and its impact on the tropical rainforest biome

Mainly agriculture (80%) but also flooding, mining, settlement, and road building have all contributed to high rates of deforestation in the tropical rainforest biome. Since 1960, around 15% of the world's largest rainforest, the Amazon rainforest, has been removed.

Deforestation: destroys habitats reducing biodiversity. When tropical rainforest is replaced by monoculture plantations, biodiversity plummets.

Road-building: impacts drainage patterns, increases runoff, soil erosion and leading to changes in soil moisture.

**Impacts of human activity**

Removal of the canopy: leads to soil erosion and loss of soil fertility; soils are quickly eroded and nutrients leached out of them.

Climate change: increases temperatures and changes precipitation patterns, with negative impacts on the highly-adapted flora and fauna of rainforests. The risks of wildfire hazards increase in tropical rainforest biomes.

# Savanna grassland biome

| Climate | |
|---|---|
| Temperature | Temperature averages range from 18 (high latitude) to 22°C (low latitude) in the wet season and 28 (high) to 32°C (low) in the dry season |
| Precipitation | 500–1500 mm during the wet season; minimal rainfall during dry season |
| Seasonality | Distinct wet and dry seasons. May to September wet season in Northern Hemisphere and November to March in Southern Hemisphere |
| **Soil** | |
| Nutrient cycle | Poor in nutrients, with most nutrients found in the vegetation |
| Soil type | Savannas often have quick-draining sandy soils or laterite soils |
| Wildfires | Ecosystem is adapted to seasonal fires, which can play a role in nutrient cycling |
| **Soil moisture budget** | |
| Water surplus | The wet season replenishes soil moisture, while the dry season leads to moisture deficits |
| **Biodiversity** | |
| Biomass | Quite low at 0.5–20 tonnes/ha |
| NPP | NPP varies between 500g/m$^2$/year and 1000g/m$^2$/year |
| Species richness | Not as biodiverse as tropical rainforests, but still a species-rich biome |

# Distribution of savanna grasslands

- Located mainly within the tropics, principally in Africa with concentrations in South America, parts of south-east Asia and Australia.
- Latitude is a determining factor, together with continentality (drier conditions in continental interiors), ocean currents and relief.
- In some areas, there is enough precipitation for this biome to be forested, but regular wildfires prevent tree species from becoming established.

# Adaptations by flora

- Grasses lie dormant during the dry season. Their deep root systems access deep groundwater, and are resilient to fire. They produce quick regrowth when the first rains arrive.
- Only a few tree species can survive the long dry season. Trees and shrubs have waxy leaves reducing moisture loss, and deep tree roots. Some are deciduous, and shedding leaves during the dry season reduces water loss through transpiration.
- Many trees and plants have adapted to be fire-resistant, such as thick bark or being able to resprout quickly after a fire, while competing vegetation is cleared after a fire. Some species even need fire for reproduction.
- Some plants store nutrients underground, such as bulbs or tubers, which they can access during the dry season.

# 6 Ecosystems under stress

## 6.4 Biomes

### Adaptations by fauna

Many animals are primarily nocturnal, when temperatures are cooler.

Many small mammals, like rodents, are highly adapted to withstand periods of drought. They are able to obtain much of their water from their diet. Some also dig burrows to stay cool and avoid predators.

Some herbivores, such as wildebeests, migrate long distances to follow seasonal changes in vegetation and water availability.

Herbivorous animals have specialised diets, adapted to consume specific types of vegetation or parts of plants that are more abundant or accessible during different seasons.

▲ **Figure 3** *Giraffes' long necks enable them to graze on parts of trees and shrubs that other herbivores cannot reach. Zebras' stripy coats make it hard for predators to distinguish individual animals from a herd.*

## Human activity and its impact on the savanna grassland biome

Humans have been influencing grassland savanna landscapes in Africa for 300,000 years, especially through fires. Today, human impacts are fragmenting and degrading savanna grasslands.

- Overgrazing removes vegetation cover that wild herbivores rely on. With no protection, soil erosion can lead to desertification, severely reducing biomass.

- Tourists use vehicles for safaris, causing soil erosion. New housing for tourists encroaches onto grassland. Land use changes as farmland is extended to feed tourists. Provision of drinking water for wild animals (for photo opportunities) may disrupt herbivore migrations. However, tourism can contribute to conservation.

- Fire suppression policies, such as allowing invasive species to outcompete grassland species that depend on fire, impact on the biome. Higher population densities reduce the extent of fires, livestock eat the fuel and humans deforest the trees, but disrupts ecosystem processes.

- Farming, such as clearing land for crops, removes the primary producers on which the ecosystems depend. Commercial agriculture may attract insect pests such as locust swarms that then also affect grassland species.

- Hunting/poaching in response to high demand for some animal products (e.g. rhino horn, ivory) threaten extinction (e.g. black rhino). Wild animals and some farmers may have conflicting interests, sometimes leading to hunting of wildlife to preserve crops.

- Climate change increases the duration/severity of droughts fragmenting savanna grasslands. Overgrazing, deforestation and urbanisation add to the pressure and disrupt herbivore communities (interrupting migration routes) and fire regimes.

## Managing human impacts

Pressure on savanna grasslands is increasing, especially in African countries with high population growth rates (e.g. Tanzania has a growth rate of around 3% per year). Managing human impacts focuses on sustainable development, regulation and conservation.

- Controlled burning: controlled burning early in the dry season regenerates grassland growth, prevents the spread of invasive species and fertilises the soil with ash. Early-season burning also reduces the risk of extensive late-season wildfires.

- Sustainable agriculture: for example, education around crop rotation to protect soils, and indigenous tree planting (not non-indigenous species) can reduce human impact.

- International agreements banning the sale of products sourced through poaching (e.g. ban on trade of products from endangered animals established by the Convention on International Trade in Endangered Species of Wild Fauna and Flora (CITES) in 1990) protect fauna.

- The establishment of national parks promotes the conservation of biodiversity, the restoration of degraded environments, education for sustainable development and sharing of revenue with local people.

- Ecotourism is a sustainable alternative to mass tourism and offers more opportunities for local people to share in revenue generation. Local populations can otherwise resent the establishment of reservations and national parks.

- Controlled (regulated) hunting of African megafauna (e.g. elephants, giraffes, lions and hyenas) brings in revenue and can help to reduce the imbalances in grassland ecosystems caused by provision of water and food to wild animals. However, it is highly controversial both in countries that supply the majority of tourists for grassland savanna safaris and in Africa.

◀ **Figure 4** Controlled burning of grassland in the Kruger National Park, South Africa

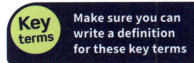

**Key terms** Make sure you can write a definition for these key terms

biome

# RETRIEVAL

Learn the answers to the questions below, then cover the answers column with a piece of paper and write down as many as you can. Check and repeat.

## Questions

**1** 'Tropical rainforests are warm and wet' – how warm and how wet, typically?

**2** What is the average NPP for the tropical rainforest biome?

**3** What percentage of light reaches the rainforest floor (in areas where the canopy layer is unbroken)?

**4** Why do rainforest plants often have drip-tips?

**5** One category of human activity is responsible for 80% of deforestation in the tropical rainforest biome: which one?

**6** Lower-latitude savanna grasslands have average temperatures of 22°C in the wet season and 28°C in the dry season. How does that compare with higher latitude savanna grasslands?

**7** Biomass in savanna grasslands varies from as low as 0.5 metric tons per hectare to 20 metric tons through the year. Why is this?

**8** Identify two ways in which some tree species have adapted to savanna grassland conditions.

**9** Name three impacts of human activities on the savanna grassland biome.

**10** Explain why afforestation might not be a suitable method of conserving savanna grassland.

*Put paper here*

## Answers

Average daytime temperature range of 28°C, 2000mm precipitation annually

An average of 2200 g/m²/year

1%

To prevent water accumulating on them, reducing the risk of fungal growth blocking leaves' ability to absorb sunlight

Agriculture

Average temperatures in higher-latitude savanna grasslands range between 18°C in the wet season and 32°C in the dry season

Because of the seasonality of the biome: vegetation dies back in the dry season, reducing biomass, then regrows rapidly in the wet season, increasing biomass

Two from: deep roots / deciduous or only have leaves during wet season / waxy leaves to reduce water loss / thick bark to withstand fire / ability to store water in trunk

Three from: overgrazing / tourism / fire suppression / farming / urbanisation / hunting and poaching / climate change

Grasses are shade-intolerant, so planting lots of trees will change the ecosystem to forested savanna, not grassland

## Previous questions

Now go back and use these questions to check your knowledge of previous topics.

## Questions

**1** Give an example of an economic cost of biodiversity decline.

**2** What are trophic levels?

**3** Define biomass.

**4** What is a plagioclimax?

*Put paper here*

## Answers

Overfishing affects livelihoods of fishing communities

The different positions within a food chain or web, indicating an organism's role in energy transfer

The total mass of all biotic organisms per unit area

A sub-climax ecosystem area where the arresting factor is human interference

# 6 Ecosystems under stress

## 6.5 Ecosystems in the British Isles over time

### Ecosystems in the British Isles

In the British Isles there are four main environments in which primary vegetation successions could progress over time to the climatic climax of temperate deciduous woodland biome: lithoseres and psammoseres (sand) on land and **hydroseres** and haloseres (saltwater) in water.

### Succession and climatic climax in lithoseres

In the British Isles, **lithoseres** have been produced by glacial retreat and isostatic rebound – the uplifting of an area after the weight of glaciation has been removed.

Pioneer species are the first organisms that colonise the bare rock (typically mosses/lichens). They will be hardy (e.g. xerophytes) and autotrophic (do not need external nutrients except sunlight). → Mosses/lichens, along with natural weathering, begin to break the rock down, creating soil and increased water retention. Decomposers break down dead mosses and lichens, releasing nutrients into the soil. → Grasses and ferns colonise the area. With further decay and nutrient release, a soil structure is created which is developed and aerated by decomposers. Flowering plants emerge, followed by shrubs. → Fast-growing pioneer trees such as birch, willow or rowan appear as the soil layer deepens and nutrient supply increases. Their decay and decomposition creates the conditions for temperate deciduous woodland, such as ash and oak – the climax community.

### Succession and climatic climax in hydroseres

Primary succession from freshwater is called a hydrosere, for example a newly constructed pond, oxbow lake or kettle lake.

At first, deep freshwater cannot support rooted plants as not enough light reaches the bottom of the pond. Pioneer species such as phytoplankton and algae, and then zooplankton (trophic level 2), are blown into the pond or carried there by insects/birds. Sediment is washed into the pond by runoff, making it shallower. → The pond becomes shallow enough for light to penetrate, and submerged rooted species grow. As they die and decay, more nutrients are built up and the pond becomes shallower. → Although floating plants are also rooted at the bottom of the pond, their leaves spread out over the pond surface, blocking sunlight. The submerged species die and decompose, adding to the nutrients available and further reducing the pond depth.

Pioneer shrubs emerge, lowering the water table and producing shade which the meadow herbs cannot tolerate. Wet woodland develops (called a carr), dominated by shrubs. The soil builds up humus and structure. Eventually, the climatic climax of temperate deciduous woodland is achieved. ← Reeds and sedges spread across the pond, creating marsh conditions that become less water-logged. As the soil dries, secondary species, such as herbs, move in and outcompete the reeds and sedges to form marsh meadow. ← In time, reedmace, sedges and rushes take root around the pond margins, binding the sediment and taking advantage of the nutrients. As leaf litter builds up, pond edges turn to marsh. Transpiration helps lower the water level. Vegetation matter fills in the pond.

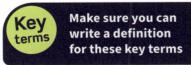 **Key terms** Make sure you can write a definition for these key terms

hydrosere    lithosere

## 6 Ecosystems under stress

### 6.5 Ecosystems in the British Isles over time

## Characteristics of the temperate deciduous woodland biome

The mild, wet temperate climate creates strong growing conditions for plants, particularly during the spring and summer months. Productivity is high and only exceeded by the tropical equatorial rainforest.

| Characteristic | Temperate deciduous woodland |
|---|---|
| **Climate** | |
| Temperature | Summer temperatures average 10–18°C, winter temperatures average 2–7°C but do dip below 0°C. |
| Precipitation | Between 800 mm and 1400 mm per year on average; wet all year round (maritime climate) but typically more precipitation in winter than summer. |
| Seasonality | Four distinct seasons (spring, summer, autumn, winter). As temperatures fall and daylight hours reduce in autumn, deciduous trees shed their leaves. |
| **Soil** | |
| Nutrient cycle | Deciduous trees have a high demand for nutrients, but the annual leaf fall and high rainfall ensure nutrients are returned to the soil, while the mild conditions are favourable to decomposers, so nutrient cycling is rapid. |
| Soil type | Brown earth soils: deep, well-drained and fertile. The large amounts of leaf litter are broken down by decomposers to form rich humus, aerated and mixed by earthworms and burrowing rodents. |
| **Soil moisture budget** | |
| Water surplus | Although dry periods and droughts occur, the high rates of precipitation and the deep structure of the soil mean that trees rarely experience water stress. |
| **Biodiversity** | |
| Biomass | Typically 150 to 300 metric tons per hectare |
| NPP | 1200 g/m²/year |
| Species richness | This is a diverse biome, with stratification of the forest into ground layer, field layer, shrub layer and canopy layer providing niches for different species to occupy. British temperate deciduous forests have up to 29 tree species, although oak, beech and sycamore dominate. There are some 24,000 insect species, 50 woodland bird species and a similar number of mammals, some of which hibernate in winter as an adaptation to the cold. |

## The effects of human activity on succession

Heather moorland is a plagioclimax vegetation community found in upland areas (above 250 m) in Britain on low-nutrient, acidic soils. The climatic climax is temperate deciduous woodland, which began to be cleared by humans around 5000 years ago to create pasture.

This clearance continues today by burning heathland, and also grazing livestock which prevents tree species from growing. Without competition from trees, shade-intolerant heather dominates the plagioclimax. Heather vegetation is adapted to grow in low-nutrient soils by having fungi in its roots, which help break down organic matter and minerals for absorption.

# RETRIEVAL

Learn the answers to the questions below, then cover the answers column with a piece of paper and write down as many as you can. Check and repeat.

## Questions | Answers

| # | Question | Answer |
|---|----------|--------|
| 1 | The four main environments in which a primary vegetation succession could take place in the British Isles are haloseres, psammoseres and which two others? | Lithosere, hydrosere |
| 2 | Pioneer species in a primary vegetation succession are often xerophytes. What is a xerophyte adapted to? | Dry conditions |
| 3 | Name two typical pioneer species for exposed rock environments. | Mosses, lichens |
| 4 | What plants are most associated with the second seral stage of a lithosere succession? | Grasses, ferns |
| 5 | Name two species of typical pioneer trees in the British Isles. | Two from: birch / willow / rowan |
| 6 | What is the climatic climax biome for the majority of the terrestrial British Isles? | Temperate deciduous woodland |
| 7 | What type of plant outcompetes submerged plants in a hydrosere, once the water is shallow enough for them to root? | Floating plants |
| 8 | What is the average winter temperature of temperate deciduous woodland biome in the British Isles? | Between 2°C and 7°C on average |
| 9 | How is heather adapted to grow in low-nutrient soils? | It has fungi in its roots which help break down organic matter and minerals for absorption |
| 10 | How does human activity maintain the heather moorland plagioclimax? Name two ways. | Burning and grazing of livestock |

*Put paper here*

## Previous questions

Now go back and use these questions to check your knowledge of previous topics.

## Questions | Answers

| # | Question | Answer |
|---|----------|--------|
| 1 | What are biotic elements of an ecosystem? | The living parts of an ecosystem |
| 2 | What is a pioneer species? | A plant species which requires very few nutrients to survive |
| 3 | What percentage of light reaches the rainforest floor (in areas where the canopy layer is unbroken)? | 1% |
| 4 | Name three impacts of human activities on the savanna grassland biome. | Three from: overgrazing / tourism / fire suppression / farming / urbanisation / hunting and poaching / climate change |
| 5 | Explain why afforestation might not be a suitable method of conserving savanna grassland. | Grasses are shade-intolerant, so planting lots of trees will change the ecosystem to forested savanna, not grassland |

*Put paper here*

# 6 Ecosystems under stress

## 6.6 Marine ecosystems

### The distribution of coral reef ecosystems

Warm coral reefs form in latitudes between 30° north and 30° south of the equator, including areas in the Caribbean Sea, the Indo-Pacific region (such as the Great Barrier Reef in Australia), the Red Sea and the Pacific Islands.

▲ *Figure 1 A coral reef, Caribbean sea*

### The main characteristics of coral reef ecosystems

Biodiversity:
Support a wide variety of species, including corals, fish, invertebrates and marine plants. Despite covering less than 1% of the ocean floor, coral reefs are home to about 25% of all marine species.

Polyps:
Primarily built by small, soft-bodied organisms called coral **polyps**, which secrete calcium carbonate (limestone) skeletons that accumulate over time, forming the hard structures associated with coral reefs.

**Coral reef ecosystems**

Symbiotic relationship:
Coral polyps have a **symbiotic** relationship with zooxanthellae (photosynthetic algae) allowing corals to thrive in nutrient-poor waters. The algae provide the coral with nutrients and oxygen; coral provides the algae with a protected environment and access to sunlight.

Vibrant colours:
The vibrant and diverse colours of coral reefs come from the pigments in both the coral polyps and the zooxanthellae.

### Environmental conditions associated with reef development

- Warm waters (23–29°C), essential for the growth and survival of the coral polyps and their symbiotic algae. Slight deviations in temperature can lead to coral bleaching, a stress response that can damage or kill coral colonies.

- Clear and sunlit waters to support the photosynthesis of zooxanthellae, which provide the corals with essential nutrients. Sedimentation can reduce the amount of light reaching the coral reef.

- Shallow waters, usually less than 45 m deep, as beyond that depth, insufficient light limits coral growth.

- Stable salinity levels, as rapid fluctuations can stress corals and disrupt their ability to maintain proper internal conditions.

- Coral reefs are often found in low-nutrient (oligotrophic) waters. Excess nutrients, such as nitrogen and phosphorus from pollution or agricultural runoff, can lead to eutrophication, which can smother and damage coral reefs.

- Moderate wave action helps to maintain water quality and prevent the accumulation of sediment and debris. Excessive wave energy can physically damage coral structures.

## The Andros Barrier Reef, Bahamas

The Andros Barrier Reef, located off Andros Island in the Bahamas, is the world's third-largest barrier reef system, covering approximately 200 km of coastline, with much of the reef at a depth of just 2.5 m. It is made up of five individual reefs that have diverse ecosystems of 164 different species of fish, including the 5 m long blue marlin – the national fish of the Bahamas.

**REVISION TIP**

You need to know a named, located example of a coral reef that you can use in answers about factors affecting the health and survival of reefs. Revise the example you have studied. This page has an example from the Bahamas.

## Natural factors affecting reef health

Water temperature: surface ocean temperatures rarely fall below 22°C , ideal for reef development. However, **coral bleaching** is an inevitable result of the predicted rise in ocean temperatures of 1.5°C by 2100.

The Bahamas experiences five times more marine heatwave events than it did 40 years ago. The reef is expected to be dead by 2040.

## Human activities and their impact

- Tourism is a significant contributor to the Bahamian economy, but unregulated tourism for recreational activities, including snorkelling and diving, can cause physical damage.

- Overfishing of particular species can cause food chain problems in the coral ecosystem, for example corals being smothered by the algae that a species of fish usually eats.

- Pollution from sewage, agricultural runoff and industrial activities can introduce harmful chemicals and contaminants into reef ecosystems. Between 1950 and 1980, logging on North Andros Island meant heavy silt pollution of the coastal waters, smothering areas of coral on the Andros Barrier Reef.

## Management and conservation efforts

- Northern and Southern Marine Parks (NSMPs) protect 20 km² of the barrier reef by limiting fishing, regulating tourism and protecting critical reef habitats; for example, there are strict bans on dumping waste in the water.

- The Andros West Side National Park (2002) covers the entire west coast of Andros Island. Development is strictly controlled, reducing risks of pollution or siltation of the reef.

- Coral restoration programmes, such as the Exuma Cays Land & Sea Park, aim to restore damaged coral populations through coral gardening and transplantation techniques.

- Crab Replenishment National Park is a protected area to allow a sustainable land crab population to be maintained for the use of the Androsian people.

## Future prospects

By declaring their reef system to be a national park, the Bahamas government has strictly controlled development that might harm the coral ecosystem. Global governance is also required to combat ocean warming if the reef is to survive. The UN's 2016 resolution on sustainable coral reef management, including the development of indicators and assessment of coral reef status and trends, is a starting point.

**Key terms** Make sure you can write a definition for these key terms

coral bleaching   polyp   symbiotic

# RETRIEVAL

Learn the answers to the questions below, then cover the answers column with a piece of paper and write down as many as you can. Check and repeat.

## Questions | Answers

| # | Question | Answer |
|---|----------|--------|
| 1 | Within which latitudes north and south of the equator do coral reefs form? | 30° |
| 2 | What percentage of the ocean floor do coral reefs cover? | 1% |
| 3 | What percentage of all marine species are coral reefs home to? | 25% |
| 4 | Coral polyps have a symbiotic relationship with photosynthetic algae called… | Zooxanthellae |
| 5 | What temperature range do warm coral polyps and their symbiotic algae require to grow and survive? | Between 23°C and 29°C |
| 6 | Coral reefs typically cannot develop below what depth of water? | Usually not deeper than 45 metres |
| 7 | Identify three threats to coral reefs. | Three from: deviations in water temperature (leads to coral bleaching) / sedimentation or pollution (reducing light reaching the coral reef) / sea level rise (deep water is too dark) / rapid fluctuations in salinity / excess nutrients (eutrophication) / excessive wave energy (physical damage to coral reefs) |
| 8 | Name three local-scale threats to the Andros Barrier Reef. | Three from: tourism / overfishing / onshore development / pollution |

*Put paper here*

## Previous questions

Now go back and use these questions to check your knowledge of previous topics.

## Questions | Answers

| # | Question | Answer |
|---|----------|--------|
| 1 | What percentage of species in the UK are threatened with extinction? | 15% |
| 2 | What is a plagioclimax? | A sub-climax ecosystem area where the arresting factor is human interference |
| 3 | What are the five main factors affecting the type and characteristics of terrestrial ecosystems? | Climate, vegetation, soils, topography, time |
| 4 | Biomass in savanna grasslands varies from as low as 0.5 metric tons per hectare to 20 metric tons through the year. Why is this? | Because of the seasonality of the biome: vegetation dies back in the dry season, reducing biomass, then regrows rapidly in the wet season, increasing biomass |
| 5 | How is heather adapted to grow in low-nutrient soils? | It has fungi in its roots, which help break down organic matter and minerals for absorption |

*Put paper here*

 **KNOWLEDGE**

## 6 Ecosystems under stress

### 6.7 Local ecosystems

## Psammoseres

The most common **psammoseres** are sand dune systems.

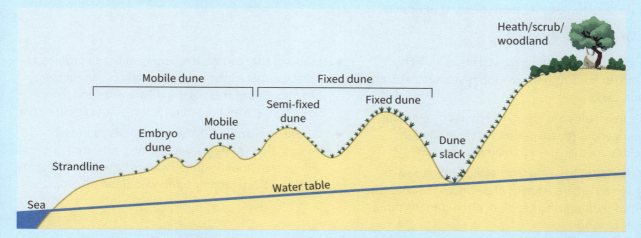

▲ *Figure 1 Sand dune succession*

## Adaptations by flora and fauna

### Marram grass

Pioneer species colonise the **embryo dunes**, which are mostly exposed sand. Marram grass is a xerophyte and is stimulated to grow by being buried by sand. Its leaves are waxy and rolled inwards to reduce evaporation. It has long roots, which grow down several metres to reach water, as well as shallow roots (rhizomes) that spread out to catch water. These extensive roots stabilise the dunes.

### Northern dune tiger beetle

This rare UK species is adapted to hunting other insects across exposed, moving sand. They dig holes in the face of dunes for shelter overnight. They also lay their eggs in holes dug into the sandy dunes, and when hatched, the larvae wait for prey, which they drag into their burrows.

### Sand lizards

England's rarest lizard needs warm, undisturbed sand to lay its eggs in, sheltered sunny places to bask, and nearby vegetation for food. Sand lizards dig burrows for shelter and hibernation, with entrances hidden by vegetation. Their numbers are declining due to loss of embryo dunes – the perfect habitat.

> **REVISION TIP**
>
> Local ecosystems can include an area of heathland, a managed parkland, a pond or a dune system. You may have done some fieldwork in a distinctive local ecosystem like this. This section focuses on a sand dune ecosystem.

**Key terms** — Make sure you can write a definition for these key terms

embryo dunes    psammosere

# KNOWLEDGE

## 6 Ecosystems under stress

### 6.7 Local ecosystems

## Ecological development and change

Sand dunes are one of the UK's most threatened ecosystems. Since the 1940s, around 80% of bare sand has been lost on the coast of Britain.

Reasons for this include:

- the expansion of housing: coastal areas are in high demand for housing and developments such as static caravan parks
- golf courses: they often stabilise dunes by planting non-indigenous grass species. Fertilisers and pesticides are then used to maintain the turf
- coastal defences: developments along the coast are protected by different forms of defences that alter the supply of sediments to beaches, often starving dune systems of sand

- sea-level rise: causes steepening of beach profiles as coastal erosion increases, removing sediment from beaches. Some areas of the UK are expected to lose 50 m of beach and coastline as a result of sea level rise
- decline in grazing: with fewer animals pastured on sand dune systems, vegetation succession is unhindered, mobile dunes with exposed sand become fixed and heathland or scrub woodland develops.

## Conservation strategies for sand dune ecosystems

Sand dune ecosystems have high biodiversity so any threats impact a significant number of rare, unusual and special species.

Common strategies for managing these impacts include:

- habitat restoration and management: for example, removing invasive species and replanting native species such as marram grass, sea holly and sea rocket
- controlled grazing: reintroducing grazing by sheep and cattle prevents some of the overgrowth of vegetation on dune systems, maintaining open sandy areas
- selective dune bulldozing: this mimics natural processes of dune blowouts and creates open sandy areas where specialised dune plants can establish themselves
- sand dune nourishment: adding sand to dune systems that are being starved of sediment by coastal defences
- boardwalks: these allow people to walk through and enjoy sand dune areas without disturbing open sand or trampling areas which have been replanted.

◀ *Figure 2*
*Indigenous species such as Marram grass can help to stabilise sand dunes*

# 6.8 Case study: A local ecosystem

## Case study: The Sefton Coast ecosystem

The Sefton Coast is an extensive dune network stretching for 17 km with an average width of 1.5 km. The ecosystem has both fixed and mobile sand dunes, together with slacks, mud flats and salt marshes.

It is a refuge for species such as the natterjack toad, sand lizard and northern dune tiger beetle, as well as rare plants such as the Isle of Man cabbage, sea bryum, matted bryum and petalwort.

It is part of the Merseyside conurbation, which has a population of over 100,000, and receives a large number of visitors throughout the year. This creates challenges for conserving the ecosystem and its rare species while meeting the high demand for leisure visits.

## Human impacts

Development within most of the Sefton Coast is currently low, because 75% of it is within greenbelt, but in the past roads, railways, housing developments and pine tree plantations have fragmented the sand dune ecosystem. Large-scale sand extraction also took place offshore until the 2000s. One recent threat has been a 60% decline in grazing land over a ten-year period, which has seen areas of the dunes become overgrown with vegetation. Coastal erosion has also removed dune habitats at Formby, while in other parts, deposition is increasing beach width. The Sefton Coast has many golf courses.

## Sustainable development

The Sefton Coast Landscape Partnership (SCLP) is a group of all the significant landowners and key partners (such as businesses and conservation groups). Their aim is to conserve and enhance the habitats and species of the ecosystem while also enabling local communities to enjoy the natural environment and promote sustainable economic growth.

> **REVISION TIP**
>
> You should revise the case study you have studied. This page uses examples from the Sefton Coast ecosystem.

The main aim is to direct visitors away from the most sensitive and vulnerable parts of the ecosystem, and towards less sensitive areas. This has involved upgrading walkways and information boards for tourists and establishing tourist-free areas to protect rare species. This is helped by some areas having a high level of protection, including sites of special scientific interest (where it is an offence to damage protected natural features).

Volunteers work with specialists to restore dune slack habitats by clearing areas of scrub and invasive species and creating or enhancing hundreds of bare sand areas to create mosaic habitats for the rare sand lizards and northern dune tiger beetles. In some areas, pine tree plantations near the coast have been cleared in order to restore the dune habitat.

There is also high demand for new housing. Sefton Council requires housing developers to contribute towards council projects that reduce impacts on the natural environment.

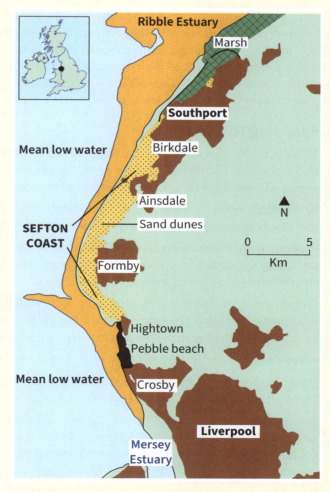

▲ *Figure 1* Location of the Sefton Coast

 **KNOWLEDGE**

## 6 Ecosystems under stress

### 6.9 Case study: Ecological change

 **Case study: Deforestation in Malaysia**

In the period 2000–2012 Malaysia had the highest rate of deforestation in the world. The rate has since reduced, from 244,300 hectares of primary forest lost in 2012 to 73,000 hectares in 2020.

**REVISION TIP**

You should revise the case study you have studied. This page has an example from Malaysia.

### Causes of rainforest deforestation

| Population pressure | Commercial agriculture | Energy development |
|---|---|---|
| During the second half of the twentieth century Malaysia's population was increasing at a rate of 3% a year. From the 1950s to the 1980s, the government resettled people in rural areas, leading to rainforest clearance for subsistence agriculture. | Clearing rainforest for oil palm cultivation has been a major cause of deforestation. However, in recent years, new plantations have been using previously cleared land. | Malaysia has developed HEP infrastructure to reduce dependence on fossil fuels. The Bakun dam and reservoir involved 700 km of rainforest being cleared. |

### Indigenous populations – the Batek

The Batek are an indigenous, semi-nomadic people of the tropical rainforest of Peninsula Malaysia. Traditionally they are expert tree-climbers, enabling them to gather rainforest fruits, hunt animals using poisoned blow-darts and catch fish. They used rainforest products to meet all their needs, but since mass deforestation and systematic dispossession of their lands, they are less able to be self-sufficient, and the government attempt to move them to permanent settlements have exacerbated this.

### Responding to change

- To reduce rainforest deforestation, Malaysia has introduced licensed selective logging and replanting, and illegal loggers are prosecuted. Replanting creates secondary succession and a subclimax community of managed forest.
- In Peninsula Malaysia, national parks and nature reserves focus on the most threatened ecosystems and protecting rainforest from development.
- Replanting programmes have focused on connecting surviving rainforest areas with secondary forest 'corridors'. A key part of the programme is to prevent young trees being smothered by fast-growing grasses, vines and ferns.
- Ecotourism is promoted in these protected areas with the income helping to keep the rainforest sustainable. For example, in the Sungai Yu Forest Reserve, Pahang, visitors can camp with the Batek people and trek across the rainforest with guides.
- At the Earth Summit in 1992 Malaysia promised to keep 50% of its land use as forest
- Malaysia's forest management practices are audited every five years by the Forest Stewardship Council (FSC) to assess the sustainability of its forestry
- Malaysia is part of the REDD Plus programme – an example of global governance that seeks to reduce emissions from deforestation and forest degradation in developing countries. For REDD Plus, Malaysia has introduced forest monitoring programmes (geospatial surveys of forest cover every two years) and a national action plan that links forest management and conservation to Malaysia's 50% forest cover pledge and its emission target reduction under the Paris Agreement.

# ⇄ RETRIEVAL

Learn the answers to the questions below, then cover the answers column with a piece of paper and write down as many as you can. Check and repeat.

## Questions | ## Answers

| # | Question | Answer |
|---|----------|--------|
| 1 | Give two examples of a distinct local ecosystem. | Two from: an area of heathland / managed parkland / a pond / a dune system |
| 2 | What is a psammosere? | A primary succession that begins on newly exposed coastal sand |
| 3 | Sand dune systems migrate towards the coast. What are the newest sand dunes closest to the strand line called? | Embryo dunes |
| 4 | Describe one adaptation of marram grass to colonising sand dunes. | One from: waxy, rolled-in leaves to reduce evaporation / long roots that grow down several metres to reach water / shallow roots called rhizomes that spread out near the surface to catch water |
| 5 | Name one example of fauna adapted to live in sand dune ecosystems. | One from: sand lizard / northern dune tiger beetle / natterjack toad (in dune slacks) |
| 6 | Name one example of flora adapted to live in sand dune ecosystems. | One from: marram grass / sea holly / sea rocket / Isle of Man cabbage / sea bryum / matted bryum / petalwort |
| 7 | Explain how golf courses are often a threat to coastal sand dune ecosystems. | Golf courses at the coast often stabilise dunes by planting non-indigenous grass species |
| 8 | Why are coastal defences often a threat to sand dune ecosystems? | If they alter the supply of sediments to beaches, they can starve dune systems of sand |
| 9 | How does selective dune bulldozering help restore sand dune ecosystems? | It mimics natural processes of dune blowouts and creates open sandy areas where specialised dune plants can establish themselves |
| 10 | What benefits to boardwalks offer as a means of combining conservation with leisure access? | They allow people to walk through and enjoy sand dune areas without disturbing open sand or trampling areas which have been replanted |

*Put paper here*

## Previous questions

Now go back and use these questions to check your knowledge of previous topics.

## Questions | ## Answers

| # | Question | Answer |
|---|----------|--------|
| 1 | What is a sub-climax in a vegetation succession? | Vegetation that has not been able to mature to the natural climatic climax vegetation for the region |
| 2 | One category of human activity is responsible for 80% of deforestation in the tropical rainforest biome: which one? | Agriculture |
| 3 | Name two typical pioneer species for exposed rock environments. | Mosses, lichens |
| 4 | What is the climatic climax biome for the majority of the terrestrial British Isles? | Temperate deciduous woodland |

*Put paper here*

## Exam-style questions

1    Outline the concept of a plagioclimax in vegetation succession.    **[4 marks]**

2    Outline the nature of trophic levels within an ecosystem.    **[4 marks]**

3    Outline one factor influencing the changing of ecosystems at a global scale.    **[4 marks]**

4    Outline the natural factors in the health and survival of reefs.    **[4 marks]**

5    Outline the characteristics of the temperate deciduous woodland biome.    **[4 marks]**

6    Outline the concept of the biome.    **[4 marks]**

> **EXAM TIP**
>
> This is section C of Paper 1. You will have a choice of either Question 5 (Hazards) or Question 6 (Ecosystems under stress). Answer questions on the topic you have studied only!

> **EXAM TIP**
>
> 4-mark questions are point-marked. That means each valid point or development of a valid point receives 1 mark, up to a maximum of 4 marks.

7    **Figure 1** shows the extent of terrestrial biomes considered wilderness and those that are considered intact or damaged/highly modified by humanity.

Analyse the data shown in **Figure 1**.    **[6 marks]**

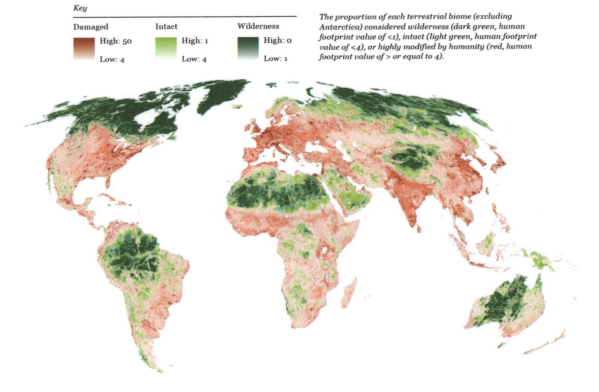

*Key*

Damaged

High: 50

Low: 4

Intact

High: 1

Low: 4

Wilderness

High: 0

Low: 1

*The proportion of each terrestrial biome (excluding Antarctica) considered wilderness (dark green, human footprint value of <1), intact (light green, human footprint value of <4), or highly modified by humanity (red, human footprint value of > or equal to 4).*

▲ *Figure 1 Status of wilderness globally (terrestrial biomes)*

8 **Figure 2** shows the estimated extent of live hard coral coverage globally, from 1980 to 2020.

Analyse the data shown in **Figure 2**.

**[6 marks]**

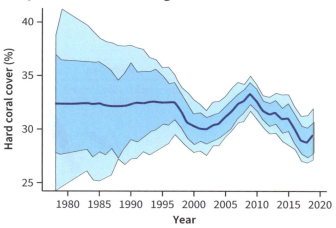

▲ **Figure 2** *Estimated extent of live hard coral coverage globally, from 1980 to 2020. The shading around the average line represents the levels of uncertainty about the data (surveying of coral reefs has improved significantly this century)*

9 Using **Figure 3** and your own knowledge, assess the development issues in the savanna grassland biome.

**[9 marks]**

Degradation spiral (LD)

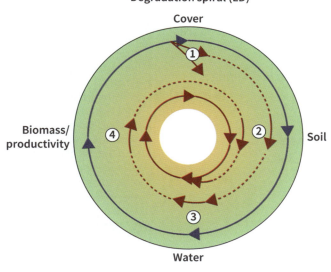

Improvement spiral (SLM)

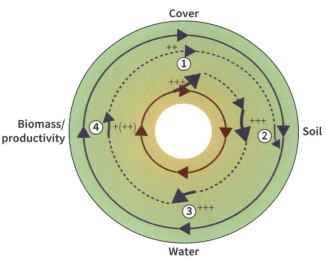

① Overuse of grass / tree cover

② Compacting/crusting, loss of soc, nutrients

③ Accelerated runoff

④ Loss of biomass and perennial grasses

① Resting, reseeding, replanting, rotational grazing....

② Surface treatment, manure

③ Water harvesting

④ Selective removal of undesirable and promotion of desirable species

Inputs and effectiveness

 +++ Very high

 ++ High

 + Medium

▲ **Figure 3** *The downward spiral of land degradation (LD) and the improvement spiral of sustainable land management (SLM). The green circle represents sustainability.*

10  **Figure 4** shows three different issues for species interactions related to climate change. (Phenology refers to relationships between plants / animals and seasons or other natural cycles.)

Using **Figure 4** and your own knowledge, assess the implications of climate change for the tropical rainforest biome.  **[9 marks]**

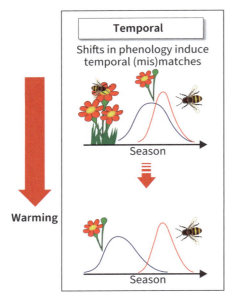 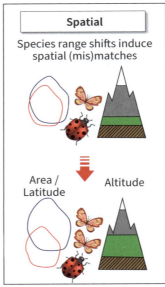 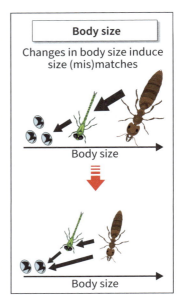

▲ **Figure 4** *Mismatches in species interactions resulting from temporal, spatial and body size responses to climate warming*

11  Analyse the interconnections between climate, vegetation and soils in the development of a lithosere.  **[9 marks]**

12  How far do you agree that global governance has a role to play in securing a viable future for the tropical rainforest biome?  **[9 marks]**

13  With reference to an ecosystem at a local scale, evaluate the extent to which causes of declining biodiversity have been successfully managed.  **[20 marks]**

14  With reference to the savanna grassland biome, assess the view that mineral nutrient cycling is the most important factor in its natural development.  **[20 marks]**

15  With reference to a specified region experiencing ecological change, assess the sustainability of the community's response/s to the change.  **[20 marks]**

16  Assess the extent to which the properties of a specified ecosystem at a local scale reflect the climatic climax community.  **[20 marks]**

> **EXAM TIP**
>
> 9-mark questions assess knowledge and understanding (AO1) for 4 marks and application (AO2) for 5 marks. Use brief examples for AO1 that back up the AO2 arguments you make. Most 9-mark questions will require a brief concluding statement.

> **EXAM TIP**
>
> 20-mark questions have 10 marks available for AO1 and 10 marks for AO2 (your argument). Some degree of balance in your answer is needed. Could you present a counterargument in your penultimate paragraph, ahead of your conclusion?

# ⚙ KNOWLEDGE

## 7 Global systems and global governance

### 7.1 Globalisation

## Dimensions of globalisation

**Globalisation** is the process of increasing interconnectivity and interdependence within our world. This involves global economic and political systems, and the world's societies and cultures.

### Flows of capital

- Money is invested by individuals, businesses and governments within a country or into a different country – **foreign direct investment (FDI)**.
- FDI can fluctuate year on year and between different regions, depending on external shocks (e.g. 2022 Russia–Ukraine conflict).

### Flows of labour

- Movement of people for work, linked to international and internal migration.
- Economic migrants move for improved job opportunities, increased wages and improved working conditions. They can be high-, semi- or low-skilled workers.
- Migrants can fill skills shortages in different regions, contribute to taxes in the host country, send money home and contribute to the global spread of different cultures and ideas.

### Flows of products

- Countries export raw materials to countries who use these products to make goods with added value.
- Location of raw materials can depend on different factors, including geology (minerals, oil, gas), climate (agricultural products), proximity to water sources (fishing).
- Direction and volume of product flows depends on supply and demand.

### Flows of services

Flows of services including financial, business and customer services, have increased due to:

- improvements in ICT – banking TNCs use the internet to take advantage of the 24/7 economy to increase their presence and profits around the world. Companies can be based anywhere and make transactions with their customers located around the world.
- deregulation – fewer rules and laws surrounding financial transactions made it easier for companies to negotiate business deals internationally.

### Flows of information

The internet has increased the speed of information transfer around the world through email, smart phone technology, social media applications and AI.

### Global marketing

- Relies on brand awareness and consumer confidence.
- Companies can adapt global brands to appeal to different markets (glocalisation).
- Faster internet connections and smart phones have increased the power of social media as a marketing tool.
- **Global marketing** drives globalisation through increased consumerism, especially of US brands.

### Patterns of production, distribution and consumption

Nineteenth and early twentieth century:

- Most manufacturing took place in Highly Developed Economies (HDE) due to the industrialisation, large markets and increasing consumer demand.

Late twentieth century:

- 'Global shift' of manufacturing from HDEs to Emerging Major Economies (EMEs) and Less Developed Economies (LDEs).
- Lower wages and investment opportunities led companies to relocate their factories to cheaper production areas. Goods are then exported, often to HDE markets.
- Former industrial areas in HDEs have experienced decline in production, causing factory closures and unemployment.

Twenty-first century:

- Rising numbers of middle-class populations with higher disposable incomes in EMEs have increased demand for consumer products in, for example, China and India.

# KNOWLEDGE

## 7 Global systems and global governance

### 7.1 Globalisation

## Factors in globalisation

Globalisation has developed due to changing technologies, systems and relationships.

### Financial systems

- Facilitate the movement of capital around the globe. FDI, individual and company investments and international loans support trade, business expansion and infrastructure development.
- Deregulation has allowed countries to trade more easily. Foreign exchange, global banking, insurance and payment networks allow cross-border transactions to be conducted.

### Communication technology

- Driven by satellite technology and fibre optic cables.
- Software applications, including social media, open source and AI are increasing availability and efficiency of communications.

### Management and information systems

- Models of production management can be used in factories in different countries, e.g. robots on assembly lines, JIT (just in time) production (vehicles), CMT (cut, make and trim) production (clothing).
- Increase efficiencies, take advantage of economies of scale, minimise costs and increase profits. Businesses benefit from global supply chains. Companies can also outsource or offshore part of their business to benefit from less strict rules or tax benefits in another country.

### Transport and communication systems

- Transport network densities differ around the world meaning some countries have better internal and international access than others.
- Key technologies have included high-speed rail, increased capacity and efficiency of shipping due to containerisation and development of deep water ports, and advancements in non-stop aircraft routes.

### Security

- Inter Governmental Organisations (IGOs) promote the globalisation of security-based decision making.
- It is also argued that global interdependence through interconnected trade systems reduces the possibility of conflicts.
- With increased and different forms of communication come an increase in the possibility and severity of cybercrime and identity theft.

### Trade agreements

- The WTO governs the world trade system, under the free trade model. It sets rules and settles disputes between member countries.
- **Trade agreements** exist between two (bilateral) or more (multilateral) countries (e.g. bilateral: UK–Japan) or as part of a larger trading bloc (EU).
- Critics say that global trade rules prevent some countries from having fair access to the global market.

**Key terms** Make sure you can write a definition for these key terms

consumption    flow of capital    foreign direct investment (FDI)
globalisation    global marketing    trade agreement

# RETRIEVAL

Learn the answers to the questions below, then cover the answers column with a piece of paper and write down as many as you can. Check and repeat.

## Questions

## Answers

| | Questions | Answers |
|---|---|---|
| 1 | What are the five dimensions of globalisation? | Flows of capital, labour, products, services and information |
| 2 | Give two examples of flows of labour. | Two from: movement of people for work / economic migrants / to fill skills shortages in the same country |
| 3 | What are flows of capital? | Movement of finance and investment between places |
| 4 | What are flows of labour? | Movement of people for work between places |
| 5 | Name three areas in which flows of services take place. | Financial, business, customer services |
| 6 | Why has the flow of services increased? | Improvements in IT, deregulation |
| 7 | What has been a major factor in increasing the flow of information around the world? | Internet |
| 8 | What is global marketing? | The promotion and advertising of goods and services around the world |
| 9 | Why did most manufacturing take place in HDEs in the nineteenth and early twentieth centuries? | Industrialisation, large markets and increasing consumer demand |
| 10 | From which type of economy has manufacturing shifted in the late twentieth century, and where has it shifted to? | From HDEs to EMEs and LDEs |
| 11 | How have improvements in transport technology helped globalisation? | Increased the speed and volume of goods and people that can travel around the world |
| 12 | How have financial systems helped develop the spread of globalisation? | Facilitate the movement of capital around the globe; deregulation |
| 13 | Name three factors in the improvement in communication technology. | Development of satellite technology, fibre optics and software applications |
| 14 | Which organisation governs the global trading system? | WTO |
| 15 | What is a bilateral trade agreement? | A trade agreement between two countries |

Put paper here

# KNOWLEDGE

## 7 Global systems and global governance

### 7.2 Global systems

## Interdependence in the contemporary world

**Economic:**
One country cannot usually be self-sufficient in all the goods and services it needs and wants, so relies on others. Trade agreements enable it to sell goods and services it produces (exports) and buy those it requires (imports). Countries may rely on other countries for FDI, aid, loans and workforce.

**Political:**
Countries collaborate to solve global problems. Transboundary issues, e.g. illegal migration, water security and climate change are often tackled by international alliances. Political IGOs are also significant players in resolving conflict and wars around the world.

**Interdependence**

**Social:**
Transport, communications and freedom to travel have increased interactions between different populations. Migration creates links between different peoples in both the source and destination areas.

**Environmental:**
The global environmental system – atmosphere, hydrosphere, cryosphere, lithosphere and biosphere – forms the basis for life on earth. All of humankind is dependent on each other to protect and manage the environment, e.g. climate change or, at regional level, drainage basin management.

> **REVISION TIP**
>
> Make notes from the spider diagram to show examples of i) economic, ii) political, iii) social and iv) environmental interdependence between countries.

## Unequal power relations

### Those benefiting more from global systems

- Countries with access to natural, human, financial and technological resources
- Countries with a locational advantage, e.g. coastline
- Some HDEs have taken advantage of global growth from historical factors such as colonialism
- HDEs have more influence than LDEs within IGOs through membership and voting rights, e.g. UN, IMF, WB
- Countries within trade blocs
- Free trade benefits some countries more than others
- HDEs and EMEs respond more effectively to unexpected external events

### Those benefiting less from global systems

- Countries with less access to natural, human, financial and technological resources
- Some landlocked countries dependent on neighbouring states for import and export access, or remote island states
- Countries that were colonised may have increased levels of dependency
- Countries vulnerable to risks and impacts of climate change
- Countries with unequal access to global markets due to trade barriers such as tariffs, quotas and subsidies
- Countries with less decision-making power within IGOs

> **REVISION TIP**
>
> Make an audio recording to explain how unequal power systems affect different countries in different ways.

# Issues associated with interdependence

|  | Promotion of stability, growth and development for people and places | Causes of inequalities, conflicts and injustices for people and places |
|---|---|---|
| Unequal flows of people | • Offsets declining birth rates and ageing population in destination country<br>• Brings in workers with different skills and experiences to fill the job gap<br>• Increased contribution to economy through taxes and disposable income<br>• Migrants contribute to the economy of their home countries through remittances<br>• Increased diversity | • 'Brain drain' – skilled workers leave for better working conditions elsewhere<br>• Social and cultural tensions between different people<br>• Increased income **inequality** between workers<br>• Challenging working conditions for lower-paid or migrant workers |
| Unequal flows of money | • Incentives to invest, save and innovate leading to economic growth<br>• Economic growth creates employment, and develops infrastructure and services<br>• IGOs, NGOs and high income individuals donate or loan money for specific projects to improve human development, e.g. foreign aid<br>• Economic mobility allows individuals and businesses to take opportunities to improve status | • HDEs account for most trade in goods and services causing global trade imbalance<br>• Income inequality can cause social and political tensions<br>• Unequal access to health care / education limits life chances and workforce productivity<br>• Less demand as less disposable income<br>• Aid can lead to dependency<br>• Debt burden slows down economies of some LDEs<br>• Corruption can distort the flow of funds<br>• TNCs cause conflicts, e.g. outcompeting local businesses, working conditions, environmental policies |
| Unequal flows of ideas | • Incentivises innovation, creativity and entrepreneurship<br>• Some argue neo-liberalism has encouraged less state, more private sector involvement and **free trade**, leading to economic growth<br>• Spread of democracy and equality<br>• Cultural exchange of ideas<br>• Encourages global networking and cross-border collaboration | • Some argue neo-liberalism has benefited the rich at the expense of the poor<br>• Some groups may feel excluded from participation<br>• Challenges in respecting human rights and equality for all groups<br>• Unequal access to and transfer of knowledge leaves some people at a disadvantage<br>• Globalisation may threaten local cultures |
| Unequal flows of technology | • R&D hubs can help with key inventions, e.g. Oxford, UK – COVID vaccine, Silicon Valley, USA – software development<br>• Attraction of FDI can increase technology levels and improve skills and knowledge of employees<br>• Increased connectivity maintains and develops global links | • Technology usually developed in HDEs and EMEs and 'trickles down' to LDEs<br>• Unequal access to technology infrastructure and services<br>• Dependency on HDEs for key technology infrastructure, e.g. satellites, fibre optics<br>• Privacy rights and technological ethics vary between places and people |

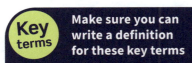

**Key terms** Make sure you can write a definition for these key terms

free trade    global system    inequality    interdependent

# RETRIEVAL

Learn the answers to the questions below, then cover the answers column with a piece of paper and write down as many as you can. Check and repeat.

## Questions | Answers

| | Questions | Answers |
|---|---|---|
| 1 | Give a definition of global systems. | Interconnections and interdependence between countries that help make and develop the world, including economic, political, societal and environmental interactions |
| 2 | What is interdependency? | People, places and countries relying on each other |
| 3 | Give two examples of how countries may rely on others economically. | Two from: FDI / aid / loans / workforce |
| 4 | Give one example of political interdependence. | One from: countries collaborate to solve global problem/ political IGOs are significant players in resolving conflicts and wars |
| 5 | Environmental interdependence: Give two parts of our global environmental system. | Two from: atmosphere / hydrosphere / cryosphere / lithosphere / biosphere |
| 6 | Give two negative impacts of unequal flows of people. | Two from: 'Brain drain' / social or cultural tensions between different groups / increased income inequality / challenging working conditions |
| 7 | Give two positive impacts of unequal flows of ideas. | Two from: incentivises innovation, creativity and entrepreneurship / encourages less state, more private sector involvement and free trade / spread of democracy and equality / cultural exchange of ideas / encourages global networking and cross-border collaboration |
| 8 | Give two examples of R&D hubs. | Oxford, Silicon Valley |
| 9 | What does FDI stand for? | Foreign direct investment |

*Put paper here*

## Previous questions

Now go back and use these questions to check your knowledge of previous topics.

## Questions | Answers

| | Questions | Answers |
|---|---|---|
| 1 | What are the five dimensions of globalisation? | Flows of capital, labour, products, services and information |
| 2 | What are flows of labour? | Movement of people for work between places |
| 3 | What is global marketing? | The promotion and advertising of goods and services around the world |
| 4 | How have improvements in transport technology helped globalisation? | Increased the speed and volume of goods and people that can travel around the world |
| 5 | Which organisation governs the global trading system? | WTO |

*Put paper here*

# KNOWLEDGE

## 7 Global systems and global governance

### 7.3 International trade

## Patterns of international trade

International trade and access to markets is an important part of the global economic system. The value of **international trade** has increased rapidly since the 1950s, reaching US$27 trillion in 2022 (80% goods and 20% services).

Since 2000, the **volume of trade** of goods internationally has increased dramatically, with China's trade volumes more than tripling, making it now a major trading partner with many countries. Trade volumes have also increased for the other major economies (US, EU) but at a slower rate.

The increase in world trade has been driven by an increase in 'South to South' trade (South America, Central America, Mexico, and the islands of the Caribbean and South and East Asia).

LDEs account for just 1 % of global exports. In 2021, over half of the world's exports of goods were made by only five economies (China, EU, USA, Japan and South Korea).

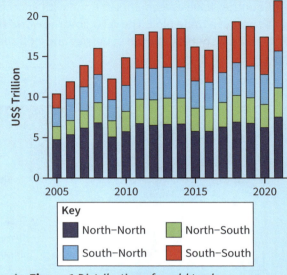

▲ **Figure 1** Distribution of world trade

## Patterns of international investment

Individuals, companies, governments or other groups use FDI to increase economic growth. FDI is either inward (investments made in a country from another country) or outward (investments made by domestic companies in a foreign economy).

- Volume of FDI has risen as globalisation has increased. In 2022, global FDI was US$1286 billion.

- In 2023, top recipients of inward FDI were USA, Brazil, Canada and Mexico. Top sources of outward FDI were USA, China and Japan.

- Since 2000, EMEs have been recipients of large amounts of FDI. EMEs such as China have invested in the 'global south' – Africa and South America, including Belt Road Initiatives.

- Sustainable and ethical investing have increased as companies seek long-term profitability from socially and environmentally responsible projects.

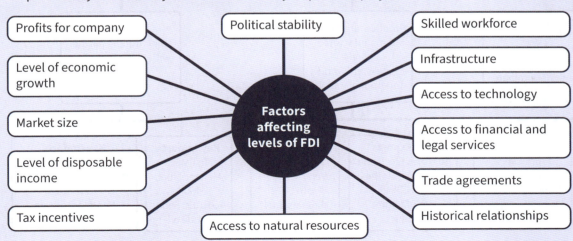

## 7 Global systems and global governance

### 7.3 International trade

## Trading patterns and relationships

Countries have agreements to trade with another country (bilateral) or as part of a trade bloc. The WTO governs world trade by setting rules and settling disputes between countries and/or trade blocs.

**REVISION TIP**

You need to know the trading patterns and relationships between HDEs, EMEs and LDEs.

Other major regional trade blocs include ASEAN and SCO

EU's major trading partners are China (goods) and the USA (services)

China is the world's largest trading nation

**Key**
- USMCA
- GCC
- Mercosur
- RCEP
- European union
- SAARC
- EAEU
- AfCFTA

USA is ranked 2nd in global imports and exports of services, 1st in global imports of goods and 2nd in global export of goods

Most trade takes place between HDEs

Increasing growth in trade between EMEs and LDEs

▲ **Figure 2** The world's regional trade blocs, 2021

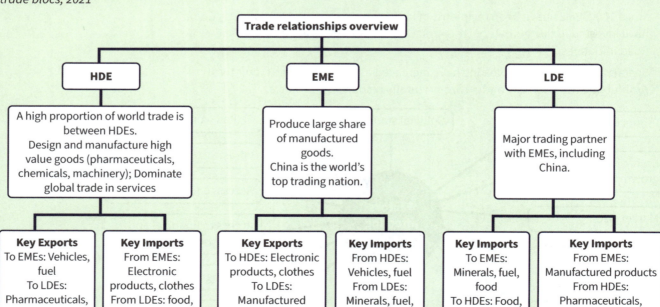

**Trade relationships overview**

**HDE**

A high proportion of world trade is between HDEs.
Design and manufacture high value goods (pharmaceuticals, chemicals, machinery); Dominate global trade in services

**Key Exports**
To EMEs: Vehicles, fuel
To LDEs: Pharmaceuticals, machines

**Key Imports**
From EMEs: Electronic products, clothes
From LDEs: food, crude oil

**EME**

Produce large share of manufactured goods.
China is the world's top trading nation.

**Key Exports**
To HDEs: Electronic products, clothes
To LDEs: Manufactured products

**Key Imports**
From HDEs: Vehicles, fuel
From LDEs: Minerals, fuel, food

**LDE**

Major trading partner with EMEs, including China.

**Key Imports**
To EMEs: Minerals, fuel, food
To HDEs: Food, crude oil

**Key Imports**
From EMEs: Manufactured products
From HDEs: Pharmaceuticals, machines

# Market accessibility

Access to markets is vital for trade.

There is **differential access to markets** which is caused by:

- WTO rules and regulations
- trade barriers (tariffs, taxes, quotas and subsidies) – HDEs can add tariffs to prevent flooding their domestic markets with imports
- trade bloc membership – members have access to internal market without barriers, reducing costs
- levels of investment in infrastructure and production – HDEs can invest highly in transport networks and factories, making production more modern and efficient
- product type – Many LDEs' economies are based on **primary products** that have lower values than manufactured goods or services. LDEs need special trade agreements to give them more access to markets, for example EU (Generalised Scheme of Preferences, Aid for Trade) and WTO (Special and Differential Treatment)
- technology – HDEs have higher levels of technological development so can innovate new products and services, maintaining intellectual property rights.

### Economic impacts of differential access to markets

- Economy becomes dependent on smaller, lower value range of primary products, whose price fluctuates on commodity markets
- Outcompeted by larger trading economies
- Low economic growth
- Less foreign exchange
- Difficult to innovate to create new products
- Difficult to withstand internal and external shocks, e.g. crop failure, rise in energy costs
- Higher prices of imported goods, e.g. manufactured products

### Social impacts of differential access to markets

- Less consumer choice
- Smaller range of job opportunities, especially in higher paid tertiary and quaternary sectors, leading to income inequality, poverty, less social mobility and more outmigration
- Dependency on primary products makes incomes seasonal, leading to underemployment

**REVISION TIP**

In order of importance, rank the reasons why some countries find it more difficult to access the global market.

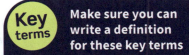

**Key terms** Make sure you can write a definition for these key terms

differential access to markets    international trade
primary product    trading pattern    volume of trade

Learn the answers to the questions below, then cover the answers column with a piece of paper and write down as many as you can. Check and repeat.

## Questions | Answers

| # | Questions | Answers |
|---|-----------|---------|
| 1 | What was the value of international trade in 2022? | US$27 trillion |
| 2 | How much do LDEs contribute to global exports | 1% |
| 3 | What is inward FDI? | Investments made in a country from another country |
| 4 | What is outward FDI? | Investments made by domestic companies in a foreign economy |
| 5 | Give FIVE factors that affect levels of FDI. | Five from: political stability / profits for company / level of economic growth / market size / level of disposable income / tax incentives / access to natural resources / skilled workforce / infrastructure / access to technology / access to financial and legal services / trade agreements / historical relationships |
| 6 | Which country is the world's largest trading nation? | China |
| 7 | Which country is the world's largest importer and exporter of services? | USA |
| 8 | From which country does the EU import fewest services? | Russia |
| 9 | State three social impacts of differential access to markets. | Less consumer choice; less range of job opportunities; seasonal employment, leading to underemployment |
| 10 | How do tariffs affect access to markets? | HDEs can add tariffs to products to prevent flooding their domestic markets with imports |

*Put paper here*

## Previous questions

Now go back and use these questions to check your knowledge of previous topics.

## Questions | Answers

| # | Questions | Answers |
|---|-----------|---------|
| 1 | Give two examples of flows of labour. | Two from: movement of people for work / economic migrants / to fill skills shortages in the same country |
| 2 | What is a bilateral trade agreement? | A trade agreement between two countries |
| 3 | Give a definition of global systems. | Interconnections and interdependence between countries that help make and develop the world, including economic, political, societal and environmental interactions |
| 4 | Environmental interdependence: Give two parts of our global environmental system. | Two from: atmosphere / hydrosphere / cryosphere / lithosphere / biosphere |
| 5 | What does FDI stand for? | Foreign direct investment |

*Put paper here*

# ⚙ KNOWLEDGE

## 7 Global systems and global governance

### 7.4 TNCs

## TNCs and globalisation

**Transnational corporations** (TNCs) play an important role in international trade and the global market. Of all global trade, 80% is linked to TNCs.

It is often more efficient and profitable for TNCs to locate parts of the business in different countries. They use a global supply chain to develop, manufacture and sell their products. This helps TNCs to spread locational risk and increase production flexibility, responding to changing commercial conditions, such as market demand.

## Spatial organisation of a TNC

Headquarters: HDE
Many of the world's TNCs started in HDE; location attracts best talent for high skilled quaternary functions e.g. finance, legal, marketing, R&D, business operations

Regional Headquarters: HDE/EME
In strategic locations closer to countries where manufacturing takes place or in key regional markets

Manufacturing: EME/LDE
Production takes place in factories (branch plants) where costs are lower than HDEs, including labour; government gives incentives to attract FDI to build factories; close to or within market for products to reduce transport costs

Market: HDE/EME/LDE (depending on product)
Products are distributed to the market e.g. container ship; TNCs want to have a large market share to increase profits

Raw materials and components: HDE/EME/LDE (depending on materials needed)
Links to resource endowment

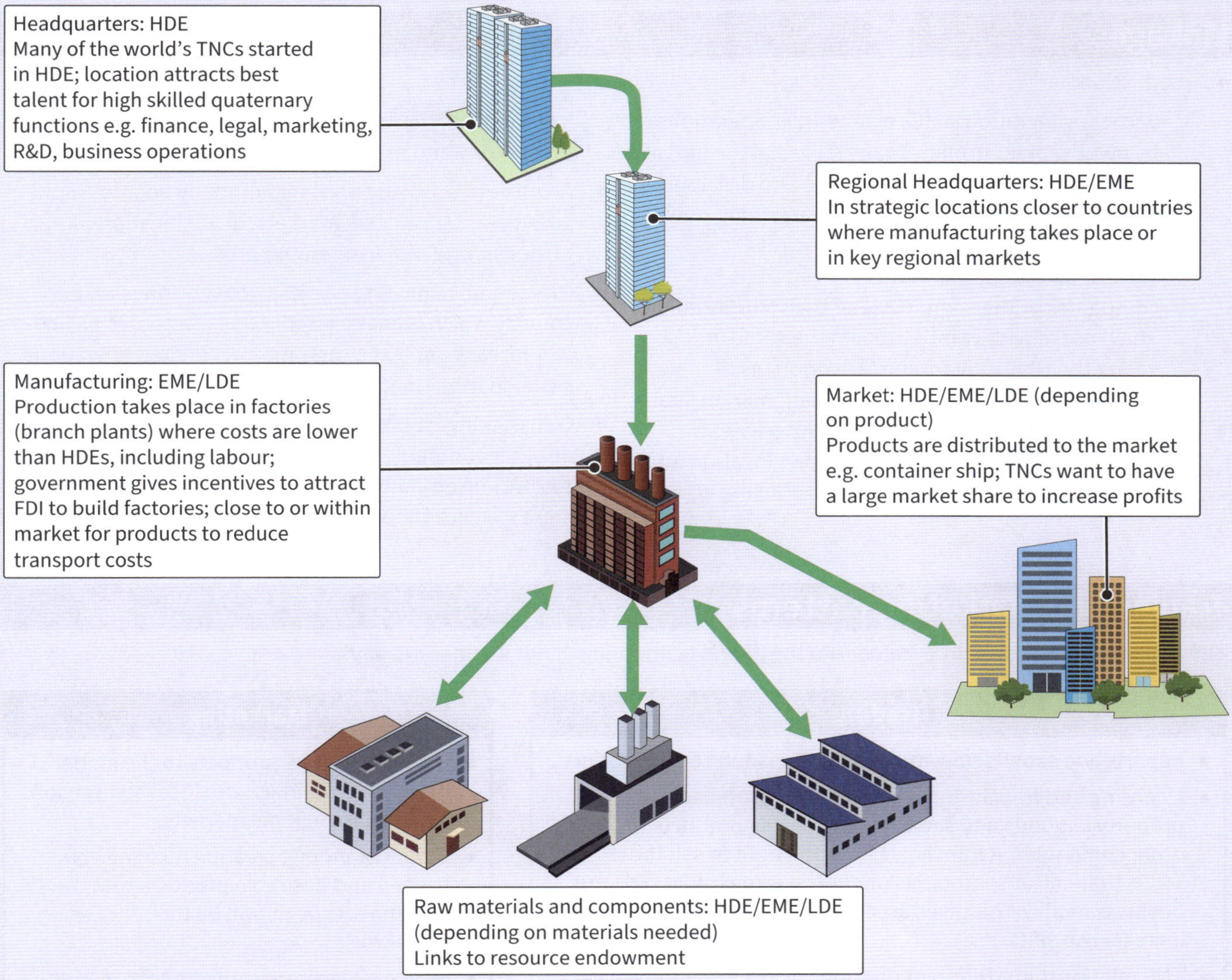

▲ **Figure 1** *Spatial organisation of a TNC*

# 7 Global systems and global governance

## 7.4 TNCs

### Production

TNCs invest in all stages of the manufacturing process.

- Primary – extraction of raw materials.
- Secondary – the raw materials are used to make other products, adding value at each stage of the manufacturing process. TNCs invest in countries where production costs are lower (labour, government incentives, cheaper land, in Special Economic Zones (SEZs)).

- Tertiary – the infrastructure to provide distribution networks to get the finished product to market and shops. This involves transport and retail services.
- Quaternary – knowledge-based services employ highly skilled workers to create new products (R&D) or to manage the business aspects of the company (financial, legal).

### Linkages

- Production:
  - Vertical integration: TNCs extend their operations by owning or controlling many stages of the global supply chain, including production and distribution. Gives TNCs advantage of greater bargaining power, reduces costs, improves quality control.
  - Horizontal integration: TNCs use different companies at the same stage of their supply chain. They merge with competitors in the same industry giving them the advantage of economies of scale, reducing competition and increasing market share.
- Mergers: joining with another company creates a larger company, which increases market location and share, gives the benefit of economies of scale, and

increases efficiency through integrating resources, technology and expertise.

- Acquisitions: buying another company enables rapid growth, access to new markets, diversification of business, and obtains new expertise.
- Offshoring: relocating part of its own business in a country which offers cost reductions, access to skilled labour, easier access to new markets (makes operations easier in those markets with different time zones).
- Outsourcing: subcontracting parts of its operation to another company, which reduces costs, gives access to specialised skills (production, services), and increases flexibility.

### Trading and marketing patterns

The activities of TNCs are integrated into the global economic system in different ways.

| Trading | Marketing |
|---|---|
| <ul><li>Imports and exports: raw materials and manufactured products.</li><li>Licensing and franchising: TNCs retain the intellectual property of goods and services by licensing products to other companies. A franchisee buys the right to sell the TNC's (franchiser's) products under the same trademark and business model. This gives access to markets with lower risk, increasing profits.</li><li>TNC relationships: Different divisions of TNCs can trade with each other (intra-firm trading). TNCs also enter into joint ventures with other TNCs to improve manufacturing processes, market access and distribution.</li><li>TNCs can stimulate local economies through the multiplier effect and the circular economy.</li></ul> | <ul><li>Retailing: selling products to the global market, sometimes with localised versions to adapt to regional markets.</li><li>Global branding: creating recognisable, trusted and desirable products that the consumer expects will be the same worldwide.</li><li>Global advertising campaigns to increase exposure of product to the market.</li><li>TNCs invest large amounts of money into marketing budgets, including using IT and social media to access growing and changing markets.</li></ul> |

# A specific TNC: Coca-Cola

Coca-Cola Company is an American TNC that produces and sells drinks.

| Revenue | Annual revenue: US$43 billion (2022) |
|---|---|
| Spatial organisation | 900 bottling partners and 950 production facilities worldwide<br>Products sold in over 200 countries and territories |
| Production and linkages | Makes concentrates and sells to authorised bottling partners, who mix concentrates with still/sparkling water, and sweeteners if required, and prepare, package and distribute the finished drinks.<br>Also has 'finished product' operations where the company owns or controls bottling, sales and distribution.<br>In 2006, company-owned bottling plants joined together to form Bottling Investments Groups (BIG).<br>Bottling plants are near to markets as transport costs increase once water is added to the concentrate.<br>Coca-Cola has bought existing bottling plants rather than building new ones to rapidly expand into new markets. |
| Trading and marketing patterns | World's largest drinks distribution system<br>One of the world's most recognisable and profitable brands<br>Huge global market – 2.2 billon servings a day<br>Diversified products to access increased market locations and market share |

**REVISION TIP**

For a TNC you have studied, draw a table to show the positive and negative impacts it has in the countries in which it operates.

## Impacts of Coca-Cola on countries in which it operates

| | Social | Economic | Environmental |
|---|---|---|---|
| **Positive** | • Coca-Cola Foundation gives grants<br>• Training and education schemes<br>• Improved standard of living for employees of Coca-Cola and its partners | • Multiplier effect from BIG and franchises<br>• Investment in R&D and bottling plants in new and expanding markets | • Circular economy with more focus on recycling<br>• Supports sustainable agriculture schemes<br>• Water replenishing schemes |
| **Negative** | • Difficulties with workers' rights in some bottling plants<br>• Health issues surrounding soft drink consumption | • Difficulties with working conditions including hours and pay in some bottling plants<br>• Profits go back to USA<br>• Changes in demand, external factors or local conditions can lead to closure of bottling plants | • Depletion of local water supplies<br>• Water pollution<br>• Litter from packaging |

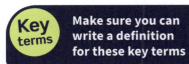 **Key terms** Make sure you can write a definition for these key terms

linkage    spatial organisation    transnational corporation

# 7 Global systems and global governance

## 7.5 Consequences of global systems

### World trade in food commodities

**Food commodities** are either raw or processed products that are sold for human consumption, such as coffee, cocoa, grains, sugar and fruit.

**Bananas**

- 90% of exported bananas are from Central and South America and the Philippines
- The largest importers are the EU, USA, China, Russia and Japan
- Demand is high in HDEs, which tend not to have the climate to grow bananas
- Technology is important in the global supply chain, including crop disease prevention and refrigerated transport
- Bananas are grown in humid, tropical regions
- Export revenue contributes towards buying food imports for producing countries
- TNCs dominate cultivation due to the long term impacts of colonisation and the large amount of investment needed, e.g. Chiquita, Dole and Del Monte
- Often sold at low prices due to short shelf life and supermarket marketing strategies
- Can comprise 75% of total monthly household income for small-scale banana farmers
- Banana crops are vulnerable to natural disasters, climate change and disease, affecting global supply chains and incomes

Workers 4–7%

Banana TNC 15–25%

Transport 20–33%

EU Tariffs 8%

Ripening 10%

Retailer 21–43%

▲ **Figure 1** *Distribution of retail price of a banana*

> **REVISION TIP**
>
> For a food commodity or manufactured product you have studied, state where it is produced and explain how it is produced, how it is distributed and where it is sold.

### Fairtrade

Fairtrade initiatives try to increase the amount of money farmers get and to encourage more sustainable practices in the global supply chain. Fairtrade labels are displayed on products that meet Fairtrade standards.

| Fairtrade benefits | Fairtrade limitations |
| --- | --- |
| • Smallholders get guaranteed minimum price and a fair income<br>• Additional money (Fairtrade premium) paid into a communal fund for workers and farmers to use<br>• Encourages cooperatives so farmers have more buying power and reduce costs through economies of scale<br>• Stable income so a reduction in poverty<br>• Training in new methods and technologies<br>• Empowerment of local communities<br>• Environmentally sustainable practices encouraged | • Fairer price of product<br>• Expensive for lower income groups<br>• Not all fruit produced is sold in Fairtrade markets<br>• Costs of certification and administration<br>• Although some farmer incomes have improved, large share of profits remain with TNCs and HDEs |

# Impacts of global trade and markets on people's lives

**Positive impacts**

- Increase in foreign exchange
- Increased employment opportunities at different skill levels
- More education and training leads to increased social mobility
- Increased consumer choice
- Competition results in lower consumer prices
- Multiplier effect leads to increased incomes
- Increased product quality
- Increase in personal wealth increases disposable income and demand for products and services increases
- Technology transfers
- Increased political stability
- Consumer awareness puts pressure on companies and IGOs to tackle ethical and environmental issues
- Improved infrastructure for trade benefits society

**Negative impacts**

- Higher paid quaternary workers brought in from other countries
- External shocks (conflict, natural disasters) cause reduced demand and closure of factories and unemployment
- Environmental issues e.g. pollution
- TNC expansion can lead to conflicts over land ownership
- Unemployment or forced to migrate if companies relocate
- Profits go to shareholders in home country so reduced 'trickle down' effect
- Many manufacturing and some service (tourism) jobs are lower income/seasonal
- Preferential treatment for TNCs (e.g. tax breaks) and workers can create political conflicts (nationally and locally)
- Time zones may make work patterns of employees less 'family orientated'
- Loss of empowerment – decisions made in the USA, China and EU have impacts on local communities
- Increased prices for, or loss of, local products that don't benefit from economies of scale or need special expertise
- Loss or adaptation of local culture and traditions
- Increased interdependence – events in other countries have a direct impact on a nation's economy

**Key terms** Make sure you can write a definition for these key terms

food commodity

**REVISION TIP**

Draw a concept map of the ways in which international trade impacts people's lives in different parts of the world.

Learn the answers to the questions below, then cover the answers column with a piece of paper and write down as many as you can. Check and repeat.

## Questions | Answers

| | Questions | | Answers |
|---|---|---|---|
| 1 | What is a transnational corporation? | Put paper here | A company that operates in more than one country |
| 2 | What are the five elements of TNC spatial organisation? | | Headquarters, regional headquarters, manufacturing, raw materials and components, market |
| 3 | What are the four stages of the manufacturing process? | | Primary, secondary, tertiary, quaternary |
| 4 | Give three examples of linkages in a TNC. | Put paper here | Three from: production / mergers / acquisitions / offshoring / outsourcing |
| 5 | What is the term used for a recognisable, trusted, desirable product the consumer expects will be the same worldwide? | | Global branding |
| 6 | Give two negative impacts of Coca-Cola on the environment. | Put paper here | Two from: overuse of local water supplies / water pollution / litter from packaging |
| 7 | Name the three main export areas for bananas. | | Central America, South America, Philippines |
| 8 | Name the five main importers of bananas. | | EU, USA, China, Russia, Japan |
| 9 | Give two aims of Fairtrade. | Put paper here | Increase the amount of money farmers get; encourage more sustainable practices |
| 10 | What is the Fairtrade premium? | | Additional money paid into a communal fund for workers and farmers to use |

## Previous questions

Now go back and use these questions to check your knowledge of previous topics.

## Questions | Answers

| | Questions | | Answers |
|---|---|---|---|
| 1 | From which type of economy has manufacturing shifted in the late twentieth century, and where has it shifted to? | Put paper here | From HDEs to EMEs and LDEs |
| 2 | Give two examples of how countries may rely on others economically. | | Two from: FDI / aid / loans / workforce |
| 3 | Why has the flow of services increased? | Put paper here | Improvements in IT, deregulation |
| 4 | What is outward FDI? | | Investments made by domestic companies in a foreign economy |
| 5 | How do tariffs affect access to markets? | | HDEs can add tariffs to products to prevent flooding their domestic markets with imports |

# 7 Global systems and global governance

## 7.6 Global governance

## What is global governance?

- **Global governance** manages cross-border issues through institutions, rules and systems, for example climate change, migration, pandemics, trade, finance, conflicts.
- There is no single international organisation responsible for the overall governance of our planet.

> **REVISION TIP**
>
> Remember, you need to know what global governance is.

## The role of norms, laws and institutions in tackling global issues

| Norms | Laws | Institutions |
|---|---|---|
| Shared behaviours and values. **Norms** are attitudes towards, and standards of, what is acceptable (or not) in society. They are often linked to human rights and may differ between nations. | Enforced system of rules and regulations. International laws are legally binding agreements between different countries. They are treaties, standards and conventions aimed at tackling global issues such as human rights, environmental management, trade laws and employment rules. | International organisations which enact laws and devise rules and policies on matters of global interest. They can enforce laws and determine whether they have been broken, while following established procedures. They may settle disputes between different countries or seek penalties for those breaking international laws. |

## The United Nations

The **United Nations** (UN) is a major international agency, set up in 1945.

The work of the UN covers five main areas:

- Maintaining international peace and security through preventing and removing threats to peace and settling international disputes peacefully
- Promoting and protecting human rights
- Delivering humanitarian aid and solving international economic, social, cultural or humanitarian problems
- Supporting sustainable development and climate action
- Upholding international law

> **REVISION TIP**
>
> Make an audio recording to explain the role of the UN.

▶ **Figure 1** *The United Nations headquarters in New York City,*

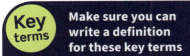 **Make sure you can write a definition for these key terms**

global governance    International Monetary Fund    norm
United Nations    World Bank    World Trade Organization

## 7 Global systems and global governance

### 7.6 Global governance

## Other important international agencies

### International Monetary Fund
- Established 1944, based in the USA
- Encourages global financial cooperation; promotes increased trade and economic growth
- Encourages governments to have policies which increase prosperity
- Offers technical assistance and training to governments
- Gives loans to countries to help improve their economies

### World Bank
- Established 1944, based in the USA
- Gives financial and technical help to countries to increase their level of development, including low interest loans and grants.

### World Trade Organization
- Established 1995, based in Switzerland
- Promotes free trade and settles trade disputes between different countries
- Puts rules in place to govern how goods, services and intellectual property are traded around the world

## Analysing global governance

### Global governance promotes growth and security
- Creates common rules which countries should follow, leading to greater security
- Common rules aim to consider all parties and to reduce differences and their impacts
- Global governance needed to tackle issues (COVID-19 pandemic, climate change, illegal migration, conflicts)
- Sharing of knowledge and expertise
- Emergency action for humanitarian crises and disaster relief
- Global projects to improve human development (e.g. UN's Millennium Development Goals (2000–2015), Sustainable Development Goals (2015–2030))
- UN Peacekeeping missions in conflict zones

### Global governance creates inequalities and injustices
- Power imbalances: domination of HDEs in decision making (e.g. no African or South American countries are permanent members, with veto powers, of the UN Security Council)
- WTO favours free trade model, which puts some LDEs at an economic disadvantage through the cost of its imports and exports
- Trade blocs may impose tariffs on imports from LDEs
- G7 and OECD are focused on economic power of members (mainly HDEs)
- Loan conditions from WB or IMF may have long term effects on LDEs' economies and societies
- UN Peacekeeping missions may not be successful

## Interactions from local to global scales

Global governance is complex and involves a range of decision making, management and enforcement at a range of scales: global (e.g. UN), International (e.g. EU), National (e.g. UK government), Regional (e.g. Welsh assembly) and local (e.g. local councils).

- Organisations at different scales of governance have varying levels of resources, decision-making powers and goals, making decisions and enforcing action difficult.
- Timeframes for decision making also fluctuates between types of governance
- Types of government (e.g. democracy) and different voting systems within international organisations affect empowerment of people in the global system

- Public sector (government) organisations work with the private (business) and voluntary (non-governmental organisation (NGO)) sectors, which can affect decision making
- Countries can alter affiliations with international agencies, changing relationships (e.g. Brexit)
- Impact and influence of media can affect global governance
- Historical ties between groups can be strong, influencing decision making today (e.g. UK and the Commonwealth)

# RETRIEVAL

Learn the answers to the questions below, then cover the answers column with a piece of paper and write down as many as you can. Check and repeat.

## Questions / Answers

| | Questions | | Answers |
|---|---|---|---|
| 1 | How does global governance work? | Put paper here | Through institutions, rules and systems that manage cross-border issues that go beyond the sphere of individual countries |
| 2 | What are norms? | | Shared behaviours and values |
| 3 | What are the five main roles of the UN? | Put paper here | International peace and security; human rights; humanitarian aid; sustainable development and climate action; international law |
| 4 | Give two examples of the type of issues international law deals with. | | Two from: human rights / environmental management / trade / employment |
| 5 | When was the United Nations founded? | | 1945 |
| 6 | When was the WTO founded? | Put paper here | 1995 |
| 7 | What does NGO stand for? | | Non-governmental organisation |
| 8 | Give two global issues that need global governance to tackle them. | Put paper here | Two from: COVID-19 pandemic / climate change / illegal migration / conflicts |
| 9 | What are the five scales of global governance? | | Global, international, national, regional, local |
| 10 | What is a public sector organisation? | Put paper here | One that is run by the government |

## Previous questions

Now go back and use these questions to check your knowledge of previous topics.

## Questions / Answers

| | Questions | | Answers |
|---|---|---|---|
| 1 | Give two positive impacts of unequal flows of ideas. | Put paper here | Two from: incentivises innovation, creativity and entrepreneurship / encourages less state, more private sector involvement and free trade / spread of democracy and equality / cultural exchange of ideas / encourages global networking and cross-border collaboration |
| 2 | State three social impacts of differential access to markets. | | Less consumer choice; less range of job opportunities; seasonal employment, leading to underemployment |
| 3 | What is a transnational corporation? | Put paper here | A company that operates in more than one country |
| 4 | What are the four stages of the manufacturing process? | | Primary, secondary, tertiary, quaternary |
| 5 | What is the Fairtrade premium? | | Additional money paid into a communal fund for workers and farmers to use |

# 7 Global systems and global governance

## 7.7 The global commons

## Location of the global commons

The **global commons** belong to everyone, and there are rights and responsibilities surrounding their use. Location of the global commons is defined by international law. They are areas of the planet that are:

- outside national jurisdictions for all nations
- accessible for all nations.

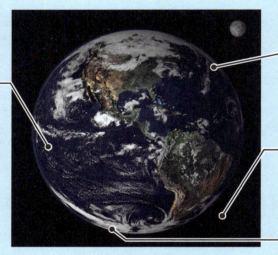

High seas: e.g. United Nations Convention on the Law of the Sea (UNCLOS)

Atmosphere: United Nations Framework, e.g. Convention on Climate Change (UNFCCC)

Outer space: e.g. Outer Space Treaty

Antarctica: e.g. Antarctic Treaty System

▶ **Figure 1** *International law is used to govern the global commons*

Other ideas of what constitutes the global commons include:

- resources of interest or value to the welfare of the community of nations, e.g. tropical rainforests and biodiversity
- science, education, information and peace.

> **REVISION TIP**
>
> Practise listing the areas considered to be the global commons.

## The rights of all to the benefits of the global commons

- Until the late twentieth century, access to the global commons was difficult due to remoteness and lack of technology.
- Technological development has increased accessibility to the global commons and their resources.
- There is increasing pressure on the space and resources of the global commons, particularly due to globalisation and climate change.
- Resources in the global commons can be exploited, regardless of the negative impact that may have on the global commons themselves or on others – 'the tragedy of the commons'.
- Countries are affected by policies made and enforced by others.
- Some argue that increased policy and decision making at a global level will help manage the global commons more effectively for all.

- The UN is the only worldwide, multilateral organisation that could develop and coordinate policies to protect and manage the global commons.
- Several treaties and conventions exist under international law to protect the global commons.
- Treaty systems linked to the global commons are complex, piecemeal and vulnerable to fast-moving changes in technology and human activities.
- Challenges remain on how agreements should be policed to ensure the sustainability of the global commons (stewardship).

> **REVISION TIP**
>
> Rank the ways in which the global commons can be protected in order of importance.

## Protection of the global commons

Some argue that increasing global interdependence requires stronger governance of the global commons, if **sustainable development** is to be achieved. Technology can help monitor the global commons to measure environmental impacts and the success of protection policies.

**REVISION TIP**

Make a memory map to show the threats to the global commons.

| | Key resources | Challenges for a sustainable future |
|---|---|---|
| **Antarctica** | • Terrestrial and marine resources<br>• Oil, gas, minerals, fish<br>• Landscape and ecosystems | • Territorial claims<br>• Fossil fuel exploration<br>• Increased tourism<br>• Protection of whales and other species<br>• Maintaining sustainable fish stocks<br>• Marine pollution |
| **High seas** | • Fish and other marine wildlife<br>• Oil, gas, sea bed exploration | • Territorial claims<br>• Ocean acidification<br>• Protection of marine life<br>• Maintaining sustainable fish stocks<br>• Bioprospecting<br>• Increased shipping<br>• Marine pollution including plastics |
| **Atmosphere** | • Essential for life on earth | • Atmospheric pollution<br>• Climate change and its impacts |
| **Outer space** | • Unique environment for scientific exploration, e.g. medical research on the International Space Station<br>• Key environment for satellite and communications technology (e.g. GPS) | • Human made space debris including out-of-life satellites<br>• Maintaining peaceful use of outer space<br>• Increasing exploration to the Moon and other celestial bodies<br>• Potential claims by countries landing on celestial bodies |

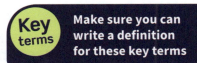 **Key terms** Make sure you can write a definition for these key terms

global commons    sustainable development

## 7 Global systems and global governance

### 7.8 Antarctica as a global common

## Geography

- **Antarctica** and the Southern Ocean, up to the **Antarctic convergence**, is part of the global commons.
- Antarctica is located in the southern hemisphere, mostly south of the Antarctic Circle (66° 33′S) and includes the South Pole at 90°S.
- It has an area of 14 million km², (larger than Europe), containing 30 million km³ of ice. In the Antarctic winter, the ocean surface freezes over, extending the size of the continent. The Antarctic ice sheet contains 60% of the world's total fresh water (90% of the world's surface fresh water). If this ice melted, the global sea level would rise by 70 m.
- Antarctic landmass is covered with glacial ice (East and West Antarctic Sheets). The Transantarctic Mountain range contains peaks over 4000 m, such as Mount Vinson.

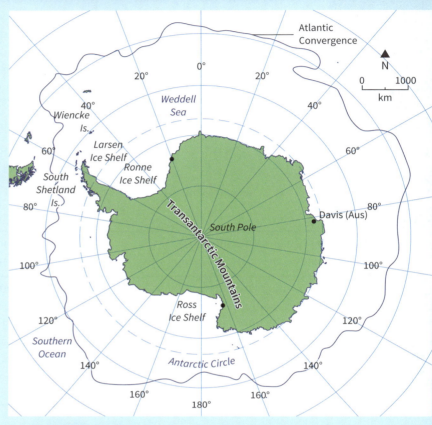

▲ **Figure 1** Antarctica

## Climate

| Climate | Cold desert climate |
|---|---|
| Temperature | Average temperature ranges from −10°C on the coast to −60°C at the highest parts of the interior. Highest temperatures between November and February |
| Precipitation | Low: snowfall equivalent to 150 mm rainfall per year |
| Winds | Very high katabatic winds (from high elevation to low under the force of gravity) which blow down from continental plateaus, exceeding 160 kph |
| Sunlight | Low sunshine levels with 24-hour darkness in the winter months |

**REVISION TIP**

Draw the outline of the Antarctic continent and label it with its key characteristics.

## Ecosystem

Less than 1% of Antarctica is ice-free so it is difficult for soils and vegetation to develop.

Terrestrial species include mosses, lichens and liverworts but no trees/shrubs.

Southern Ocean is one of the most productive parts of the world's oceans due to nutrient availability. Nutrients are brought up from the seabed around the Antarctic Convergence.

**Antarctic ecosystem**

Coastal areas are used as breeding areas for species (e.g. Weddell Seal, Emperor penguins).

Ecosystem structure and species behaviour are vulnerable to climate change.

Terrestrial and marine ecosystems are vulnerable to climate change.

Phytoplankton (producers) are at the base of the food chain and provide food for krill.

## Threats

### Climate change

Greatest long-term threat to the continent. Uneven temperature changes with temperatures rising rapidly (over 3 °C) since 1950. Global warming causes ice shelf collapse and increased ocean acidification, which leads to reduction of krill population, compromising the Antarctic food chain.

### Fishing

Despite legal limits fishing is reducing global fish stocks (including krill). Illegal fishing also creates problems, threatening stocks for predators, e.g. seabirds and/ or marine mammals. Harmful fishing practices affect seals, whales, seabirds. Difficult to monitor and police.

### Whaling

Ban on commercial whaling by IWC in 1982 has led to recovery in whale populations. Whaling is still done by some countries, under the guise of scientific research.

### Mineral exploration

Climatic conditions and remoteness make exploration expensive and dangerous. Commercial mining in Antarctica is banned. Increased demand may put pressure on the exploitation of mineral deposits thought to be in the Transantarctic Mountains, as well as offshore oil and gas in the Southern Ocean.

### Tourism

There has been a recent growth in tourism, which can lead to more support and investment in the protection of the Antarctic. Antarctica's remoteness leads to a high carbon footprint for travellers, and boats create pollution. Litter and waste can cause environmental damage at landing sites, and tourists can disrupt species' behaviour. Management is mainly self regulated by the International Association of Antarctica Tour Operators (IAATO).

### Scientific research

There are over 50 permanent research stations with 1200 scientists in winter and 4800 in summer. Antarctica is a unique place for research as it is relatively untouched – the Antarctic Treaty lays down laws about the activities that can take place. Research has helped understanding of climate change, ozone depletion, sea level rise and ecosystems. Researchers must ensure their activities do not harm the fragile ecosystem. All non-indigenous species, including huskies, were banned in the 1990s.

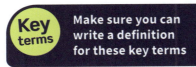
**Key terms** — Make sure you can write a definition for these key terms

Antarctica    Antarctic convergence

# 7 Global systems and global governance

## 7.9 Global governance in Antarctica

### Territorial claims to Antarctica

Antarctica is part of the global commons and the last remaining 'wilderness', a unique place on the planet. It is not governed by one country, nor does it have its own government. Some countries have territorial claims, or reserve the right to claim, parts of Antarctica.

Other countries have research bases on the Antarctic continent (e.g. India, China) or on islands off the Antarctic Peninsula, such as King George Island (e.g. Brazil, Russia). The USA has not made a claim but has three year-round research bases.

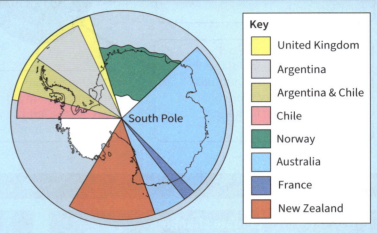

**Key**
- United Kingdom
- Argentina
- Argentina & Chile
- Chile
- Norway
- Australia
- France
- New Zealand

▲ **Figure 1** *Territorial claims to Antarctica*

### Intergovernmental organisations (IGOs)

A range of treaties, agreements and organisations aim to sustainably manage Antarctica.

| | |
|---|---|
| **United Nations Environment Programme (UNEP)** | The leading global authority on the environment focusing on climate, nature and pollution. This includes the protection and sustainable management of marine and coastal environments, as well as rights and governance. |
| **International Whaling Commission (1946)** | Responsible for the management of whaling, and conservation of whales, with 88 member countries.<br><br>1994: designated a sanctuary for whales in the Southern Ocean; prohibiting commercial whaling.<br><br>2009: IWC Southern Ocean Research Partnership to improve conservation of large whale species through research. |
| **The Antarctic Treaty (1959)** | Originally signed by 12 countries, now 56. Key elements: Antarctica used only for peaceful purposes; freedom and cooperation and exchange of scientific investigation; not to recognise, dispute, establish or allow future claims; ban of military activities; nuclear testing and radioactive waste disposal prohibited.<br><br>Evolved into the Antarctic Treaty System and includes the Commission for the Conservation of Antarctic Marine Living Resources (CCAMLR). |
| **Protocol on Environmental Protection to the Antarctic Treaty (1991)** | Designates Antarctica as a 'natural reserve, devoted to peace and science'. Prohibits all mineral resource activities, except for scientific research. Assesses and monitors human activities including: environmental impact assessments, protecting wildlife, waste and pollution management, designating protected areas and guidance on environmental emergencies. |
| **IWC Whaling Moratorium (1982)** | Pause on commercial whaling to allow for species recovery. (Norway and Iceland set their own catch limits in the North and Arctic Seas.) Japan left the IWC in 2019 and is not bound by the moratorium. Catches it makes are reported to the IWC. |

## The role of NGOs

**Non-governmental organisations** (NGOs) have an important role to play in monitoring threats and enhancing protection of Antarctica.

- Independently monitor activities.
- Report on how successful management strategies are at dealing with threats to the Antarctic.
- Raise international awareness of evolving environmental issues such as climate change.

## Key NGOs

- WWF: Lobbied for new regulations in 2011 to stop ships carrying or using heavy fuel in the Antarctic Ocean.
- Antarctic and Southern Ocean Coalition (ASOC): Protects vulnerable ecosystems by providing a unified voice for Antarctic conservation. Only environmental NGO to observe Antarctic Treaty meetings, representing the Antarctic conservation community where key decisions about the continent's future are made. Through its Antarctic Ocean Alliance (AOA) promotes special designation of marine protected areas (MPAs).
- Greenpeace: Campaigns for environmental protection (Antarctic World Park – 1983). Ships travel to Antarctica each summer to document and raise awareness of environmental degradation. Monitors whales and illegal whaling activity in Antarctic waters.

> **REVISION TIP**
>
> Choose one NGO and explain how its work is protecting Antarctica.

## Consequences of global governance

### How Antarctica is governed affects people worldwide

- The Antarctic Treaty is one of the most successful international agreements, with peaceful cooperation and scientific research between countries.
- Research on melting ice raises awareness of the impacts climate change has on Antarctica and beyond.
- Global research collaboration benefits science worldwide.
- Protection of resources focuses on long-term sustainability for many rather than short-term economic gain for the few.
- Tourist activities are managed for the long-term future of the ecosystem.

### Criticisms of the developing governance of Antarctica

- It is difficult to enforce agreements – there is no 'Antarctic police force'.
- There are no legal penalties for breaking agreements.
- Countries have to negotiate disputes or take them to the International Court of Justice.
- Complex systems make decision making more difficult and slower.
- Globalisation has increased interest in the continent by many countries who may have different perspectives on what should be managed and how.

> **REVISION TIP**
>
> Practice explaining to a friend why the protection of the Antarctic continent matters to the global population.

 **Key terms** — Make sure you can write a definition for these key terms

United Nations Environment Programme
International Whaling Commission (IWC)     Antarctic Treaty
Protocol on Environmental Protection to the Antarctic Treaty
IWC Whaling Moratorium
non-governmental organisation (NGO)
intergovernmental organisation (IGO)

# RETRIEVAL

Learn the answers to the questions below, then cover the answers column with a piece of paper and write down as many as you can. Check and repeat.

## Questions | Answers

| # | Question | Answer |
|---|----------|--------|
| 1 | What are the global commons? | Areas of the planet that are outside national jurisdictions and accessible for all nations |
| 2 | What are the four locations of the global commons under international law? | Antarctica, high seas, atmosphere, outer space |
| 3 | Give two challenges for a sustainable future for the atmosphere. | Atmospheric pollution, climate change and its impacts |
| 4 | What is the Antarctic Convergence? | Boundary between the cold water from the Southern Ocean and warmer waters to the north |
| 5 | Which species is the primary producer in the Antarctic food web? | Phytoplankton |
| 6 | Name three ways tourists can cause environmental damage to Antarctica. | Litter, waste, disrupt species' behaviour |
| 7 | Does Antarctica belong to any country? | No |
| 8 | What is the role of the IWC? | Global organisation responsible for the management of whaling, and conservation of whales |
| 9 | Give the five key elements of the Antarctic Treaty. | Peaceful purposes; freedom, cooperation and exchange of scientific investigation; not to recognise, dispute, establish or allow future claims; banning military activities; nuclear explosions and radioactive waste disposal prohibited |
| 10 | Name one NGO active in Antarctica. | One from: WWF / ASOC / Greenpeace |

*Put paper here*

## Previous questions

Now go back and use these questions to check your knowledge of previous topics.

## Questions | Answers

| # | Question | Answer |
|---|----------|--------|
| 1 | Name three areas in which flows of services take place. | Financial, business, customer services |
| 2 | What is inward FDI? | Investments made in a country from another country |
| 3 | Give three examples of linkages in a TNC. | Three from: production / mergers / acquisitions / offshoring / outsourcing |
| 4 | How does global governance work? | Through institutions, rules and systems that manage cross-border issues that go beyond the sphere of individual countries |
| 5 | What are the five scales of global governance? | Global, international, national, regional, local |

*Put paper here*

# KNOWLEDGE

## 7 Global systems and global governance

### 7.10 Globalisation critique

## Analysing and evaluating key issues

Critiquing globalisation requires knowledge and understanding of the key issues and the ability to analyse and evaluate those issues. When assessing different impacts of globalisation, consideration should be given to each of the following:

| Location | Where is the impact? | Local / regional / national / international / global |
|---|---|---|
| Severity | How significant is the impact? | Low / medium / high |
| Nature | What is the impact? | Social / economic / political / environmental |
| Timescale | How long does the impact last for? | Short term / medium term / long term |
| Positive/negative | How much has the impact been positive or negative? | Positive / negative |
| People | Who is involved? | Individuals / national governments / IGOs / trade blocs / TNCs / small businesses |
| Direct/indirect | How much does one impact lead to another? | Interactions between impacts, e.g. economic impact leads to physical change |

## Benefits of globalisation

| Growth | Global economic growth |
|---|---|
| | Trade of goods, services, intellectual property |
| | FDI encourages growth and regional development through the multiplier effect |
| | Investment in infrastructure and services, e.g. education and health |
| | LDEs can enter the global market |
| | Increased employment opportunities, including through TNC investment |
| Development | Extreme poverty reduction |
| | Increased standard of living |
| | Increased empowerment and social mobility |
| | Increased consumer choice |
| | Lower prices for consumers |
| | Increased awareness of environmental issues, including recycling and carbon footprint |
| | Working towards Sustainable Development Goal targets |
| Integration | Free trade removes barriers |
| | Shared decision making |
| | Multiculturalism and diversity |
| | Sharing of technology |
| | Connectivity via the internet |
| | IGOs help global community with financial resources and humanitarian support |
| Stability | More global stability through interdependency |
| | Cooperation with environmental issues |

**REVISION TIP**

Make a judgement on how far globalisation benefits the global population, and practise writing the justification for your viewpoint.

## 7 Global systems and global governance

### 7.10 Globalisation critique

## Costs of globalisation

| Inequalities | Increased inequalities between low and high incomes within and between countries |
| --- | --- |
| | Highest income 20% consumes 86% of world's resources; lowest income 80% consumes 14% |
| | Barriers to free trade still exist, leaving some countries disadvantaged |
| | Inequalities in defence and military spending between countries |
| | HDEs dominate decision making in many IGOs |
| | Increased power of TNCs, usually based in HDEs |
| | Deindustrialisation of HDEs to lower cost production in EMEs and LDEs, causing unemployment and dereliction |
| Injustice | Increase in illegal migration |
| | Increase in smuggling including human trafficking and the illegal wildlife trade |
| | IGOs not always able to enforce their rules and penalise those who break them |
| | Reduction in quality of working conditions for some workers |
| | Currency manipulation to gain cost advantages |
| | 'Westernisation' leads to loss of local traditions, culture and languages |
| | Loss, or perception of loss, of sovereignty |
| Conflict | Need for natural resources by countries, and TNCs could be conflicting with needs of local community and the environment |
| | IGOs not always able to solve economic, social, political and environmental issues, at a range of scales |
| | Use of internet for increasing speed, type and amount of illegal activities |
| Environmental impact | Resource exploitation affects habitats and biodiversity |
| | Increased transportation and carbon emissions, threatening green space |
| | Increased consumer waste |
| | Increased air, water, land, noise and visual pollution |
| | Threats to the use and management of the global commons |
| | Difficult to visualise the long-term negative environmental impacts |

 **REVISION TIP**

Make an audio or video recording in which you outline the benefits and costs of globalisation for different groups.

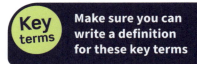 **Key terms**

**Make sure you can write a definition for these key terms**

integration

# RETRIEVAL

Learn the answers to the questions below, then cover the answers column with a piece of paper and write down as many as you can. Check and repeat.

## Questions

## Answers

| | Questions | Answers |
|---|---|---|
| 1 | Name four factors to consider when examining the impacts of globalisation. | Four from: location / severity / nature / timescale / positive or negative / people / direct or indirect |
| 2 | Name the four key benefits of globalisation. | Economic growth, development, integration, stability |
| 3 | Name the four key costs of globalisation. | Inequalities, injustice, conflict, environmental impact |
| 4 | How can globalisation benefit growth? Name four ways. | Four from: global economic growth / trade of goods, services, intellectual property / FDI encourages growth and regional development through the multiplier effect / investment in infrastructure and services / LDEs can enter the global market / increased employment opportunities, including through TNC investment |
| 5 | How can globalisation benefit integration? Name four ways. | Four from: free trade removes barriers / shared decision making / multiculturalism and diversity / sharing of technology / connectivity via the internet / IGOs help global community with financial resources and humanitarian support |
| 6 | How can globalisation increase injustice? Name four ways. | Four from: increase in illegal migration / increase in smuggling / IGOs not always able to enforce their rules and penalise those who break them / reduction in quality of working conditions for some workers / currency manipulation to gain cost advantages / 'westernisation' leads to loss of local traditions, culture and languages / loss, or perception of loss, of sovereignty |
| 7 | Give four environmental impacts of globalisation. | Four from: resource exploitation affects habitats and biodiversity / increased transportation and carbon emissions / increased consumer waste / increased air, water, land, noise and visual pollution / threats to the use and management of the global commons |

*Put paper here*

## Previous questions

Now go back and use these questions to check your knowledge of previous topics.

## Questions

## Answers

| | Questions | Answers |
|---|---|---|
| 1 | Name the three main export areas for bananas. | Central America, South America, Philippines |
| 2 | Give two examples of the type of issues international law deals with. | Two from: human rights / environmental management / trade / employment |
| 3 | What does NGO stand for? | Non-governmental organisation |
| 4 | What is the Antarctic Convergence? | Boundary between the cold water from the Southern Ocean and warmer waters to the north |
| 5 | Name three ways tourists can cause environmental damage to Antarctica. | Litter, waste, disrupt species' behaviour |

*Put paper here*

# PRACTICE

1    Outline how global marketing reflects globalisation.                                    **[4 marks]**

2    Explain how unequal power relations enable some states to drive
     global systems to their own advantage.                                                 **[4 marks]**

> **EXAM TIP**
>
> 4-mark questions are point-marked. That means each valid point or development of a valid point receives one mark, up to a maximum of 4 marks.

3    Explain features and trends in the volume and pattern of investment
     associated with globalisation.                                                         **[4 marks]**

4    Outline the spatial organisation of one transnational corporation (TNC)
     you have studied.                                                                      **[4 marks]**

5    Outline how the role of norms and laws differ within global
     governance.                                                                            **[4 marks]**

6    Explain why Antarctica is considered a global common.                                  **[4 marks]**

7    Explain how globalisation can have environmental costs.                                **[4 marks]**

8    Using **Figure 1** and your own knowledge, assess the extent to which
     global systems can promote growth for places.                                          **[6 marks]**

▲ **Figure 1** *Infosys Headquarters, an Indian information
technology TNC in Bengaluru, India*

**9** Analyse the data shown in **Figures 2a**, **2b** and **2c**. **[6 marks]**

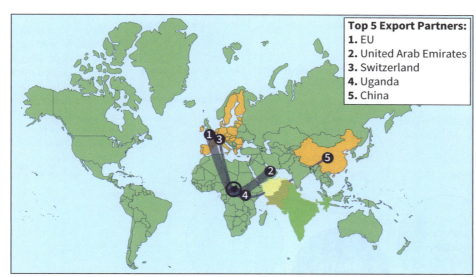

**Top 5 Export Partners:**
**1.** EU
**2.** United Arab Emirates
**3.** Switzerland
**4.** Uganda
**5.** China

| Central African Republic | $US million (2022) |
|---|---|
| Goods exports | 143 |
| Goods imports | 496 |
| Service exports | 38 |
| Service imports | 353 |

▲ **Figure 2b** *Value of Central African Republic exports and imports in 2022*

▲ **Figure 2a** *Top 5 countries to which Central African Republic exports*

In 2021, out of 226 world economies, CAR was 194th for exports and 199th for imports.

CAR Top 5 exports: gold, rough wood, diamonds, sawn wood and vehicles.

CAR Top 5 imports: medicines, broadcasting equipment, vaccines, cars and poultry.

◀ **Figure 2c** *Central African Republic imports and exports in 2021*

**10** Study **Figure 3**. Analyse the data presented about the world's largest transnational corporations by revenue. **[6 marks]**

> **EXAM TIP**
>
> In 6-mark 'Analyse' questions using a resource, all 6 marks are for AO3 (using geographical skills). In this case, you will be using geographical skills to interpret and analyse data.

| 2003 | | | 2023 | | |
|---|---|---|---|---|---|
| Rank | Company (Country, Sector) | Revenue ($US million) | Rank | Company (Country, Sector) | Revenue ($US million) |
| 1 | Walmart (US, Retail) | 246,525 | 1 | Walmart (US, Retail) | 611,289 |
| 2 | General Motors Corporation (US, Vehicles) | 186,763 | 2 | Saudi Aramco (Saudi Arabia, Energy) | 603,651 |
| 3 | Exxon Mobile (US, Energy) | 182,466 | 3 | State Grid (China, Energy) | 530,008 |
| 4 | Royal Dutch/Shell (Netherlands/UK, Energy) | 179,431 | 4 | Amazon (US, Retail) | 513, 983 |
| 5 | BP (UK, Energy) | 178,721 | 5 | China National Petroleum (China, Energy) | 483,019 |
| 6 | Ford Motor Company (US, Vehicles) | 163,871 | 6 | Sinopec Group (China, Energy) | 471,154 |
| 7 | DaimlerChrysler AG (Germany/USA, Vehicles) | 141,421 | 7 | Exxon Mobile (US, Energy) | 413,680 |
| 8 | Toyota Motor Corporation (Japan, Vehicles) | 131,754 | 8 | Apple (US, Technology) | 394,328 |
| 9 | General Electric Company (USA, Energy) | 131,698 | 9 | Shell (UK, Energy) | 386,201 |
| 10 | Mitsubishi Corporation (Japan, Vehicles) | 109,386 | 10 | United Health Group (US, Health) | 324,162 |

▲ **Figure 3** *Top TNCs by revenue ($US million) in 2003 and 2023*

# Exam-style questions

**11**  Using **Figure 4**, showing aspects of globalisation put forward by US and UK survey participants, and your own knowledge, assess the extent to which it reflects different aspects of globalisation.

**[6 marks]**

**EXAM TIP**

PEEL is a good way to structure your answers for 6-mark questions. Make three developed points (P) for your answer, include evidence (E) from the resource for each point, explain (E) a reason for each point you make, and link back (L) to the question.

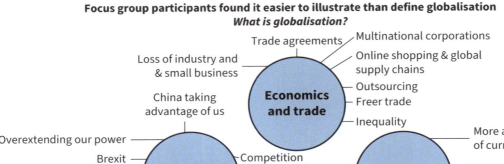

Note: This graphical representation reflects the breadth of ideas shared by focus group participants, not the frequency with which these ideas came up.
Source: Focus groups conducted Aug. 19–Nov. 20, 2019.
"In U.S. and UK, Globalisation Leaves Some Feeling 'Left Behind' or 'Swept Up'"

▲ *Figure 4*

**12**  Using **Figures 5a** and **5b** and your own knowledge, assess the issues associated with attempts at global governance by UN agencies.    **[6 marks]**

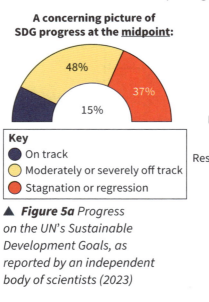

▲ *Figure 5a Progress on the UN's Sustainable Development Goals, as reported by an independent body of scientists (2023)*

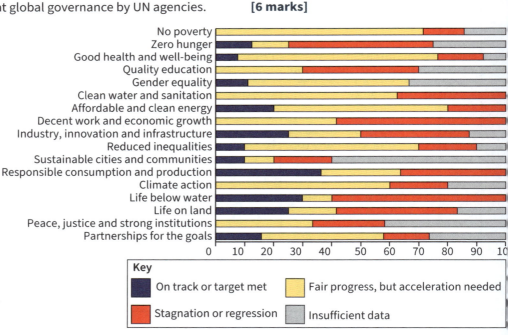

▲ *Figure 5b Progress assessment for the 17 Goals based on assessed targets, 2023 or latest data (percentage)*

13    Using **Figures 6a, 6b** and your own knowledge, discuss the threat to Antarctica arising from fishing.        **[6 marks]**

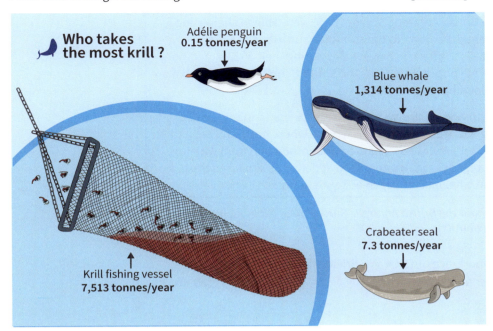

▲ **Figure 6a** *Who takes the most krill in Antarctica?*

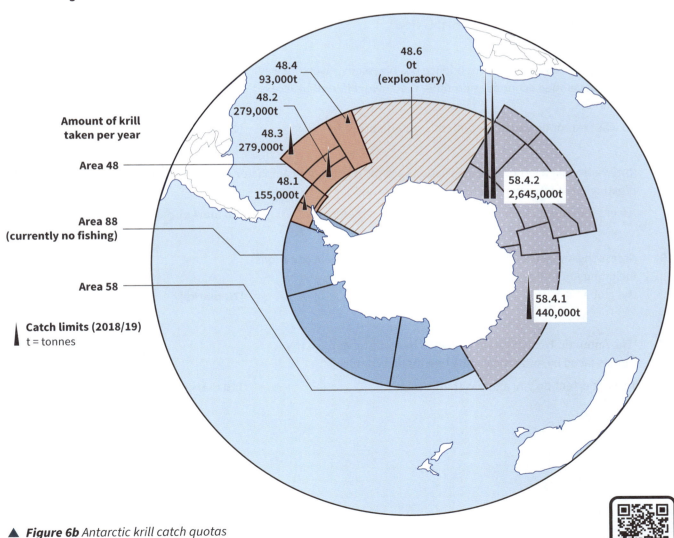

▲ **Figure 6b** *Antarctic krill catch quotas*

**14** Using **Figure 7** and your own knowledge, critically assess the impact of the global governance of whaling in Antarctica. **[6 marks]**

> Scientists led by the British Antarctic Survey (BAS) were able to share good news from their recent expedition to the sub-Antarctic island of South Georgia. They counted 55 Antarctic blue whales during their 2020 expedition, which they describe as "unprecedented". The South Georgia Waters remain an important summer feeding ground.
>
> After three years of surveys, scientists were excited to see so many whales visiting South Georgia to feed again. This is a place where both whaling and sealing were carried out extensively. It is clear that protection from whaling has worked, with humpback whales now seen at densities similar to those a century earlier, when whaling first began at South Georgia.

▲ **Figure 7** *A report about the British Antarctic Survey (BAS) expedition to the sub-Antarctic island of South Georgia in 2020.*

**15** Assess the relative importance of TNCs as a driving force in the globalisation of the economy and society. **[20 marks]**

**16** 'Global systems are more likely to cause conflicts than promote development.'

To what extent do you agree with this view? **[20 marks]**

**17** 'Our ever increasing interdependence has big issues, given the way highly developed economies promote their own differential access to markets.'

To what extent do you agree with this statement? **[20 marks]**

**18** 'International government organisations are the most significant factor in creating unequal power relations between states.'

Assess the extent to which you agree with this statement. **[20 marks]**

> **EXAM TIP**
>
> Use examples you have studied to back up your points.
>
> Reread the question and make an overall judgement in your conclusion.

**19** Assess the impacts of the global governance of the global commons, including Antarctica, on your life and the lives of people across the globe. **[20 marks]**

**20** 'The Antarctic Treaty System is an effective way to manage the threats faced by Antarctica in the twenty-first century.'

To what extent do you agree with this view? **[20 marks]**

> **EXAM TIP**
>
> A PEEL approach works well for 20-mark answers – backing up points with evidence provides AO1 and explaining points and linking them back to the question provides AO2.

# ⚙ KNOWLEDGE

## 8 Changing places

### 8.1 The concept of place

## The concept of place

**Place** is location plus meaning: a specific location that possesses unique physical, cultural and social attributes that distinguish it from other locations. A place can have different meanings to different people. **Placelessness** is the concept that a location lacks uniqueness – the location carries the same meanings as other locations.

**The importance of place**

- Identity and belonging: people form strong connections to places; places shape individuals' identities and provide a sense of belonging
- Cultural significance: places can be associated with important events and traditions
- Social interaction: places are where people gather, work, relax and build relationships: they are where communities connect
- Cultural diversity: regions and locations often have distinct languages, cuisines, art forms and ways of life which enrich human experiences
- Economic activities: location is critical for economic activities
- Environmental interaction: places differ because of their different environments and environmental factors shape human activities
- Sense of security: factors such as crime rates, natural disaster risks and the availability of emergency services can affect people's sense of security and peace of mind

## Insider and outsider perspectives on place

- An **insider** perspective is the perspective of someone who lives in the place.
- An **outsider** perspective on a place comes from someone who is visiting, or has never been there. This perspective may be derived from media representations of a place. It may also come from someone who feels they don't belong in a place or feels unwelcome in a place.

## Categories of place

|  | Near places | Distant places |
|---|---|---|
| **Proximity** | Geographically close | Geographically far |
| **Familiarity** | High: direct impact on daily life | Lower: little / less impact on daily life |
| **Spatial interaction** | Frequent and face-to-face, which can lead to shared culture | Less frequent (e.g. through trade, migration, communication); though globalisation has increased interactions in recent years |
| **Sense of place** | Strong sense of local identity linked to place | May be less strong. A sense of identity may be formed in opposition to distant places: the 'other'. |
| **Cultural homogeneity** | Greater homogeneity. Potentially limited variety of language, traditions and way of life, but this is dependent on location | Greater heterogeneity. Potentially more variety in language, culture and ways of life |
| **Perceptions** | More accurate: based on experience | Often based on limited information, may be stereotypical |

## 8 Changing places

### 8.1 The concept of place

## Experienced places and media places

**Experienced places** are places that you have lived in or visited, which allows a deeper understanding of the place.

**Media places** are places that we only know through media such as the Internet, literature, TV, songs and art. Perceptions of media places come from what is presented to us. Media presentations may focus on extreme negatives or positives of a place. Media places can become experienced places, but not vice versa.

## Factors contributing to the character of places

### Endogenous factors

**Endogenous factors** originate internally and relate to the site of a place, such as location, **topography**, physical geography, land use, built environment and infrastructure, demographic and economic characteristics.

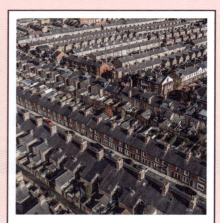

- Topography: flat land, lots of space for building
- Land use: urban, residential (terraced housing)
- Built environment: densely-packed, uniform
- Infrastructure: buildings along long, narrow, parallel roads

© Crown copyright

- Topography: hills enclose the place
- Land use: rural, forested, wind turbines on hill tops
- Physical geography: steep relief, dissected by small rivers, waterfalls
- Location: isolated? At the end of a road

- Economic characteristics: busy shopping arcade
- Built environment: city market or arcade
- Demographic characteristics: mostly families, ethnically diverse community?

### Exogenous factors

**Exogenous factors** are those that originate externally and relate to links to or influences from other places, and relationships with other places shown by movement or flow across space, e.g. the movement of people, the flow of money.

**Key terms**

Make sure you can write a definition for these key terms

distant place    endogenous factor    exogenous factor
experienced place    insider    media place    near place    outsider
place    placelessness    topography

# RETRIEVAL

Learn the answers to the questions below, then cover the answers column with a piece of paper and write down as many as you can. Check and repeat.

## Questions

**1** What is the concept of place in geography?

**2** What does the concept of placelessness refer to?

**3** Name three aspects of the importance of place in human life and experience.

**4** In what ways are places important for social interaction?

**5** Name two factors which might influence people's sense of security and peace of mind about a place?

**6** Why might a place have cultural significance?

**7** Location often affects access to resources, markets and infrastructure. What category of importance would these aspects of a place fit with?

**8** What is an insider perspective on a place?

**9** What is an outsider perspective on a place?

**10** What distinguishes experienced places from media places?

**11** What are exogenous factors?

**12** What are endogenous factors?

**13** Is topography an exogenous or endogenous factor?

**14** Are demographic characteristics exogenous or endogenous?

**15** Is migration to a place exogenous or endogenous?

*Put paper here*

## Answers

Place is location plus meaning – a specific location with unique physical, cultural and social attributes

Placelessness refers to a location lacking uniqueness and carrying the same meanings as other locations

Three from: identity / cultural significance / social interaction / economic activities / environmental interactions / sense of security / cultural diversity

Places are where people gather, work, relax and build relationships: they are where communities connect

Two from: crime rates / natural disaster risks / the availability of emergency services

A place could be associated with an important event or a tradition

Economic importance of place

An insider perspective comes from someone who lives in the place and has direct experience of it

An outsider perspective may come from someone visiting or from media representations and may indicate feeling excluded or unwelcome

Experienced places are those visited or lived in, providing a deeper understanding, while media places are known only through media representations

Factors that originate externally

Factors that originate internally

Endogenous

Endogenous

Endogenous

 **KNOWLEDGE**

## 8 Changing places

### 8.2 Relationships and connections

## Connections with other places

Relationships and connections with other places, both near and distant, are essential drivers of change.

```
                    ┌─────────────────────────────┐
                    │ Relationships and connections │
                    └─────────────────────────────┘
```

| Cultural exchange: connections can result in the adoption of new traditions, languages, foods and technologies, leading to the evolution of local cultures. | Economic development: the flow of resources, money (investment) and workers can lead to the growth or decline of industries. | Social dynamics: migration patterns and demographic change can reshape or influence demographic, socio-economic and cultural characteristics of a place. | Globalisation: a greater interconnectedness between places and an increased flow of ideas, goods and information can reduce the uniqueness of places. |

What should we look at to understand how a place has changed or stayed the same?

- One way is looking at the demographic and cultural characteristics of a place, and how those have changed.

- Another is to consider economic change and socio-economic changes such as changes in inequality.

> **REVISION TIP**
>
> In your local (near) place study or your contrasting (distant) place study, you will have looked at *either* changing demographic and cultural characteristics, *or* economic change and social inequalities. Exam questions will not require you to answer for both.

## Economic change

Economic changes can produce a wide range of changes in places; for example changes in levels of employment and unemployment; in the balance of sectors of employment (primary; secondary; etc.); in how much disposable income people have; income differentials within the community; access to services for different economic groups (including health; education; transport) and economic provision for accessibility.

## Deindustrialisation

Deindustrialisation is a key economic change affecting places, linked to globalisation.

- Before deindustrialisation, many places were closely linked to a major heavy industry or primary industry employer, and this meant a strong link to identity.

- Deindustrialisation stripped this common identity away, causing very high unemployment. Economic decline led to negative representations of such places.

- Social inequalities intensified between areas heavily impacted by deindustrialisation, and areas less affected – such as London.

- New economic activities developed, some of which continued a connection in manufacturing. For example, the north-east of England became a centre of car making for foreign-owned companies such as Nissan.

# External forces

Government policies, the decisions of multinational corporations and the impacts of international or global institutions are external forces that all operate from local to global scales.

Government regeneration policies are often responses to economic changes at a regional scale, such as deindustrialisation, globalisation and urban decline.

Regeneration policies often seek to make places more attractive to multinational corporations at a local scale. Governments may reduce barriers to investment in specific places with tax breaks and infrastructure provision.

Decisions by multinational corporations, for example, to invest in a new factory in a place, or to stop investing in a steel-making plant unless more government support is provided, can have local and regional-scale impacts.

Government policies may target social or environmental issues, for example policies to increase housebuilding, to insulate more homes, or to reduce carbon emissions or air pollution. These can be national and regional.

Government policies respond to political issues, for example to reduce or increase immigration, or increase or decrease connections with other places (e.g. trade deals, Brexit), or invest in/halt major infrastructure projects (e.g. HS2).

International organisations provide investment for regional development, for example the EU invested in the Heads of the Valleys dual carriageway in south Wales.

▲ **Figure 1** *Government policies can influence house builders*

**REVISION TIP**

In your local place study and your contrasting place study, you will look at *either* government policies or the decisions of multinational corporations *or* the impacts of international or global institutions.

# Changing demographic, socio-economic and cultural characteristics

- These characteristics of places are shaped by changing flows of people, resources, money and investment, and ideas.
- This happens at all scales, from local to global.

# 8 Changing places

## 8.2 Relationships and connections

### Scarborough – past and present connections

Past and present connections can have a strong influence on continuity and change. Scarborough is a seaside town in North Yorkshire that has seen less change than other places because seaside tourism is still a dominant influence.

**REVISION TIP**

You should revise the place you have studied. This page focuses on the example of Scarborough.

Some factors may not change: Scarborough's topography and physical geography is key to why the settlement is where it is, and influences what it means to people as a place.

Scarborough's 'cultural landscape' is tied to its history and economic function as a tourist resort, spa and fair (there is a folk song called 'Scarborough Fair').

Scarborough's residents may actively resist changes to the physical and cultural landscape. For example, there was strong resistance to the proposed demolition of a local theatre.

Scarborough is still the second most popular holiday destination for British holidaymakers. Its cultural characteristics remain strongly connected to seaside holidays, although far fewer British people holiday at the British seaside than was the case before the 1970s.

Scarborough is a popular place for retirement. In 2023 almost one in three residents were aged 65 or older. Often retirees are looking for quality of life, not job opportunities. Retirees may tend to not increase cultural diversity.

In the 2021 census, 97.5% of people living in Scarborough described themselves as white compared to 85% in England. Ethnically diverse people may feel excluded as they do not see themselves represented in the population.

Just 3% of all visitors come from abroad. A preference for city-based holidays, for example, could be a factor.

▲ **Figure 2** *Seaside tourism is still a dominant influence in Scarborough*

## 8.3 Meaning and representation

### Place meanings

As well as objective factors that characterise places, such as the physical geography of a place or the amount of money invested in it, people also attach meanings to places. These are called **place meanings**.

- This can happen as a result of people's own experiences of living in a place (insider perspectives) or visiting it for themselves.

- People also attach meanings to places because of how others represent these places to them – **place representations**.

Place meanings and place representations can be negative as well as positive (or indifferent). Place meanings and place representations can also change. And, of course, places can have different meanings for different people.

▲ **Figure 1** The place meaning for a place you visit may be different from what that place means for people living there, for example in cities such as Paris

```
                    Meaning and connections
```

| Shaping identity: | Representation: | Cultural significance: | Place branding and |
|---|---|---|---|
| The meaning attributed to a place is closely tied to its history, culture and representation. How a place is portrayed in literature, media, art and popular culture can shape its identity and the way it is perceived by both insiders and new arrivals. Places can mean different things to different people. | How a place is represented can significantly influence place meanings. Media representations can create stereotypes or highlight specific aspects of a place, impacting on how people understand and engage with it. | The meaning of a place can change over time as its cultural mix changes. Historical events or shifts in societal values can alter the way a place is perceived. | reimagining: Branding strategies that aim to shape the image of a place, for example to attract tourism or investment, can start to construct new place meanings. |

# 8 Changing places

## 8.3 Meaning and representation

### How are place meanings formed?

- Cultural geographers have different theories about how place meanings are formed.
- Yi-Fu Tuan sees the emotional investment that people make in a place as critical to how space becomes place. That emotional investment comes from living in a place and associating with it.
- Tuan also sees a person's position in society as influencing how meanings are made.
- We have different perspectives on space depending on our perspectives and our identities. For example:
  - Some spaces are not open to everyone: the privileged may get to experience them directly while others are locked out.
  - Urban spaces at night can feel threatening to an individual, but feel safe and enjoyable for a group.
  - Communities can create places where they feel at home in contrast to places where they are made to feel unwelcome.

### Engagement and identity

Children make particularly strong place meanings through their active engagement with places like playgrounds and parks. As we grow up, we add and deepen everyday place meanings through further engagements with our local space – the route to school or college, places we meet with friends or places we avoid. We find things about our local place that strengthen our own identities and link us to the communities we are members of.

### How people present and represent the world to others

Representations of the world often seem neutral, such as maps. But they are in fact **social constructs**, which means they are created by people in society through shared interpretations and assumptions. One sign of this is that people in different cultures may not share the same representations.

#### Maps

Colonial maps often placed Europe at the centre, reinforcing the construct of European importance and superiority. This Eurocentric view has influenced perceptions of global importance and power dynamics. Maps made from another perspective (for example, that of Japan) may look very different from maps made from a European perspective.

#### Art and literature

Representations of places in art and in literature are social constructs. For example, in the nineteenth century, the French artist Gustave Doré's drawings or Charles Dickens' novels have had a lasting influence on representations of Victorian London, though both were criticised at the time for exaggerating poverty.

#### Place names

Toponyms – names given to places on Earth – are indications of how people represent the world to others. The names given to places reveal a lot about their physical geography, the people who have lived there, their original functions and their traditions. For example:

- Chiswick – from the Old English for cheese dairy
- Swindon – from the Old English for pig valley
- Birmingham – the home of Beorma's people

Many places in Australia had their original names erased by colonisers who 'renamed' them. More recently the original names given by Aboriginal and Torres Strait Islander peoples have been acknowledged and used more regularly to recognise the first peoples and true owners from whom land was stolen.

## Different forms of representation

- **Formal representations** of places are created by official institutions such as government, big businesses, advertising agencies or heritage organisations. An Ordnance Survey map of a place or census data about a place are formal representations.

- **Informal representations** are made by individuals and groups. A social media post, poem, song, painting, story, film or joke about a place would generally be informal representations.

## Differences between representations

Formal and informal representations of places can produce conflicting representations. One reason is that the purpose behind formal representations may require them to be accurate, to be acceptable to large groups of people and to promote positive feelings, while informal representations do not need to be accountable in the same way. For example:

- Census data is collected in a standard way, so that all areas covered are comparable. Planning each census takes at least five years; all the questions are trialled, a team of people follow up households that have not responded, and statistical measures and checks are applied to ensure the data collected is fully representative of places.

- The song 'I Predict a Riot' was written by the Kaiser Chiefs about Leeds, their home city. It was based on the experiences of Nick Hodgson, who DJ'd at a nightclub in the city. The lyrics reference actual experiences, references to famous people from Leeds and also, as Hodgson has stated, words that rhymed.

## Influencing and creating place meanings

Regeneration policies often seek to change or create new meanings for places, as well as provide the infrastructure and financial incentives that attract new investment. The aim is to replace negative perspectives of a place with positive ones, or change a less attractive place identity into a more attractive one. Two key strategies are rebranding and reimagining.

- **Rebranding** places refers to changing or influencing the image or perception of a place to highlight what makes it more attractive or appealing, usually for economic or social objectives. It's often driven by a need to shift place identity away from negative perspectives or to create a place meaning that is in line with developing economic trends or opportunities.

- **Reimagining** places goes beyond the reshaping of rebranding. It involves creating new place meanings by the rethinking of place functions, meanings and purposes. Reimagining places can involve a wider range of stakeholders than rebranding. It can be about an attempt to change underlying structural issues leading to inequalities.

- While this is not always the case, rebranding is generally top-down, involving local government and marketing agencies; reimagining can be more bottom-up, sometimes even driven by the grass roots of a community.

- While rebranding can be limited to a new slogan, new advertising and new signage around a place, reimagining can involve changes to place function.

## 8 Changing places

### 8.3 Meaning and representation

## Examples of rebranding

### Detroit rebranding

Detroit, a city in the USA once synonymous with urban decline, has been rebranding itself as a hub for innovation, art and culture. This is seen in initiatives promoting its thriving arts scene, tech start-ups and efforts to attract businesses and tourists. The aim is to shift perceptions from a 'declining industrial city' to an 'emerging, vibrant urban centre', with the recognition that Detroit's hard times have given a special resilience and strength to the people of 'Motor City'.

### Liverpool rebranding

Liverpool underwent significant rebranding in the early 21st century. After decades of economic decline that resulted in negative place identities and a declining population, the city's significant cultural heritage was recognised by the

▲ **Figure 2** *A museum celebrating the Motown record label in Detroit's Historic District*

EU granting it European Capital of Culture status for 2008. The Liverpool Culture Company created and marketed thousands of cultural events related to the city's status that reached a large audience globally as well as in the UK. 15 million visitors came to Liverpool in 2008.

## Examples of reimagining

- The transformation of the High Line in New York City from a derelict railway track to a thriving urban park is a case of reimagining. This wasn't just about altering the image of the space, but about repurposing and re-envisioning its role in urban life. The impetus for reimagining was informal – a local group called Friends of the High Line – rather than formal: city government wanted to demolish the old elevated railway line.

- Sabarmati Riverfront, Ahmedabad, India: over time, the Sabarmati riverfront area had degraded into a polluted and neglected area. Launched by the Ahmedabad Municipal Corporation, the Sabarmati Riverfront project reimagined this place, turning it into a vibrant urban area with parks, promenades and cultural spaces.

▲ **Figure 3** *The High Line, New York City*

## Impacts of rebranding and reimagining

Influencing place meaning away from negative perspectives or creating new place meanings can have transformational economic and social impacts on places.

- The rebranding of Liverpool in 2008 was part of a successful regeneration for the city that attracted private sector investment of £1 billion in the Liverpool ONE shopping centre.

- The US$115 million that the New York City government spent on the High Line Park reimagining has attracted five million visitors a year, and the areas that the park runs through have seen US$2 billion in private investment and the creation of 12,000 jobs.

> However, rebranding is a competitive business, and does not always meet all its goals. For example, in 2018 Detroit lost its bid to be the location of a new headquarters for Amazon because the company did not think it would be able to attract the people with the skills it needed to 'Motor City'. (Arlington, Virginia won the bid.)

## Influences of the past on present place meaning

Most rebranding and reimagining efforts attempt to weave aspects of the past into the identity they are trying to create. Heritage offers unique characteristics that make a place distinctive and can also tap into deep-seated values, memories and emotions associated with the place. A strong association with heritage can be very effective in attracting visitors.

- Restoring old buildings linking new developments with a place's past; for example, nineteenth- and twentieth-century warehouses in Liverpool's Albert Docks were restored and converted into apartments.

- Investing in museums and other visitor attractions highlighting unique aspects of the place's past; for example, the Beatles museum in 1990.

- Naming and branding new developments after historic connections; for example, Speke airport was renamed Liverpool John Lennon Airport.

- Setting up heritage trails; for example, the heritage trail at Liverpool ONE guides visitors on a tour of places with links to Liverpool's maritime history.

▲ *Figure 4* *Liverpool's Albert Docks: warehouses converted into residential blocks*

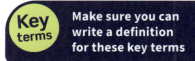

**Key terms** Make sure you can write a definition for these key terms

formal representation    informal representation    place meaning
place representation    rebranding    reimagining    social construct

# RETRIEVAL

Learn the answers to the questions below, then cover the answers column with a piece of paper and write down as many as you can. Check and repeat.

## Questions

**1** Name two ways in which relationships and connections with other places can be drivers of change in a place.

**2** Name two changes in place that can result from economic changes.

**3** What does Yi-Fu Tuan suggest influences how place meanings are formed?

**4** How are everyday place meanings deepened over time?

**5** An OS map of a place or census data about a place are examples of what type of representation?

**6** Define rebranding of a place.

**7** Define reimagining of a place.

## Answers

Two from: cultural exchange / economic development / social dynamics / globalisation

Two from: changes in levels of employment / in the balance of sectors of employment / in disposable income / in income differentials / in access to services

A person's position in society

Though active engagements with local place

Formal representation

Changing or influencing the image or perception of a place to highlight what makes it more attractive or appealing, usually for economic or social objectives

Creating new place meanings by the rethinking of place functions, meanings and purposes

*Put paper here*

## Previous questions

Now go back and use these questions to check your knowledge of previous topics.

## Questions

**1** What is the concept of place in geography?

**2** What is an outsider perspective?

**3** What is an insider perspective on a place?

**4** What are exogenous factors?

**5** Is migration to a place exogenous or endogenous?

## Answers

Place is location plus meaning; a specific location with unique physical, cultural and social attributes

An outsider perspective may come from someone visiting or from media representations and may indicate feeling excluded or unwelcome

An insider perspective comes from someone who lives in the place and has direct experience of it

Exogenous factors originate externally

Exogenous

*Put paper here*

 **KNOWLEDGE**

## 8 Changing places

### 8.4 Quantitative and qualitative data

## The usefulness of different sources

| Source | Strengths | Limitations |
|---|---|---|
| Photographs | An accurate representation of change: before and after photos help us visualise change | Can be subjective; may not represent the overall situation – what is included and what excluded? |
| Text | Gives personal and emotional context; offers historical perspectives | Subjective; based on personal experiences and not always generalisable |
| Audio-visual media | Combines visuals and sounds for a comprehensive understanding of place | Subjective or biased; production quality may affect interpretation |
| Artistic representations | Paintings can be used to show historical change; they capture place meaning in a way we can interpret | Art is subjective, not objective / accurate; who was the artwork created for? What or who might have been excluded? |
| Oral sources (interviews, memoirs, songs, etc.) | Captures personal narratives; offers insights into personal experiences | Memories can be selective and biased; historically, diaries and memoirs from lower income people are rare |
| Statistics, such as census data | Allow objective representations and comparisons of different demographics and employment types, levels of education | Statistics can be manipulated; they can give a skewed perspective if only a narrow range of results is viewed; not everyone completes the census; data can be vague |
| Maps | Accurate representations of land-use such as agriculture or retail areas; maps record the scale of changes in a settlement such as new housing estates | Old hand-drawn maps are often subjective, not accurate. OS maps don't indicate what buildings are used for; don't always identify land use. Indigenous names and ownership may not be included. |
| Geo-located data | Insider and outsider perspectives on place characteristics can be gathered by using hashtags and geotags | Might not be comprehensive; depends on user input and participation |
| Geospatial data, including GIS applications | Accurate and precise data collection; can be used for advanced spatial analysis | Requires specific tools and expertise to analyse; data can be complex and overwhelming |

## The Index of Multiple Deprivation (IMD)

The **IMD** ranks over 32,000 neighbourhoods or Lower-layer Super Output Areas (LSOAs) across England according to a combination of seven domains of deprivation, each based on a 'basket' of indicators. The data is organised in deciles, produced by ranking the 32,844 LSOAs and dividing them into ten equal-sized groups. Decile one represents the most deprived 10% of LSOAs nationally and decile ten the least deprived 10%.

When interpreting maps based on the IMD, you should keep the following in mind:

- Deciles show that one area was more or less deprived than another, but not by how much.
- LSOAs give data about places but not about the individuals who live there.
- The IMD measures deprivation but not affluence.

 **Make sure you can write a definition for these key terms**

Index of Multiple Deprivation (IMD)
qualitative data    quantitative data

# RETRIEVAL

Learn the answers to the questions below, then cover the answers column with a piece of paper and write down as many as you can. Check and repeat.

## Questions

## Answers

1. Are paintings, oral sources and audio-media qualitative or quantitative sources for places?

   Qualitative

2. Suggest two limitations of census records as a source for place studies.

   Two from: statistics can be manipulated / they can give a skewed perspective if only a narrow range of results is viewed / not everyone completes the census

3. Suggest two limitations of maps as a source for place studies.

   Two from: old hand-drawn maps are often subjective, not accurate / OS maps don't indicate what buildings are used for / OS maps don't always identify land use

4. What does IMD stand for?

   Index of Multiple Deprivation

5. What does decile one of the IMD represent?

   The most deprived 10% of LSOAs nationally

6. Suggest one limitation of the IMD as a measure of place characteristics.

   One from: deciles show that one area was more or less deprived than another, but not by how much / LSOAs give data about places but not about the individuals who live there / the IMD measures deprivation but not affluence

7. Name three of the domains (or indicators) of deprivation measured by the IMD.

   Three from: income / employment / education / health / crime / barriers to housing and services / living environment

*Put paper here*

## Previous questions

Now go back and use these questions to check your knowledge of previous topics.

## Questions

## Answers

1. What does the concept of placelessness refer to?

   Placelessness refers to a location lacking uniqueness and carrying the same meanings as other locations

2. Why might a place have cultural significance?

   For example, a place could be associated with an important event or a tradition

3. What distinguishes experienced places from media places?

   Experienced places are those visited or lived in, providing a deeper understanding, while media places are known only through media representations

4. Name two changes in place that can result from economic changes.

   Two from: changes in levels of employment / in the balance of sectors of employment / in disposable income / in income differentials / in access to services

5. How are everyday place meanings deepened over time?

   Though active engagements with local place

6. Define reimagining of a place.

   Creating new place meanings by the rethinking of place functions, meanings and purposes

*Put paper here*

## 8 Changing places

### 8.5 Local place study

## Gentrification in Withington, Manchester

**Gentrification** refers to the transformation of an urban area as a result of the influx of more affluent individuals, often resulting in increased property values and a shift in its demographic, cultural and economic characteristics.

While gentrification can result in economic growth and reduced crime rates, it also raises concerns about displacement of longer-term residents, who may no longer be able to afford to live in the area, and the loss of historical and cultural identity.

 **REVISION TIP**

You should revise the local place study you have studied, which will have investigated how people's lived experience of the place, and its characteristics, have changed over time, looking at *either* the changing demographic and cultural characteristics *or* economic change and social inequalities.

**Change over time in Withington**

| |
|---|
| Manchester grew rapidly in the nineteenth century as a result of the industrial revolution. As early as the 1850s, wealthier Mancunians moved away from the dirt and noise of the inner-city factories to villages on the outskirts, such as Withington. From just over 1200 inhabitants in 1841, by the 1890s its population had grown to over 14,000. |
| Development continued through the twentieth century, but by the 1980s, Manchester was in economic decline, with a 22% unemployment rate. The large private homes of Withington were often divided into affordable flats for city workers and their families. |
| Regeneration projects for the CBD of Manchester in the 2000s proved to be successful, and as the city centre became unaffordable, people once again began to move to nearby areas, such as Withington. |

## How did Withington attract gentrification?

- Proximity: four miles south of Manchester's city centre, making it an attractive proposition for those who work or study in the city.

- Education: many students from the University of Manchester and Manchester Metropolitan University share houses in Withington (studentification). Affluent parents are attracted to the good schools in the area, with six primaries, two independent girls' schools and Manchester Grammar School.

- Transport: excellent bus services link to the city centre and other parts of Manchester.

- Affordability: property was more affordable than in the city centre or more established suburbs, attracting professionals and initiating the gentrification process.

- Local amenities: more affluent residents demanded high-quality restaurants, cafes and boutique shops

which are a pull factor for middle-class individuals and families.

- Cultural shifts: Greater Manchester has seen an upswing in cultural events, music and art over the past few decades. Withington has an increasingly vibrant community atmosphere, fuelling gentrification.

- Safety and crime rates: an increase in more affluent residents has corresponded with a drop in crime rates, making the area even more desirable.

- Community initiatives: efforts by local community groups to revitalise the area, through initiatives like local festivals, community gardens and clean-up events, have enhanced Withington's appeal. A mural of Marcus Rashford was commissioned by community street art group Withington Walls.

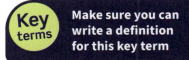 **Key terms** **Make sure you can write a definition for this key term**

gentrification

# 8 Changing places

## 8.5 Local place study

## Withington demographics

Withington now has an uneven population structure, dominated by young adults.

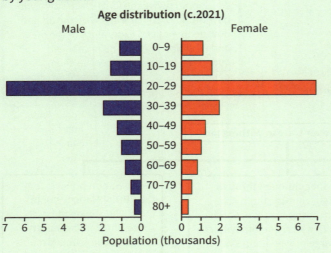

**Age distribution (c.2021)**

▲ *Figure 1 Withington population age structure (census, 2021)*

## Withington socio-economics

Maps of the Index of Multiple Deprivation outputs for Withington indicate a mix of relatively deprived (e.g. the Old Moat council estate) and non-deprived areas, common for many parts of Manchester. Crime rates are very low, although there is a problem with opportunistic thefts, such as phones being snatched from students.

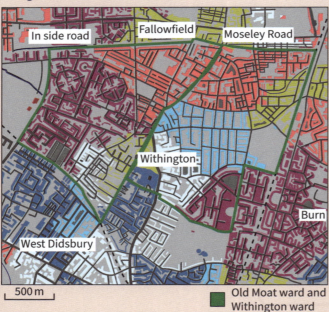

▲ *Figure 2 Map of the Index of Multiple Deprivation for Withington ward (green). The red shading represents the most deprived decile.*

## Suitable data sources for a local place study

- statistics, such as census data
- maps
- geo-located data
- geospatial data, including geographic information systems (GIS) applications
- photographs
- text, from varied media
- audio-visual media
- artistic representations
- oral sources, such as interviews, reminiscences, songs, etc.

## Social media representations of Withington

### Clairebear273

Hi everyone: We're moving to Manchester – I will be working in the city centre. I'm expecting a baby and we've seen an adorable renovated Victorian house in Withington. Withington looks great with some nice cafes and the commute to the centre is really quick (20mins), but is it safe there? I've heard there's a lot of crime? Thanks ☺

### Fairymum88

Withington is 100% safe: I've lived here for years. It is quite mixed though with some roads that are nicer than others. Try and rent/buy near the tramline as those are the best areas.

### Rashford4PM

Withington is a great place to live: really vibrant. We've got a lot of community projects too which was really helpful when my kids were toddlers. But you can't get to the city centre from Withington in 20mins – more like 45 mins in rush hour.

# 8.6 Contrasting place study

## Change in Boston, Lincolnshire

### Economic changes in the medieval period

In the medieval period, Boston was England's second most important port, one of only a few ports that were allowed to export wool to Europe. But in the fifteenth century, changes to the wool trade and the silting up of Boston's port were factors in economic and political problems that hit Boston very hard.

### Land-use changes in the eighteenth century

In the eighteenth century, there was conflict between wealthy landowners who wanted to drain the surrounding Fens (marshland) and those who made a living from the Fens. However, the drained Fens left fertile land, and Boston became a prosperous farming area. Food processing industries developed in the nineteenth and twentieth centuries.

### Labour changes in the twentieth century

The agricultural and food processing sector in Lincolnshire had always relied on migrant labour (e.g. for flower and fruit picking). After Eastern European countries joined the EU in 2004, Boston became a popular destination for migrants from countries such as Poland and Lithuania.

### Brexit and social change

Boston residents voted strongly (74.9%) in favour of leaving the EU in the 2016 referendum on EU membership. Boston residents felt that immigration had been uncontrolled.

- Concern about the lack of sufficient local infrastructure and services (e.g. NHS services or education).

- A change in the culture, with familiar shops being replaced by shops catering for Eastern European and Baltic customers.

- Perception of a lack of integration by many immigrants, and a fear that they were taking the jobs that could have been done by local people.

- An increase in violent crime primarily between migrants. In 2016, Boston was named as the 'most murderous' place in England and Wales.

**REVISION TIP**

You should revise the contrasting place study you have studied. It should contrast with your local study, for example a rural place if your local place is part of a city. Remember, the focus is on how people's lived experience of the place, and its characteristics, have changed over time, looking at *either* the changing demographic and cultural characteristics *or* economic change and social inequalities.

# KNOWLEDGE

## 8 Changing places

### 8.6 Contrasting place study

## Immigration in Boston

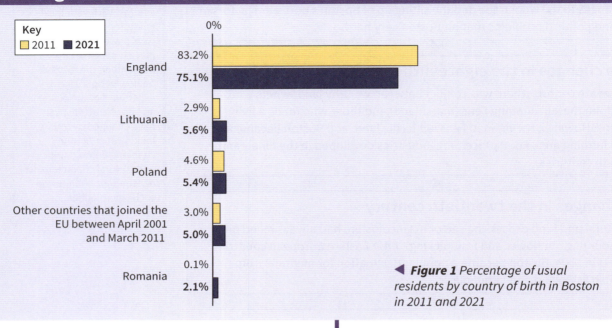

**Key**
□ 2011  ■ 2021

0%

England
83.2%
75.1%

Lithuania
2.9%
5.6%

Poland
4.6%
5.4%

Other countries that joined the EU between April 2001 and March 2011
3.0%
5.0%

Romania
0.1%
2.1%

◀ **Figure 1** *Percentage of usual residents by country of birth in Boston in 2011 and 2021*

## Post-Brexit Boston

Outsider perspectives on Boston are often influenced by its Brexit referendum vote, although Brexit also still resonates in some insider perspectives. While many still support the decision to leave the EU, criticisms that lower immigration has not been delivered has resulted in the Conservative local administration being largely replaced by the Boston Independent Group. Their aim is to restore civic pride and to 'put Boston back on the map where it belongs'.

## Socio-economic indicators

- Ranked in the bottom 10% of local authority areas in England for health (2021), with high levels of obesity in adults.
- Of the 36 neighbourhoods in Boston, six were among the most income-deprived in England, while two were in the 20% of least income-deprived neighbourhoods in England.
- In 2023, 4.1% of working-age people in Boston were unemployed (3.6% in the East Midlands region).
- Boston has 20% fewer higher and intermediate professional households than the national average.
- 25% of Boston residents have no qualifications (18.08% for England).
- Average hourly pay in 2023 was £12.23 in Boston (£15.49 for England).
- The crime rate in Boston in 2022 was 116 crimes per 1000 people (85 crimes per 1000 people for Lincolnshire).

# RETRIEVAL

Learn the answers to the questions below, then cover the answers column with a piece of paper and write down as many as you can. Check and repeat.

## Questions

## Answers

1. Suggest two positive changes resulting from the gentrification of an urban area.

Two from: economic growth / reduced crime rates / improved infrastructure / improved services

2. Suggest two negative changes resulting from the gentrification of an urban area.

Two from: social displacement (people no longer being able to afford to live in an area) / loss of historic or changes to cultural identity of a place / resentment from surrounding places about services, opportunities, etc that they are not receiving

3. Suggest three reasons why Withington attracted gentrification.

Three from: proximity to city centre / good educational opportunities / good transport links / affordability (initially) / good housing stock / improving local amenities / lower crime rates / community culture and initiatives

4. What is distinctive about the population structure of Withington?

It is dominated by 20–29-year-olds

5. What exogenous factor reduced Boston's prosperity in the fifteenth century?

Changes to the wool trade

6. What endogenous factor in the eighteenth century caused a change in the lived experience of Bostonians who made a living from the Fens?

The draining of the Fens

7. In what year did the countries of Poland and Lithuania join the EU?

2004

8. What is distinctive about Boston's 74.9% vote to leave the EU in the 2016 referendum?

It was the highest percentage for Leave in the UK

*Put paper here*

## Previous questions

Now go back and use these questions to check your knowledge of previous topics.

## Questions

## Answers

1. In what ways are places important for social interaction?

Places are where people gather, work, relax and build relationships: they are where communities connect

2. What distinguishes experienced places from media places?

Experienced places are those visited or lived in, providing a deeper understanding, while media places are known only through media representations

3. What are endogenous factors?

Factors that originate internally

4. Name two ways in which relationships and connections with other places can be drivers of change in a place.

Two from: cultural exchange / economic development / social dynamics / globalisation

5. Are paintings, oral sources and audio-media qualitative or quantitative sources for places?

Qualitative

*Put paper here*

## Exam-style questions

1   Explain the difference between insider and outsider perspectives on place.   **[4 marks]**

2   Outline how economic characteristics can contribute to the character of a place.   **[4 marks]**

**EXAM TIP**

Remember that 4-mark questions are point-marked, so focus on making four points or two points with development to pick up the available marks – and then move on!

3   Outline how maps can be used to present place characteristics.   **[4 marks]**

4   Outline how endogenous factors contribute to the character of a place.   **[4 marks]**

5   **Figure 1a** shows a painting of the Sankey Valley in Cheshire by Thomas Talbot Bury. **Figure 1b** is a modern photograph of the same place.

Evaluate the usefulness of **Figure 1a** and **Figure 1b** for investigating the changing characteristics of the Sankey Valley in Cheshire.   **[6 marks]**

**EXAM TIP**

These 6-mark questions have 2 marks for AO1 and 4 marks for AO2, which means demonstrating your knowledge and understanding of the concept and then applying that to the resource.

▲ *Figure 1a* *The Sankey Valley in Cheshire in 1831 by Thomas Talbot Bury*

▲ *Figure 1b* *Nine Arches Viaduct, Newton-le-Willows in 2020*

**6**   **Figure 2a** is a map of Credenhill, a village in Herefordshire, in 1938.
**Figure 2b** is a map of Credenhill today, presented at the same scale.

Analyse the changes shown between **Figure 2a** and **Figure 2b**.   **[6 marks]**

**EXAM TIP**

6-mark questions that ask you to analyse require you to describe what the data shows (in this case the differences between the two maps), identifying the differences.

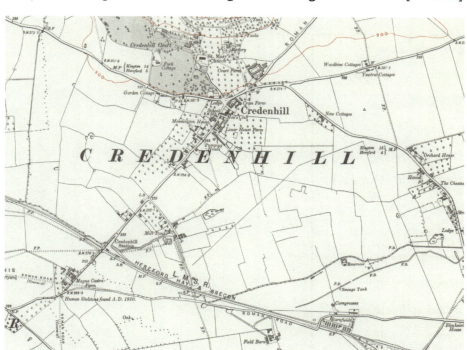

▲ **Figure 2a** *A map of Credenhill, a village in Herefordshire, in 1938*

▲ **Figure 2b** *Map of Credenhill in 2023*

7   **Figure 3a** shows a poem about the view from Westminster Bridge in London written in 1803. **Figure 3b** shows a protest on Westminster Bridge in 2019.

Using **Figure 3a**, **Figure 3b** and your own knowledge, to what extent do you agree that place-meaning is shaped by different perspectives?   **[6 marks]**

> Earth has not any thing to show more fair:
>
> Dull would he be of soul who could pass by
>
> A sight so touching in its majesty:
>
> This City now doth, like a garment, wear
>
> The beauty of the morning; silent, bare,
>
> Ships, towers, domes, theatres, and temples lie
>
> Open unto the fields, and to the sky;
>
> All bright and glittering in the smokeless air.
>
> Never did sun more beautifully steep
>
> In his first splendour, valley, rock, or hill;
>
> Ne'er saw I, never felt, a calm so deep!
>
> The river glideth at his own sweet will:
>
> Dear God! the very houses seem asleep;
>
> And all that mighty heart is lying still!

▲ **Figure 3a** *'Composed upon Westminster Bridge, September 3, 1802', by William Wordsworth*

▲ **Figure 3b** *A protest on Westminster Bridge in 2019*

**8** **Figure 4** shows information published online by Carmarthenshire County Council in 2020 about the development of a new suburb west of the town of Llanelli, Wales.

Using **Figure 4** and your own knowledge, evaluate attempts by external agencies to influence or create specific place meanings.   **[6 marks]**

> **Pentre Awel**
>
> Pentre Awel is the first development of its scope and size in Wales providing world-class medical research and health care delivery and supporting and encouraging people to lead active and healthy lives.
>
> As well as improving health and wellbeing, the project will create over 1,800 jobs and training/apprenticeship opportunities, and is expected to boost the local economy by £467million over the next 15 years.
>
> All set within the natural landscape, around a freshwater lake and within walking distance of the Millennium Coastal Park, Pentre Awel will feature landscaped outdoor public spaces for recreation, with walking and cycling paths and stunning coastal views.
>
> Set within an 83-acre site at Delta Lakes, the council-led project will be developed in phases across four zones.

▲ **Figure 4** *Information from Carmarthenshire County Council*

> **EXAM TIP**
>
> This 6-mark question also has 2 marks available for AO1 and 4 marks for AO2: 2 marks for knowledge and understanding of how external agencies influence/create place meanings, and then 4 marks for applying that understanding to the source.

**9** 'Government policies have the most impact on the connections that people have with a place.'

With reference to your distant place, critically assess this statement.   **[20 marks]**

**10** With reference to your local place, assess the extent to which different sources reflect change and/or continuity in this place.   **[20 marks]**

**11** 'Differing perceptions of the same place can be explained by people's socio-economic status.'

With reference to either your local or distant place, to what extent do you agree with this statement?   **[20 marks]**

> **EXAM TIP**
>
> 20-mark questions that ask 'to what extent do you agree?' need to have a brief introduction that states the extent to which you agree (to a large/lesser extent) and a conclusion that summarises how you reached this judgement.

**12** You have studied a local and a distant place.
Assess the extent to which the demographic characteristics or patterns of social inequality are influenced by shifting flows of people in both your local and your distant place.   **[20 marks]**

**13** 'Maps are the least effective way of representing the changing character of a place over time.'

With reference to either your local or distant place, critically assess the extent to which you agree with this statement.   **[20 marks]**

**14** How far do you agree that the most important process of change for local places in the future will be in relation to mitigating the impacts of climate change?   **[20 marks]**

**15** Evaluate the usefulness of different sources in the study of your local place.   **[20 marks]**

## 9 Contemporary urban environments

### 9.1 Urbanisation

## Urbanisation and its importance in human affairs

Economic centres for production (including services) and consumption

Some have economies (GDP) larger than some countries, increasing their global power and influence

**Urban areas are important**

Hubs for social, cultural and educational exchange

Centres for political power and decision making

High population densities, so any incident will impact many people

**REVISION TIP**

Think about why urbanisation is important in the world today and in the future.

## Global patterns of urbanisation

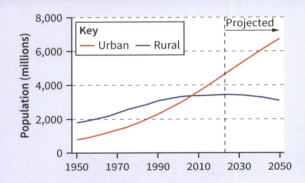

◀ **Figure 1** *The urban and rural population of the world. Around 55% of the world's population live in urban areas.*

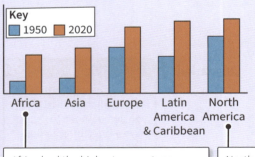

◀ **Figure 2** *The share of urban dwellers has increased in every world region since 1950*

Africa had the highest percentage increase (204%) but still has the lowest proportion of urban residents

North America, South America, Central America, Mexico, and the islands of the Caribbean and Europe have the highest percentage of urban dwellers

## Future growth

- The UN predicts that 68% of the global population will live in urban areas by 2050. India, China and Nigeria will account for 35% of this projected growth between 2018 and 2050. By 2050, India will have 416 million urban residents, China 255 million and Nigeria 189 million.

- In many cities in HDEs, the growth of urban populations has slowed down and in some cases people are moving out of urban areas.

- In EMEs and LDEs, people are moving from rural to urban areas and from smaller towns and cities areas to larger urban areas.

# Cycle of urbanisation

**Urbanisation** has taken place at different rates in different regions of the world. HDEs urbanised rapidly in the nineteenth century due to the industrial revolution. EMEs experienced rapid urban growth in the twentieth century.

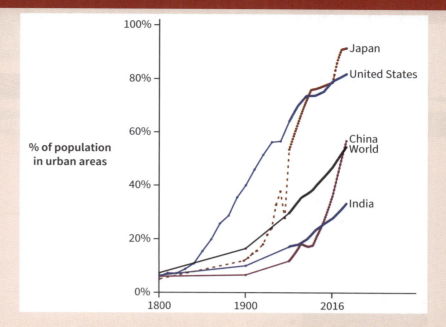

▶ *Figure 3 Urbanisation from 1800 to 2016*

There are four cycles of urbanisation, which can take place at the same time, especially in HDEs. Urbanisation is still the main process in LDEs although **suburbanisation** and even **counter-urbanisation** are taking place in some Asian cities.

**Urbanisation**: increase in the proportion of the population living in urban areas. Mid 1800s industrial revolution

**Suburbanisation**: outward growth of the urban population from the centre towards its edge. Interwar: 1920s–1930s and post WWII 1950s–1970s

**Counter-urbanisation**: movement of people out of an urban area to smaller towns and cities or rural areas. 1980s–1990s

**Urban resurgence**: movement of people from rural areas back into urban areas. 2000s onwards

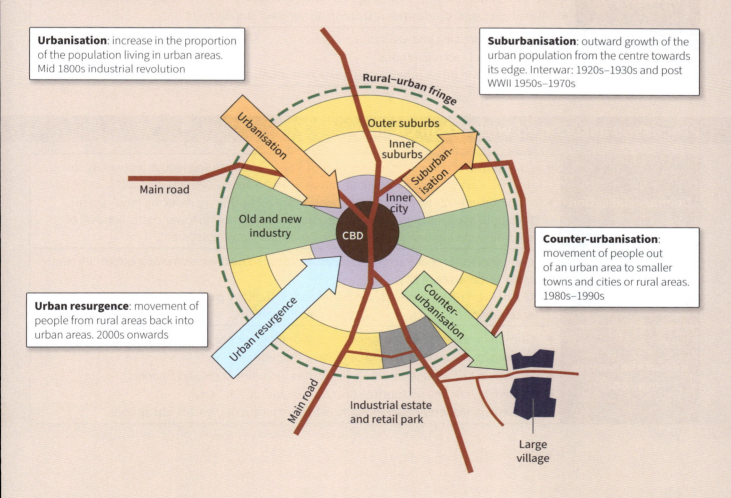

▲ *Figure 4 The four processes of urbanisation with UK timeline*

## 9 Contemporary urban environments

### 9.1 Urbanisation

## Processes associated with urbanisation and urban growth

| Demographic | • Rural to urban migration<br>• Natural increase<br>• International migration |
|---|---|
| Economic | • FDI investment, including through TNCs<br>• Legislation and financial products to support home ownership<br>• Increased job opportunities at different skill levels<br>• Decline in agriculture in rural areas |
| Social | • Intra-urban migration (age/family life cycle)<br>• Inward migration of groups from different communities<br>• Increased income levels<br>• More housing availability |
| Technological | • Transport infrastructure: e.g. roads, metro systems, suburban rail<br>• Utilities infrastructure: electricity, gas, water, broadband<br>• Rise of quaternary industries attracting higher-paid employment |
| Political | • Reforms on housing and sanitation<br>• Planning laws and regulations<br>• Policies expand urban areas or designate new urban areas |

## Processes associated with urban change

| Deindustrialisation | • Long-term decrease of economic output from manufacturing industries leading to decline, unemployment and deprivation in urban areas.<br>• Common in HDEs post-Second World War due to machinery reducing requirement for labour, globalisation of manufacturing to countries with lower costs, removal of subsidies and increased environmental protection and legislation. |
|---|---|
| Decentralisation | • Competition for space and need for lower costs, so businesses moving out of city centre towards the edge of city.<br>• Out of town retail, business, industrial and technology parks enabling increased accessibility needed for cars and large goods vehicles. |
| Rise of the service economy | • Growth in research and development including science and technology.<br>• Growth in office space for management functions.<br>• Growth in knowledge-intensive industries including legal and finance.<br>• Growth in leisure and tourism economy including 24/7 city.<br>• Growth in property developments in former industrial/waterfront spaces. |

# Urban policy and regeneration in Britain since 1979

- Different political parties have different ideas and priorities of how issues in towns and cities should be approached and how urban policies are funded.
- **Urban regeneration** is often government-led but requires partnerships with the private and voluntary sectors to implement strategies.
- Between 1979 and 2020, the European Union also provided finance for some urban projects in the UK.

| UK Government | Year | Schemes |
|---|---|---|
| 1979–97: Conservative | 1981 | Urban Development Corporations (UDCs) |
| | 1981 | Enterprise Zones |
| | 1986 | The Inner Cities Initiative<br>- Inner City Task Forces<br>- City Action Teams) |
| | 1991 | City Challenge |
| | 1992 | European Regional Development Fund |
| | 1997 | Single Regeneration Budget |
| 1997–2010: Labour | 1997 | Single Regeneration Budget (Challenge Fund) |
| | 1997 | Regional Development Agencies (RDAs) |
| | 1999 | English Partnerships (EPs) |
| | 1999 | Urban Regeneration Companies (URCs) |
| | 1997–2010 | A range of government and QUANGO initiatives, e.g. Education Action Zones and Millennium schemes |
| 2010–2015: Coalition | 2010 | Local Enterprise Partnerships (LEPs) |
| | 2010 | New Homes Bonus |
| | 2010 | Community Infrastructure Levy |
| | 2011 | Localism Act and Tax Increment Financing (TIF) |
| 2015: Conservative | 2017 | Metro Mayors (outside of London) |
| | 2019 | Levelling Up |
| | 2023 | Revised National Planning Policy Framework (NPPF) |

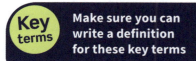

**Key terms** Make sure you can write a definition for these key terms

counter-urbanisation   decentralisation   deindustrialisation
suburbanisation   urbanisation   urban regeneration

### 9.2 Megacities and world cities

## The emergence of megacities

- A **megacity** is an urban area with a population of over 10 million people.
- In 1950 there were two megacities (Tokyo and New York), in 2023 there were 34, with most in Asia (21), South America, Central America, Mexico, and the islands of the Caribbean (6) or Africa (3), and Tokyo was the largest with a population of 37.2 million.
- In 2030 there are expected to be 43 megacities, with Delhi the largest at 38.9 million.
- They are not always a country's capital city, e.g. Sao Paulo, Shanghai, Osaka.

## Growth of megacities

Factors leading to growth of megacities:

- rural to urban migration in EMEs and LDEs
- high natural increase and high fertility rates in LDEs
- widening of urban boundaries
- globalisation and economic growth.

> **REVISION TIP**
>
> Learning key facts and statistics about megacities and world cities will help you discuss and compare them in detail in an exam.

## The role of megacities in global and regional economies

Megacities:

- dominate regional economies
- are home of regional headquarters for TNCs
- are regional centres for employment and services
- are top of the urban hierarchy
- influence the global economy – some megacities have a larger economy than a country (Tokyo's economy is larger than Canada's; Paris's is larger than South Africa's).

## Characteristics of a megacity

Land use:
Varied and dynamic, developed over time; usually unplanned (except China); growth of large informal settlements in LDEs

Function:
Multiple functions, which creates employment in different sectors (government, administration, business, services, education, art and culture, sport)

Location:
Large metropolitan area, some are agglomerations (two or more cities joined together); flat, well drained land; access to water supply

**Characteristics of a megacity**

Environmental issues:
Air and water quality; access to green space; noise and visual pollution; overcrowding; waste disposal

Connectivity:
Transport hubs, often with coastal or river ports for trade and in-migration

Social:
Inequality (income extremes); urban expansion may not keep up with basic needs, e.g. infrastructure, water, food; districts with a concentration of ethnic/foreign communities

Demographic: High population and population density

# World cities

A **world city** is one that is disproportionately important to the world economy compared to other cities. The Globalization and World Cities Research Network (GaWC) studies the economic interconnectedness of cities.

| GaWC Level | Characteristics, number and examples (2020) |
|---|---|
| Alpha++ | Cities that are significantly more integrated than all other cities. *2: London, New York* |
| Alpha+ | Highly integrated cities, mainly in Asia-Pacific. *7: e.g. Hong Kong, Singapore, Shanghai, Beijing, Dubai, Paris* |
| Alpha and Alpha− | Very important world cities linking major regions to the world economy. *Alpha 15: e.g. Sydney, Brussels, Mumbai* <br> *Alpha− 26: e.g. Seoul, Buenos Aires, Bengaluru* |
| Beta | Link region or state to world economy, *e.g. Cape Town* |
| Gamma | Link smaller regions or states to the world economy, *e.g. Dakar* |

# Characteristics of a world city

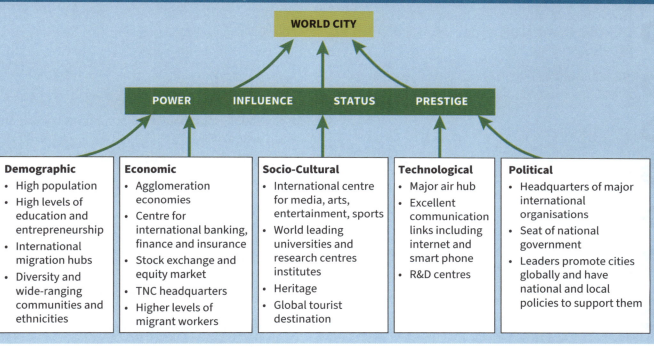

**WORLD CITY**

POWER  INFLUENCE  STATUS  PRESTIGE

**Demographic**
- High population
- High levels of education and entrepreneurship
- International migration hubs
- Diversity and wide-ranging communities and ethnicities

**Economic**
- Agglomeration economies
- Centre for international banking, finance and insurance
- Stock exchange and equity market
- TNC headquarters
- Higher levels of migrant workers

**Socio-Cultural**
- International centre for media, arts, entertainment, sports
- World leading universities and research centres institutes
- Heritage
- Global tourist destination

**Technological**
- Major air hub
- Excellent communication links including internet and smart phone
- R&D centres

**Political**
- Headquarters of major international organisations
- Seat of national government
- Leaders promote cities globally and have national and local policies to support them

# Social segregation

- Isolation of communities
- lack of understanding can lead to hostilities
- sense of injustice
- lack of investment in some areas
- increased stress and anxiety leading to mental health issues.

| Strategies to manage these issues – London | Strategies to manage these issues – Rio de Janeiro |
|---|---|
| • Local councils improving services and practices to tackle ethnic inequality; community group initiatives to improve integration | • Local authorities, NGOs and community groups working together on projects to improve housing, infrastructure and services |

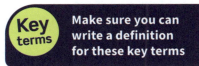

**Key terms** Make sure you can write a definition for these key terms

megacity    world city

# RETRIEVAL

Learn the answers to the questions below, then cover the answers column with a piece of paper and write down as many as you can. Check and repeat.

## Questions | Answers

| | Questions | Answers |
|---|---|---|
| 1 | What is urbanisation? | Increase in the proportion of the population living in urban areas |
| 2 | Give three reasons why urban areas are important. | Three from: an incident will impact many people / economic centres for production and consumption / centres for political power and decision making / hubs for social, cultural and educational exchange / some have economies larger than some countries, increasing their global power and influence |
| 3 | What percentage of the global population live in urban areas? | 55% |
| 4 | Which three countries will account for 35% of projected growth of the world's urban population between 2018 and 2050? | India, China and Nigeria |
| 5 | Why did HDEs experience rapid urbanisation in the nineteenth century? | The industrial revolution |
| 6 | Name the four stages in the cycle of urbanisation. | Urbanisation, suburbanisation, counter-urbanisation, urban resurgence |
| 7 | Name the three processes of urban change. | Deindustrialisation, decentralisation, rise of service economy |
| 8 | What is a megacity? | An urban area with a population of over 10 million people |
| 9 | Name the megacity with the largest population in 2023, and the one expected to be largest in 2030. | Tokyo and Delhi |
| 10 | Name three factors leading to megacity growth. | Three from: rural to urban migration / high natural increase / high fertility rates in LDEs / widening of urban boundaries / globalisation and economic growth |
| 11 | How many megacities were there in 2023? | 34 |
| 12 | Give three roles of megacities. | Three from: dominate regional economies / home of regional headquarters for TNCs / regional centres for employment and services / top of the urban hierarchy / influence the global economy |
| 13 | What is a world city? | A city that has significant global influence |
| 14 | Which organisation classifies world cities? | Global and World Cities Research Network (GaWC) |
| 15 | How many Alpha+ cities are there? | 7 |

*Put paper here*

# 9 Contemporary urban environments

## 9.3 Urban forms

### Physical geography and demography

The **Urban form** of a city is its physical characteristics – its shape, size, layout and density. It can be examined at different scales – street, neighbourhood or the whole urban region. Urban areas in contrasting places have many shared characteristics.

| Physical Factors | Human factors |
|---|---|
| • Water supply: Close to water source but avoiding regular flooding<br><br>• Topography: Flat land allows for faster, cheaper urban expansion; high points for defence in early urban centres; coasts, rivers, valleys restrict expansion<br><br>• Aspect and shelter: Building on valley sides with most sunlight and sheltered from winds<br><br>• Natural resources: early urban growth linked to resource exploitation | • Demographics: Population change (natural increase/decrease, migration) affects housing, services, employment<br><br>• Political: National and local planning policy (housing, infrastructure, transport, economic activities, green and blue space)<br><br>• Economic: market forces (supply and demand) or government intervention affect value of land. High land value in city centres affects location of retail/industry/housing<br><br>• Transport: Road and rail networks encourage growth along routes (linear) and by stations<br><br>• Communication: Location of high speed broadband and mobile phone masts |

> **REVISION TIP**
>
> Think about ranking the factors in order of importance. This will help you remember them.

### Factors that influence spatial patterns

#### Land Use

| Spatial pattern | Influencing factors |
|---|---|
| HDEs, e.g. UK:<br>• Focused around CBD<br>• Older inner city areas next to city centre<br>• Residential suburbs<br>• Edge of town developments for retail, services, leisure, transport<br><br>EMEs/LDEs<br>• CBDs with rapidly expanding urban areas (new settlements or former colonial towns)<br>• High value housing in city centres<br>• Informal settlements on edge of towns | • Slow growing, or decreasing, populations<br>• Land values and building density decreases with increased distance from CBD (highest land value)<br>• Outside city centre, housing prices increase with distance<br>• Industries in redeveloped inner city have decentralised<br>• Older housing renovated (gentrification)<br>• Population booms and changing lifestyles lead to housing expansion<br>• Radial growth along transport networks, with later infilling<br>• Planning and redevelopment within city<br>• Renovated city centre housing can be expensive<br>• Fast-growing populations<br>• Planning regulation and enforcement depends on government<br>• Housing prices decrease with distance from CBD<br>• Facilities in informal settlements developed by residents |

## 9 Contemporary urban environments

### Economic inequality

| Spatial pattern | Influencing factors |
|---|---|
| Income inequality leads to richer and poorer areas of the city | Differences in:<br>• evolving economic activities leading to changes in employment opportunities<br>• urban poverty rates<br>• house prices<br>• level of 'desirability' of local areas within a city<br>• level of government support including the welfare state<br>• presence of informal settlements<br>• crime rates |

### Social segregation

| Spatial pattern | Influencing factors |
|---|---|
| Different groups are separated from each other within the city, e.g. high- and low-income residents, ethnic and community groups | • Government policy and planning, e.g. townships under apartheid in South Africa<br>• Large post-Second World War local authority housing estates in UK<br>• People from ethnic, religious or cultural background choose to join existing communities in the city<br>• Location of places of worship or community groups<br>• Affordability of housing costs |

### Cultural diversity

| Spatial pattern | Influencing factors |
|---|---|
| • Many cities are culturally diverse<br>• Internal and international migrants to the city can come from different ethnic groups<br>• Central areas may be more culturally diverse than suburbs | • People from the same ethnic background may be concentrated in the same area<br>• Availability of education and cultural institutions<br>• Access to culturally diverse shops, food and entertainment in city centres |

# Characteristics of new urban landscapes

## Town centre mixed developments

Mixed city centre land use (residential, leisure, entertainment, offices, retail, education, administration). Funded by government and private investors. More accessible land use to encourage people back into city centres e.g. BedZED in Hackbridge, London

Mixed waterfront redevelopments e.g. Gunwharf Quays, Portsmouth

## Fortress developments

Uses defensible space design and security (gates, CCTV, AI). Usually for retail and residential. Increases safety for users but can polarise and discourages social mixing, e.g. Barra da Tijuca, Brazil

## Cultural and heritage quarters

Focus on historical background and architecture, cultural experience (arts, creative industries, entertainment) e.g. Birmingham

## Gentrified areas

Public and private investment housing schemes to regenerate run down inner cities. Older buildings and streetscapes are renovated. Wealthy, upwardly mobile professionals often buy properties. Other businesses and services are attracted to the area (multiplier effect). House prices may become unaffordable for existing residents, e.g. Hackney, East London

## Edge cities

Located next to towns and cities, close to major road junctions. Mixed use developments include housing, schools, offices, health services, retail and entertainment. Often reliant on car ownership e.g. Whiteley between Portsmouth and Southampton, Hampshire

# The concept of the post-modern 'Western' city

Post-modern 'Western' cities emerged at the end of the twentieth century in HDEs with evolving urban structure, architecture and planning caused by dynamic economies and societies. They are characterised by fragmentation rather than traditional urban forms.

- Multi-centred
- Mix of architectural styles, including historical designs and unique buildings constructed for how they look (aesthetics) as much as what they do (purpose)
- Diverse and multicultural populations with different ethnicities and communities concentrated in different parts of the city
- Fragmentation affects social and economic equality
- Employment in tertiary and quaternary sectors, 24/7 and gig economy present

- 'Outward facing' and globalised
- Focus on consumer culture, commercialism and technology
- High-tech corridors
- Market orientated, with international investment encouraged
- Range of stakeholder involvement
- Environmental sustainability through green space, energy efficient buildings and sustainable transport networks.

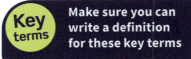

**Key terms** Make sure you can write a definition for these key terms

edge city   fortress developments   heritage quarter   mixed development   urban form

## 9 Contemporary urban environments

### 9.4 Urban issues

## Two contrasting urban areas

### London, UK

- London, capital of the UK, is a world city, with a population of 8.8 million (2021).
- It is a multicultural city with over 1.1 million residents not born in the UK, mostly from India, Poland, Bangladesh, Pakistan and Romania.
- White groups made up 54% of the population, with Asian groups 21% and Black groups 14%.
- Proportions of the largest religious groups were Christian, no religion, Muslim and Hindu.

### Rio de Janeiro, Brazil

- Rio de Janeiro, a coastal city in eastern Brazil, has a population of 6 million (16 million in greater metropolitan area).
- White groups make up 50% of the population, with concentrations of different groups throughout the city.
- Lagoa, Rio's richest neighbourhood, is 90% white and there are high concentrations of other ethnicities in the poorer north and west zones, including the favelas (informal settlements).

## Social segregation

- Isolation of communities
- lack of understanding can lead to hostilities
- sense of injustice

- lack of investment in some areas
- increased stress and anxiety leading to mental health issues.

### Strategies to manage these issues – London

- Local councils improving services and practices to tackle ethnic inequality; community group initiatives to improve integration

### Strategies to manage these issues – Rio de Janeiro

- Local authorities, NGOs and community groups working together on projects to improve housing, infrastructure and services

# Economic inequality

- Poverty
- housing costs and quality
- educational attainment
- health
- crime
- social and political unrest
- transportation.

### Strategies to manage these issues – London

- Increase local wages: UK National Living Wage £11.42 per hour (2024); Real Living Wage (£13.15 per hour in London)
- Schools for success programme; improved interventions and increased funding resulting in the 'London effect' – higher attainment of pupils in London
- Mayor of London's Police and Crime plan to improve safety and security
- Affordable Homes Programme London; shared ownership; help to buy; 'First Dibs'; Living Rent schemes; integrating more affordable housing into new residential developments
- Health inequalities strategies to address physical and mental health issues
- Transport strategy to transform London's streets, improve public transport and create opportunities for new homes and jobs

### Strategies to manage these issues – Rio de Janeiro

- Increase local wages: Brazil Minimum Wage (2023) R$1320 per month (£214)
- NGOs: Project Favela developing schools, teachers and online education
- NGOs and local community groups tackle social issues; more police stations in favelas with Pacifying Police Units (UPP)
- Increasing building quality through self-help schemes to improve property in the favelas; local government integration of favelas into established local neighbourhoods
- Expansion of primary health care facilities
- Development of sustainable transport including light rail system, electric buses and bike lane expansion

# Cultural diversity

- Tensions between different cultural groups
- discrimination
- lack of access to services due to language barriers.

### Strategies to manage these issues – London

- 'Inclusive London Strategy' focusing on working with local people and community groups to improve neighbourhoods, increase opportunities for young people, increase job opportunities and business, improve transport and promote inclusivity

> **REVISION TIP**
>
> Think about how you will use case study statistics within developed explanations, to explain their impacts. Don't just list facts.

### Strategies to manage these issues – Rio de Janeiro

- Access to green and coastal space to improve physical and mental health

**Key terms** Make sure you can write a definition for these key terms

cultural diversity   economic inequality   social segregation

# RETRIEVAL

Learn the answers to the questions below, then cover the answers column with a piece of paper and write down as many as you can. Check and repeat.

## Questions

## Answers

| # | Question | | Answer |
|---|---|---|---|
| 1 | What does the 'urban form' of a city mean? | Put paper here | Its physical characteristics – its shape, size, layout, density |
| 2 | Name three physical factors that can affect urban forms. | | Water supply, topography, natural resources |
| 3 | Name three human factors that can affect urban forms. | | Three from: demographics / political / economic / transport / communication |
| 4 | What are the factors that influence the spatial pattern relating to cultural diversity? | | People from same ethnic background may cluster in same area; availability of education and cultural institutions; access to culturally diverse shops, food and entertainment in city centres |
| 5 | Name three types of new urban landscapes. | Put paper here | Three from: town centre mixed developments / cultural and heritage quarters / fortress developments / gentrified areas / edge cities |
| 6 | Is the post-modern 'Western' city based around a city centre or is it multi-centred? | | Multi-centred |
| 7 | Name two schemes in London helping to increase affordable housing. | | Two from: Affordable Homes Programme, London / shared ownership / help to buy / 'First Dibs' / Living Rent schemes |
| 8 | How does Project Favela help education in Rio de Janeiro? | Put paper here | Develops schools, teachers and online education |
| 9 | What strategies are used in Rio de Janeiro to improve transportation? | | Development of sustainable transport including light rail system, electric buses and bike lane expansion |
| 10 | What are the aims of the 'Inclusive London Strategy'? | | Improve neighbourhoods, increase opportunities for young people, increase job opportunities and business, improve transport, promote inclusivity |

## Previous questions

Now go back and use these questions to check your knowledge of previous topics.

## Questions

## Answers

| # | Question | | Answer |
|---|---|---|---|
| 1 | What is urbanisation? | Put paper here | Increase in the proportion of the population living in urban areas |
| 2 | Why did HDEs experience rapid urbanisation in the nineteenth century? | | The industrial revolution |
| 3 | Name the three processes of urban change. | | Deindustrialisation, decentralisation, rise of service economy |
| 4 | What is a megacity? | Put paper here | An urban area with a population of over 10 million people |
| 5 | Name three factors leading to megacity growth. | | Three from: rural to urban migration / high natural increase / high fertility rates in LDEs / widening of urban boundaries / globalisation and economic growth |

## 9.5 Urban climate

## Urban microclimates

Urban areas can create their own microclimates by affecting local weather and climate.

Urban microclimates are affected by:

- structure and form of the urban area
- human activities within it
- physical location of the urban area, e.g. inland, coastal, valley, hills
- diurnal/nocturnal (daytime/ night-time) temperatures and seasonal changes.

Urban areas can create a 'climatic dome' that affects different atmospheric variables:

- temperature
- precipitation
- relative humidity
- wind speed and direction
- air quality and visibility.

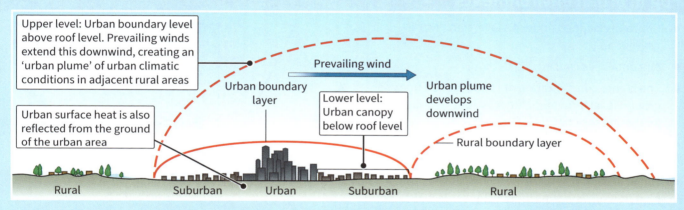

▲ *Figure 1 The urban climate dome has two levels – lower and upper levels*

## The urban heat island effect (UHI)

**UHI** occurs when the urban area has significantly higher temperatures than surrounding rural areas.

### Reasons for the UHI:

#### Ground surfaces and buildings

- Urban areas have a lower albedo as dark surfaces and materials absorb heat energy during the day and release it slowly at night.
- Steel, glass and light colour paint reflect heat energy more readily back into the atmosphere.
- Heat is leaked into atmosphere due to poor building insulation and air conditioning units.

#### Evaporation

- Less evapotranspiration as there is less vegetation than in rural areas.
- Drainage systems remove water quickly so less heat energy used to evaporate surface water.

#### Human activities

- Heat created by residents, industry, transport and power stations.
- Rising heat, water vapour and particulate matter from air pollution, acting as condensation nuclei, cause increased chance of convectional rainfall in urban areas.

#### London's UHI effect

- Thermal gradient decreases in temperature from urban centre (11°C) to rural areas (5°C); steeper over east as lower building density.
- Thermal gradients are:
  — lower in the winter
  — stronger at night as heat energy is trapped by cloud and particulate matter, creating a localised greenhouse effect.

## 9 Contemporary urban environments

### 9.5 Urban climate

## Precipitation in urban environments

- Urban areas have up to 15% more precipitation and more days with rainfall than surrounding rural areas. Cloud bursts cause intense rainfall events.
- Convection currents from the UHI, and air currents caused by air movement between buildings of different heights, cause uplift. Air rises and cools below the dew point causing water vapour to condense, increasing the chance of rainfall.
- Condensation nuclei from particulate matter allow cloud droplets to form, increasing the chance of rainfall.
- Fewer frost days in urban areas compared with surrounding rural areas.

## Fog in urban environments

- Cool air is trapped under an inversion layer of warmer air.
- Air condenses causing fog to form.
- The higher concentration of condensation nuclei at ground level from air pollutants adds to this effect, increasing the chance of fog.
- Higher chance of fog during winter when temperatures are cooler.
- Fog in urban areas disperses more slowly during high pressure systems (anticyclones) as winds are less strong.

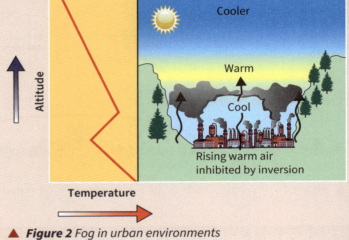

▲ **Figure 2** Fog in urban environments

## Thunderstorms in urban environments

Thunderstorms are more likely to occur in summer when the heat causes stronger uplifts of air by convection. Air cools and condenses quickly. Thunderclouds form, leading to thunder, lightning and intense periods of rainfall. Particulate matter in the atmosphere acts as condensation nuclei and causes more intense thunderstorms.

**Key terms** Make sure you can write a definition for these key terms

particulate pollution   photochemical smog
urban heat island effect (UHI)

# Wind in urban environments

Urban structures have a significant effect on air movement in the city.

- Friction: buildings create friction, slowing down air movement.
- Urban canyon effect: tall buildings on either side of narrow streets funnel winds.
- Venturi effect: pressure difference between two buildings in close proximity causes violent gusts of wind.
- Turbulence: winds are deflected down, around and over the top of buildings.

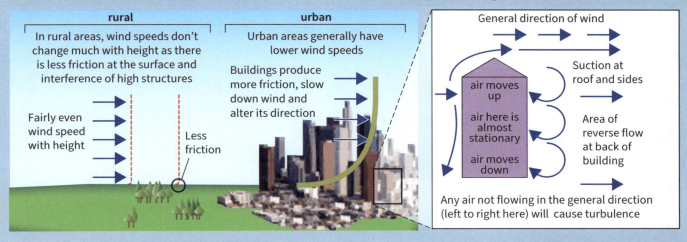

**rural**

In rural areas, wind speeds don't change much with height as there is less friction at the surface and interference of high structures

Fairly even wind speed with height

Less friction

**urban**

Urban areas generally have lower wind speeds

Buildings produce more friction, slow down wind and alter its direction

General direction of wind

Suction at roof and sides

air moves up

air here is almost stationary

air moves down

Area of reverse flow at back of building

Any air not flowing in the general direction (left to right here) will cause turbulence

▲ **Figure 3** *Tall buildings interfere with airflow in urban areas*

# Air quality in urban environments

Air quality is generally lower in urban areas than in rural areas due to vehicle emissions, burning fossil fuels for heating, and industrial processes.

**Particulate pollution** includes chemical compounds, some toxic, which can enter the bloodstream causing serious health impacts including respiratory problems (asthma, bronchitis, lung cancer).

**Photochemical smog** is a mixture of pollutants formed when nitrogen oxides and organic compounds react to sunlight. This causes a brown haze above urban areas. Cities in warm, dry, sunny climates or within a valley or 'bowl' are more susceptible to it as the surrounding topography makes dispersion more difficult.

# Pollution reduction policies

All urban areas experience air pollution issues. Rapidly growing cities in EMEs and LDEs face particular challenges. Government intervention and technology are the two main strategies used to reduce air pollution in urban areas.

| Government intervention |
| --- |
| - Clean air acts |
| - Industrial pollution controls |
| - Vehicle emissions limits |
| - Air quality targets, monitoring and warnings |
| - Congestion charge to enter urban areas |
| - Limiting the number of vehicles in urban areas |
| - Low emission zones in urban areas |
| - Promoting sustainable transport, including public transport systems, park and ride, car sharing, bikes and improving pedestrian access |

| Technology |
| --- |
| - Hybrid electric and electric vehicles |
| - Solar panels |
| - Heat pumps |
| - Low emitting stoves and heaters to reduce indoor air pollution |

# RETRIEVAL

Learn the answers to the questions below, then cover the answers column with a piece of paper and write down as many as you can. Check and repeat.

## Questions

## Answers

| | Question | | Answer |
|---|---|---|---|
| 1 | What four factors affect urban microclimates? | | Structure and form; human activities; physical location; diurnal/nocturnal temperatures and seasonal changes |
| 2 | A climatic dome can affect which atmospheric variables? | | Temperature; precipitation; relative humidity; wind speed and direction; air quality and visibility |
| 3 | What is an urban heat island? | | When the urban area has significantly higher temperatures than surrounding rural areas |
| 4 | What two reasons linked to evaporation contribute to UHI? | | Less evapotranspiration as there is less vegetation than in rural areas; drainage systems remove water quickly so less heat energy used to evaporate surface water |
| 5 | When are thermal gradients lower? | | Winter |
| 6 | How much more precipitation do urban areas have than surrounding rural areas? | | 15% |
| 7 | Why does fog in urban areas disperse more slowly during high pressure systems (anticyclones)? | | Winds are less strong |
| 8 | What is the urban canyon effect? | | Tall buildings on either side of narrow streets funnelling winds |
| 9 | Name three serious health issues caused by particulate matter. | | Asthma, bronchitis, lung cancer |
| 10 | Name three ways that technology can help to reduce pollution. | | Three from: hybrid electric and electric vehicles / solar panels / heat pumps / low emitting stoves and heaters to reduce indoor air pollution |

*Put paper here*

## Previous questions

Now go back and use these questions to check your knowledge of previous topics.

## Questions

## Answers

| | Question | | Answer |
|---|---|---|---|
| 1 | Name the four stages in the cycle of urbanisation. | | Urbanisation, suburbanisation, counter-urbanisation, urban resurgence |
| 2 | Which organisation classifies world cities? | | Global and World Cities Research Network (GaWC) |
| 3 | What does the 'urban form' of a city mean? | | Its physical characteristics – its shape, size, layout, density |
| 4 | Name two schemes in London helping to increase affordable housing. | | Two from: Affordable Homes Programme, London / Shared Ownership / Help to Buy / 'First Dibs' / Living Rent schemes |
| 5 | What strategies are used in Rio de Janeiro to improve transportation? | | Development of sustainable transport including light rail system, electric buses and bike lane expansion |

*Put paper here*

## 9 Contemporary urban environments

### 9.6 Urban drainage

## Surfaces and catchment characteristics

Urban areas have up to 15% more precipitation than surrounding rural areas because:

- higher air temperatures hold more moisture
- more condensation nuclei formed from particulate matter.

Hard surfaces (e.g. tarmac, roof tiles) in urban areas increase surface runoff / overland flow and reduce infiltration during a rainfall event. Urban areas have drainage systems to remove surface water quickly. Cities in HDEs and EMEs are likely to have more investment in their drainage systems.

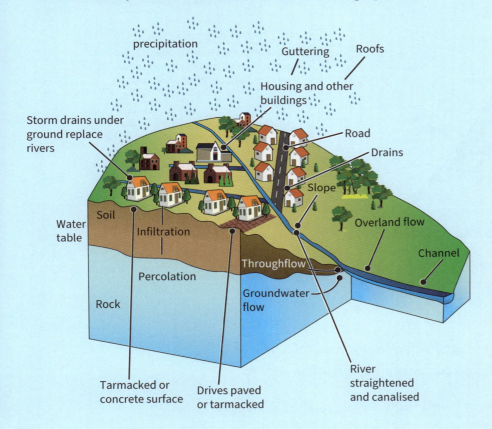

Vegetation and soil surface cover is reduced, reducing soil infiltration rates

Increased area of **impermeable** surfaces (roofs, roads, guttering) putting pressure on drainage systems

Natural streams may be replaced by underground channels (storm drains, culverts)

▲ *Figure 1 Urban areas change the surface of a drainage basin*

## Impacts on drainage basin storage areas

- Urban river storage capacity is increased by channelisation, dredging and building embankments.
- Urban areas include natural (e.g. lakes, ponds) and human-made (e.g. reservoir, swimming pools) water stores.
- Temporary water storage in puddles and depressions.
- Less vegetation decreases infiltration.
- More impermeable surfaces so less soil moisture storage.

**REVISION TIP**

Make a spider diagram of the ways in which urban areas can affect drainage basin flows and stores.

# 9 Contemporary urban environments

## 9.6 Urban drainage

### Urban water cycle

Flood hydrographs show the response of a river to a rainfall event. Discharge to urban river channels over time is measured using a flood hydrograph.

In urban areas, hydrographs are 'flashy', meaning:

- a high proportion of rainwater flows into urban river channels, increasing peak discharge
- lag times are reduced as rainwater flows into the urban river channel more quickly.

Figure 2 shows the different response rates of a river in urban and rural areas to a rainfall event.

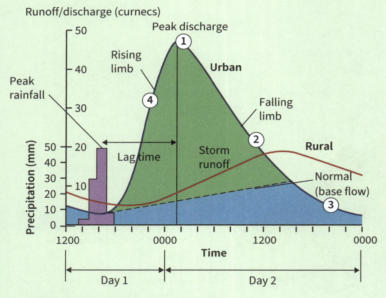

① Discharge rises quickly due to rapid surface runoff and reaches its peak around 10 hours after peak rainfall

② Discharge falls at a slower rate due to base flow increasing. Base flow is the normal flow of water in the river derived from throughflow and groundwater flow.

③ Base flow slowly declines as throughflow declines

④ Precipitation causes the discharge in the river to rise

▲ *Figure 2 A 'flashy' flood hydrograph*

Humans adapt the inputs, flows, stores and outputs of the hydrological cycle.

- More impermeable surfaces, so high proportion of precipitation flows quickly into rivers.
- Ground water pumped to the surface for urban activities (e.g. residential, industry, business, leisure).
- Wastewater (including sewerage) flows to water treatment plants.

Flash floods occur in urban areas when there is:

- too much surface water for drainage systems to cope with
- debris blocking drainage systems causing overflow
- building on flood plains.

# Catchment management

Catchment management is managing the interaction between water and land throughout the **catchment area** of the drainage basin system. Catchment management is important when tackling issues such as:

- flooding and flood prevention
- water pollution
- water shortages
- erosion.

Effective catchment management can improve urban drainage. Urban drainage is important as it rapidly removes surface water during and after a rainfall event, reducing the risk of flooding and flood damage.

# Hard engineering

Constructing and maintaining artificial structures to control rivers.

Aims include:

- increase the volume of water that can flow through the drainage basin
- increase the speed at which water can be discharged from the river system.

Strategies include:

- channel widening
- channel dredging
- raising embankment height
- river straightening
- channelisation of existing river
- flood relief channels (diversion spillways)
- drainage clearance.

# Soft engineering

Using the natural environment and natural processes to reduce river flooding.

Aims include:

- increase infiltration rates, reducing and slowing down surface flow
- preventative measures to reduce flood risk.

Strategies include:

- afforestation in upper catchment
- restoring vegetation on riverbanks
- river restoration
- floodplain zoning to restrict building development
- flood warning systems.

# Sustainable Urban Drainage Systems (SUDS)

**Sustainable Urban Drainage Systems (SUDS)** copy natural drainage systems in order to:

- reduce or slow down surface runoff, reducing flood risk
- limit wastewater and sewerage flooding during intense rainfall periods
- reduce water pollution
- replenish groundwater sources
- increase green spaces (habitat diversity, recreational area).

SUDS strategies:

- Swales – shallow, wide, vegetated channels to store or increase water flow, and remove pollutants during rainfall events
- Infiltration trenches
- Permeable surfaces to increase infiltration rates
- Balancing ponds and bioretention basins
- Green roofs
- Greywater capture units to collect rainwater on roofs.

## 9 Contemporary urban environments

### 9.6 Urban drainage

## River restoration: Sheffield's Blue Loop, UK

**River restoration** aims to remove artificial drainage including hard engineering structures and replace them with management strategies based on natural processes and the natural environment.

Sheffield's Blue Loop is a nearly 13 km stretch of waterway including the Tinsley Canal and the River Don from Sheffield City Centre to Meadowhall.

| | |
|---|---|
| **Reasons for and aims of the project** | • River Don and Tinsley Canal were key parts of the city of Sheffield's industrial landscape in the nineteenth and twentieth centuries<br>• 1970s globalisation caused key industries to close<br>• Both water courses were heavily polluted, with limited wildlife<br>• Tinsley Canal and towpath were derelict and in need of urgent repair |
| **Contributions of parties involved** | • Funding: National organisations – Sheffield Development Corporation, British Waterways, Natural England, National Lottery<br>• Local organisations – Blue Loop Community Project (2013 onwards: Friends of the Blue Loop) and other local NGOs |
| **Project activities** | • Footpath maintenance<br>• Cutting and pruning vegetation<br>• Treating invasive species<br>• Removing litter and debris from the footpaths and river channel<br>• Local schools, community groups, businesses' and individuals' engagement, including 'Trout in the Classroom' project |
| **Evaluation of project outcomes** | • Increasing accessibility for cyclists, walkers and runners along nearly 13 km of river and canal front<br>• Restoring natural habitats and species, e.g. kingfishers, fish<br>• Removing non-native species, e.g. Japanese knotweed<br>• SUDS strategies adopted by local businesses<br>• Increasing wetland areas and vegetation on the floodplain |

> **REVISION TIP**
>
> You should revise the river restoration project you have studied. This page gives the example of Sheffield's Blue Loop.

> **REVISION TIP**
>
> For your example of a river restoration and conservation project, write down why it was needed, and how successful it was. Say why you made this judgement.

---

**Key terms** Make sure you can write a definition for these key terms

catchment area    impermeable    river restoration
Sustainable Urban Drainage Systems (SUDS)

# ⇄ RETRIEVAL

Learn the answers to the questions below, then cover the answers column with a piece of paper and write down as many as you can. Check and repeat.

## Questions | Answers

| Questions | Answers |
| --- | --- |
| **1** Why do urban areas have more rainfall than surrounding rural areas? | Higher air temperatures hold more moisture; more condensation nuclei formed from particulate matter |
| **2** Name three ways in which urban areas alter drainage basin surfaces. | Reduction in vegetation and soil surface cover, increase in impermeable surfaces, replacement of natural streams by underground channels |
| **3** Give two reasons why urban hydrographs are 'flashy'. | Increased peak discharge because higher proportion of rainwater flowing into urban river channels; lag times are reduced |
| **4** Why do flash floods occur in urban areas? | Too much surface water for drainage systems to cope with; debris blocking drainage systems causing overflow; buildings on flood plains |
| **5** What is a catchment area? | The area drained by a river and all its tributaries |
| **6** What is the difference between 'hard engineering' and 'soft engineering' river management? | Hard engineering: constructing and maintaining artificial structures to control rivers; soft engineering: using the natural environment and natural processes to reduce river flooding |
| **7** Why do SUDS copy natural drainage systems? | To reduce or slow down surface runoff, reducing flood risk; limit wastewater and sewerage flooding during intense rainfall periods; reduce water pollution; replenish groundwater sources; increase green spaces |
| **8** Name two SUDS strategies. | Two from: swales / infiltration trenches / permeable surfaces / balancing ponds and bioretention basins / green roofs / greywater capture units |

*Put paper here*

## Previous questions

Now go back and use these questions to check your knowledge of previous topics.

## Questions | Answers

| Questions | Answers |
| --- | --- |
| **1** Give three roles of megacities. | Three from: dominate regional economies / home of regional headquarters for TNCs / regional centres for employment and services / top of the urban hierarchy / influence the global economy |
| **2** Name three types of new urban landscapes. | Three from: town centre mixed developments / cultural and heritage quarters / fortress developments / gentrified areas / edge cities |
| **3** A climatic dome can affect which atmospheric variables? | Temperature; precipitation; relative humidity; wind speed and direction; air quality and visibility |
| **4** Why does fog in urban areas disperse more slowly during high pressure systems (anticyclones)? | Winds are less strong |

*Put paper here*

# ⚙ KNOWLEDGE

## 9 Contemporary urban environments

### 9.7 Urban waste

## Urban waste generation

The amount of global waste is estimated to increase from just over 2 billion tonnes in 2016 to 3.4 billion tonnes by 2050.

| Urban waste source | Characteristics |
|---|---|
| Industrial. | Includes toxic and hazardous materials (solvents, chemicals) requiring special disposal and solid waste from energy plants, construction work and manufacturing industries |
| Commercial | Waste from businesses, e.g. offices, retail, hospitality. Includes paper, cardboard, food and plastic |
| Personal consumption | Domestic waste from households and individuals. Includes paper, cardboard, food, plastic, electrical goods, furniture |

Some tertiary and quaternary activities may produce waste hazardous to health and requiring special disposal, e.g. medical services (pharmaceutical or hospital waste).

**Municipal solid waste (MSW)** is found in all three urban waste sources. It includes packaging, plastic, food, street and market waste.

## Waste streams

A **waste stream** is the flow of waste from its source (industrial, commercial, personal consumption) to its recovery, **recycling** or final disposal. The type, amount and disposal method of waste varies.

### Economic characteristics

- Increase in wealth leads to higher levels of consumption and more waste.
- The amount of different types of waste is related to the income level of the country.
- Food and green waste make up the largest share of waste in all countries.
- HDEs produce more paper-based and plastic waste than LDEs.
- HDEs invest more money into waste recovery, recycling and disposal.

### Lifestyles

- HDEs have higher recycling rates than LDEs.
- Recycling facilities are more accessible in HDEs.
- Higher proportions of people live in urban areas in HDEs so higher volumes of waste produced.
- Urban waste contains more waste from manufactured and consumer goods (plastic, packaging, paper).
- HDEs have more packaging of ready meals, fast food and processed food to dispose of.

### Attitudes

- Some people have a positive attitude towards minimising waste and recycling.
- Some people have a 'throw away' culture where goods are used for a short time before being disposed of.
- Food safety regulations are stricter in HDEs. Food may still be edible after the use by date and so may be thrown away unnecessarily.

# The environmental impacts of waste disposal

| Approach to waste disposal | Environmental impacts |
|---|---|
| Unregulated | Waste dumped in streets or unofficial sites – an eyesore and risk to health<br>Hazardous substances damage ecosystems<br>Endangers wildlife (swallowing, entanglement) |
| Recycling | Waste cleaned and used again (e.g. glass bottles) or broken down and reprocessed into new products<br>Using recycled products decreases demand for raw materials and reduces energy use and carbon emissions<br>Collection and recycling centres required |
| Recovery | Waste is used instead of raw materials, reducing need for resource exploitation |
| Incineration | Burning of combustible waste reduces its volume by up to 90%<br>Disposal of ash still required<br>Carbon emissions during the combustion process<br>Other toxic pollutants released when some materials burned<br>Not all types of waste can be burned safely |
| **Landfill** (burial) | Waste is disposed in specific land-based sites, mines or quarries and left to decompose over time<br>Site is lined to reduce leaching of chemicals into the ground<br>Increased risk of contamination of soils and ground water<br>Methane builds up during decomposition, contributing to the human-enhanced greenhouse effect – some countries use this to produce energy<br>Visual pollution and malodorous |
| Submergence | Some waste is illegally dumped at sea<br>Leaks increase toxic or radioactive levels in oceans |
| Trade | Waste can be imported and exported by countries<br>HDEs send waste to LDEs for further processing and recycling, including electronic products<br>Final disposal may harm the environment through air, water or land pollution |

> **REVISION TIP**
>
> Make an audio recording to explain different approaches to waste disposal. Listen to the recording to help you revise.

## Collection of waste

The percentage of waste that is collected and disposed of by the local authority or municipality varies between high-income and low-income areas. The waste that is not collected is dealt with by the household by dumping, burning or composting.

Globally, urban areas have higher collection rates than rural areas. HDEs have higher rates of collection than LDEs in both urban and rural areas.

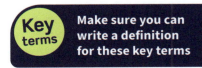

**Key terms** Make sure you can write a definition for these key terms

landfill    municipal solid waste    recycling    waste stream

## 9 Contemporary urban environments

### 9.7 Urban waste

## Comparison of incineration and landfill approaches

Governments at a local level have to decide the best way to dispose of waste that cannot be recycled – incineration and landfill are two ways they do this.

| | Advantages | Disadvantages |
|---|---|---|
| Incineration | • Heat energy produced, contributing to the energy mix<br>• Compact site<br>• Medical waste disposed of more safely<br>• Once operational, incinerator lasts a long time<br>• Ash can be used as a resource<br>• Can be located in urban and rural areas | • Large initial investment needed<br>• Carbon dioxide and particulate matter emissions<br>• Not all waste can be burned |
| Landfill | • Uses abandoned or disused land (edge of city or rural areas)<br>• Easy to manage and relatively cost effective<br>• Methane can be used for energy<br>• When full can be landscaped for leisure purposes<br>• Safe if well managed | • Visual pollution<br>• Malodorous<br>• Attracts wildlife (scavengers and decomposers, e.g. rats)<br>• Risk of subsidence<br>• Methane (greenhouse gas) produced<br>• Risk of toxic leaks into soil and ground water<br>• Lorry traffic to and from site |

**REVISION TIP**

Without looking at this book, write as many advantages and disadvantages of incineration and landfill as you can remember. Check your answers and repeat after a few days to help you remember.

## Waste disposal in Singapore

Singapore is an HDE in southern Asia. It has a population of 5.6 million which is 100% urban. It is an island country, making space for landfill limited and kept to an absolute minimum.

- Singapore's solid waste management system focuses on waste minimisation and recycling.
- Solid waste has increased from 1260 tonnes/day (1970) to 8741 tonnes/day (2021) because of Singapore's growing population and economy.
- Solid waste management is segregated into different types of products for reuse and recycling in homes and businesses.

- Waste that cannot be segregated is collected and sent to 'waste to energy' plants to be incinerated.
- Incineration reduced solid waste volume by 90% and the energy is used for electricity.
- Waste that cannot be burned and ash from incineration is sent to Semakau, an island south of Singapore used for landfill.

**REVISION TIP**

You will have studied incineration and landfill approaches to waste disposal in relation to a specific urban area. This page gives the example of Singapore.

# RETRIEVAL

Learn the answers to the questions below, then cover the answers column with a piece of paper and write down as many as you can. Check and repeat.

## Questions | Answers

| # | Question | Answer |
|---|----------|--------|
| 1 | How much waste is expected to be generated globally by 2050? | 3.4 billion tonnes |
| 2 | What are the three main sources of urban waste? | Industrial, commercial, personal consumption |
| 3 | What is Municipal Solid Waste? | It includes packaging, plastic, food, street and market waste and is found in all three urban waste sources |
| 4 | Which category has the largest share of global waste? | Food and green waste (44%) |
| 5 | Name three lifestyle factors that influence the global waste stream. | Three from: HDEs have higher recycling rates than LDEs / recycling facilities are more accessible in HDEs / higher proportion of people living in urban areas in HDEs so higher volumes of waste produced / urban waste contains more waste from manufactured and consumer goods / HDEs have more packaging of ready meals, fast food and processed food to dispose of |
| 6 | Do HDEs, EMEs or LDEs have the highest rates of recycling? | HDEs |
| 7 | Name five alternative types of waste disposal. | Five from: unregulated / recycling / recovery / incineration / burial / submergence / trade |
| 8 | What are Singapore's incinerators called? | Waste to energy plants |

*Put paper here*

## Previous questions

Now go back and use these questions to check your knowledge of previous topics.

## Questions | Answers

| # | Question | Answer |
|---|----------|--------|
| 1 | Name three physical factors that can affect urban forms. | Water supply, topography, natural resources |
| 2 | Is the post-modern 'Western' city based around a city centre or is it multi-centred? | Multi-centred |
| 3 | What four factors affect urban microclimates? | Structure and form; human activities; physical location; diurnal/nocturnal temperatures and seasonal changes |
| 4 | What is the urban canyon effect? | Tall buildings on either side of narrow streets funnelling winds |
| 5 | Why do urban areas have more rainfall than surrounding rural areas? | Higher air temperatures hold more moisture; more condensation nuclei formed from particulate matter |

*Put paper here*

## 9 Contemporary urban environments

### 9.8 Urban environmental issues

## Environmental problems in urban areas

| Environmental issue | Causes and effects |
|---|---|
| **Atmospheric pollution**<br> | • Urban areas have greater levels of atmospheric pollution than rural areas<br>• Industrial activity, energy use and vehicle emissions increase gas emissions and particulate matter into the atmosphere<br>• EMEs and LDEs often have higher levels of atmospheric pollution than HDEs because of industrial activity, and environmental standards may be lower. Car ownership is also increasing.<br>• Toxic air (gases: carbon monoxide, carbon dioxide, sulphur dioxide, nitrogen dioxide, and particulates) contributes to premature deaths (respiratory, lung cancer and cardiovascular diseases) – around 7 million worldwide annually, and 4000 in London in 2019.<br>• Visibility is also restricted, leading to road accidents and loss of education as schools are shut |
| **Water pollution**<br> | • Surface runoff and wastewater from human activities carry pollutants<br>• Domestic wastewater from households, especially where no water treatment is possible<br>• Rubber, metal and oil from vehicles are washed into the drainage system<br>• Industrial waste from factories<br>• Toxic leakage from unregulated dumps and landfill sites<br>• Solid waste, including litter and plastics, in urban watercourses<br>• EMEs and LDEs often have higher levels of water pollution than HDEs as environmental standards may be lower and less enforced<br>• Waterborne diseases, such as cholera, dysentery and diarrhoea cause health issues, particularly in young people<br>• Malaria and the zika virus can be prevalent near stagnant water<br>• Long-term damage to habitats and wildlife as pollutants stay in ecosystems |
| **Dereliction**<br> | • Built environments decay through time due to weathering and use<br>• Industries and businesses shut down through globalisation and/or deindustrialisation<br>• Building materials declared unsafe after extensive use and not cost effective to repair, e.g. asbestos, RAAC, some forms of cladding<br>• 'Broken windows theory' – visible signs of neglect leads to more buildings falling into disrepair in an area<br>• 'Planning blight' – delays in plans for redevelopment leave buildings empty and becoming derelict over time<br>• Crime – vandalism and graffiti of built structures |

# Strategies to manage environmental problems

| Environmental issue | Management strategies |
|---|---|
| Atmospheric pollution  | • Congestion charging: drivers pay to drive into urban areas<br>• Ultra Low Emission Zones: charges for vehicles that do not meet emission standards within designated areas. ULEZ in central and inner London is projected to cut air pollution-related hospital admissions by one million by 2050 and save £5billion in health care costs<br>• Improved public transport: networks, frequency, cashless payments<br>• Cycling and walking: bike lanes, bike hire and improved access for pedestrians<br>• Clean air legislation and environmental permits reducing air pollution from industry<br>• Improved technology: electric vehicles, developing more sustainable aviation fuel<br>• Raising awareness of impacts of poor air quality |
| Water pollution  | • Increased efficiency for urban waste water treatment systems: water from sewerage system is filtered using different processes before clean water is put back into local rivers<br>• Decrease amount of nutrients, nitrates, pathogen, ammonia and pharmaceuticals in waste to reduce impact on ecosystems<br>• Legislation: laws and penalties to reduce mismanagement of industrial waste<br>• NGOs, e.g. WaterAid, working with local partners to improve water and sanitation in LDEs<br>• IGO-supported schemes to improve water, sanitation and hygiene (WASH) in urban areas<br>• SUDS to remove pollutants |
| Dereliction  | • Sustainable urban redevelopment projects, including brownfield sites<br>• Local community involvement in urban planning<br>• Government grants and loans for residents to improve their homes and to encourage local businesses to set up<br>• Conservation area status to protect heritage areas<br>• Improving safety and security in urban areas<br>• Repurposing older buildings for new land uses<br>• Decontamination of land |

**REVISION TIP**

For your two examples of contrasting urban areas:
• think about the causes of pollution and dereliction
• list the strategies used to manage these issues that have been the most effective.

**Key terms** Make sure you can write a definition for these key terms

atmospheric pollution    dereliction    water pollution

## 9 Contemporary urban environments

### 9.9 Sustainable urban development

## The impact of urban areas on the environment

| Scale | Environmental impact |
|-------|---------------------|
| Local | • Competition for land and urban sprawl<br>• Consumption of water, food and energy<br>• Air quality<br>• Vehicle use and traffic congestion<br>• Consumption of goods<br>• Disposal of waste |
| Global | • Urban areas are a major contributor to climate change, producing around 75% of global greenhouse gases (UNEP)<br>• Major cities are also hubs for global trade, migration and tourism, increasing the impact of transport and transport networks around the world |

### Ecological footprint

- An **ecological footprint** of an urban area relates to the environmental resources necessary to produce its goods and services and manage its waste.
- Ecological footprints are higher in cities in HDEs than in LDEs.

## Sustainability in the urban environment

- Natural: Management of the environment, resources and waste

- Physical: Provision of resources to support urban residents, including adequate, affordable, safe housing, food, water and energy

- Social: Quality of life of urban residents: including access to health and education services; inclusive, respectful communities; safe, secure and politically stable

- Economic: Capacity for economic growth without long-term negative effects. Available employment, low inequality and opportunities for social mobility

## Nature and features of sustainable cities

- Urban Boundary to prevent urban sprawl and encourage brownfield development
- Integrated transport increasing networks and capacity to meet demand and reduce car use; increased cycling and walking
- Mixed land use of commercial and residential
- Pedestrian friendly with small street blocks and public green space
- Energy efficient buildings
- Renewable energy use
- Green waste disposal
- Increased efficiency of water collection and use

### Liveability

A sustainable city should improve the quality of life of those who live there. The combination of factors that do this is called **liveability**. These factors include:

- quality of the built and natural environment
- economic prosperity
- social stability and equality
- educational opportunities
- quality of health
- access to local services
- cultural, entertainment and leisure opportunities.

# Development of sustainable cities

| Opportunities | Challenges |
|---|---|
| • More efficient and cost-effective services and utility provision in more densely populated areas<br>• Increased awareness of value of a sustainable approach<br>• Technological innovation and implementation | • Rapid urban population growth (EMEs, LDEs)<br>• Decline of urban centres (HDEs)<br>• Complex planning regulations<br>• Updating infrastructure in older or rapidly growing cities<br>• Long-term political strategies needed<br>• Involvement of range of stakeholders (different scales of government, businesses, communities, NGOs)<br>• Rise in economic and social inequality<br>• Increased impact of globalisation: migration and influence of TNCs<br>• Climate change<br>• Natural disaster risks |

## Strategies for developing more sustainable cities

| Issue | Management strategy |
|---|---|
| Reducing vehicle traffic | Vehicle users pay to enter urban areas, e.g. London congestion charge |
| Increasing energy efficiency in buildings | Improving insulation and reducing heating and air conditioning needs, e.g. BedZED development, Hackbridge, London |
| Increasing renewable energy | Increasing use of solar, wind, biomass, e.g. London Community Energy Fund |
| Increasing water use efficiency | Reducing water usage and increasing greywater capacity, e.g. Oceanhamnen sustainable district; Helsingborg – new circular sewer system |
| Improving recycling and waste disposal | Increasing amount of products that can be reused and recycled, and reducing waste incineration and landfill, e.g. Singapore waste minimisation and recycling programmes |
| Increasing green space | Accessibility to open, biodiverse spaces, e.g. Queen Elizabeth Olympic Park, London |
| Liveable communities | All community facilities and services accessible within walking or cycling distance, e.g. 10-minute town concept |
| Risk of damage from natural hazards and climate change | Reducing impact of natural hazards, e.g. Hamburg green roofs |
| Urban heat island effect | Reducing urban temperatures, e.g. Melbourne Urban Forest Fund |
| Improving food security and reducing food miles | Increasing food production within urban spaces, e.g. New York roof and urban farms |

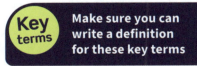

**Key terms** Make sure you can write a definition for these key terms

sustainable city   ecological footprint   liveability

# RETRIEVAL

Learn the answers to the questions below, then cover the answers column with a piece of paper and write down as many as you can. Check and repeat.

## Questions

| | Questions | Answers |
|---|---|---|
| 1 | Name three gases that contribute to atmospheric pollution. | Three from: carbon monoxide / carbon dioxide / sulphur dioxide / nitrogen dioxide |
| 2 | What substances are washed from cars into drainage systems? | Rubber, metal and oil |
| 3 | What is planning blight? | Delays in plans for redevelopment leaves buildings empty and derelict over time |
| 4 | What is an Ultra Low Emission Zone? | Vehicle exhausts must meet emission standards or drivers have to pay to use their vehicles within designated areas |
| 5 | Name one NGO improving water quality and sanitation in urban areas in LDEs. | WaterAid |
| 6 | What is sustainable urban development? | Concept of living and working in an urban area without compromising the needs of future generations |
| 7 | What is the ecological footprint of an urban area? | The environmental resources necessary to produce its goods and services and manage its waste |
| 8 | What are the four dimensions of urban sustainability? | Natural, physical, social, economic |
| 9 | Give three elements included in the concept of liveability. | Three from: quality of the built and natural environment / economic prosperity / social stability and equality / educational opportunities / quality of health / access to local services / cultural, entertainment and leisure opportunities |
| 10 | Are the ecological footprints of cities higher in HDEs or LDEs? | HDEs |

*Put paper here*

## Previous questions

Now go back and use these questions to check your knowledge of previous topics.

## Questions

| | Questions | Answers |
|---|---|---|
| 1 | What is an urban heat island? | The urban area has significantly higher temperatures than surrounding rural areas |
| 2 | Give two reasons why urban hydrographs are 'flashy'. | Increased peak discharge because higher proportion of rainwater flowing into urban river channels; lag times are reduced |
| 3 | What are the three main sources of urban waste? | Industrial, commercial, personal consumption |
| 4 | What is Municipal Solid Waste? | It includes packaging, plastic, food, street and market waste and is found in all three urban waste sources |
| 5 | Name five alternative types of waste disposal. | Five from: unregulated / recycling / recovery / incineration / burial / submergence / trade |

*Put paper here*

## 9 Contemporary urban environments

### 9.10 Case study: An urban area in the UK

## Case study: London, UK

- London, capital of the UK, is a world city with a population of 8.8 million.

- GDP is around £500 billion per year, larger than many countries.

- London's population declined after the Second World War because of decrease in employment opportunities (dock closure and deindustrialisation) and counter-urbanisation.

- Since 1981, London's population has increased due to a focus on investment in the tertiary and quaternary economy: tourism and knowledge-based industries. Brownfield sites have been used for property redevelopment.

- Between 2011 and 2021 there was a positive percentage population change in most London boroughs.

- London is the most ethnically diverse region in the UK. In 2021, 63.2% identified as an ethnic minority group and 40% were born outside the UK.

England ▲ 6.6 %   London ▲ 7.7 %   City of London ▲ 16.4 %

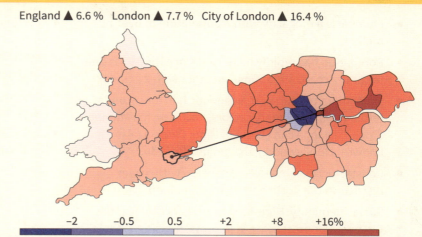

| −2 | −0.5 | 0.5 | +2 | +8 | +16% |

▲ **Figure 1** *Percentage population change in London and surrounding areas, 2011 to 2021*

## Urban regeneration

| | Area | Strategies |
|---|---|---|
| **Business-led and community-led** | London Docklands | 1980s, 1990s: London Docklands Development Corporation and Enterprise Zone<br>Focus on economic regeneration through a new financial district, property development and transport links (Canary Wharf)<br>Criticised for lack of affordable housing and appropriate of employment opportunities |
| **Community-led** | East End | 2000, 2010s: London Thames Gateway Development Corporation (LTGDC) worked with public and private partners to develop area with focus on social needs and sustainability. 35% affordable homes, employment, commercial, education and green space created |
| **Sports-led** | Stratford and Newham | Legacy Corporation set up to develop the 2012 London Olympic site after the games. Olympic Stadium: home ground of West Ham United and used for other sporting events and concerts. Aquatics Centre and Velodrome. Retail developments: Westfield Stratford and cultural and educational developments. New transport infrastructure. Greenspace enhanced through tree planting and river and canal restoration. Olympic Village used to create 2800 new homes with 1200 'affordable homes' |
| **Sustainability-led** | Hackbridge | 2000s: BedZED – First London zero-carbon community<br>52% of construction materials sourced within 35 miles, with 15% of materials used reclaimed or recycled; biomass boilers, energy efficient designed homes; car sharing, mixed housing stock, including affordable housing; greenspace; water saving toilets and appliances; community engagement |

**KNOWLEDGE**

# 9 Contemporary urban environments

## 9.11 Case study: A contrasting urban area

### Case study: Mumbai, India

- Mumbai is a megacity with a population of over 21 million.
- Population has rapidly increased since the 1950s due to rural–urban migration from other states in India.
- Major Indian Ocean port, economic hub for secondary, tertiary and quaternary sectors, including the Mumbai Stock Exchange and centre of Hindi Cinema (also known as 'Bollywood'), the Indian film industry.
- Coastal location restricts expansion with newer districts in the north and on the mainland.
- Social inequality: High-income earners live near the CBD, with poorest living in informal housing. Dharavi is India's largest informal settlement – unplanned settlement on low lying land, dense housing with high occupancy rates, poor sanitation and limited security of tenure. Strong sense of community with high local employment rates.

Mumbai is located in a high flood risk area due to:

- monsoon climate – intense rainfall in June-September and risk of thunderstorms. Urban drainage systems may not cope with downpours

- cyclone and tropical storm risks – damage from intense wind and rain caused by low pressure weather systems developing in the Indian Ocean. Low pressure and high winds increases risk of coastal flooding
- river drainage patterns – Mumbai has several rivers and creeks that can overflow
- tsunami risk – tectonically active region so risk of tsunamis affecting Mumbai due to its coastal location.

▶ **Figure 1**
*Mumbai is the captial of Maharashtra state in India*

Overcrowding: high density housing, with average of over four people per household

Lack of effective sanitation: lack of toilet and sewerage facilities

Traffic congestion: rising vehicle ownership and increased economic activity leads to more congestion

Air pollution: air quality is reduced due to vehicle and industrial emissions

**Key urban issues**

Waste disposal: more goods consumed as population increases, putting pressure on existing waste disposal systems

Water shortages: drought season reduces water availability leading to water shortages and rationing

Pressure on medical services: need to treat health conditions caused by air pollutants and waterborne diseases

## Urban management strategies

| Sanitation | Traffic management | Public transport | Waste |
|---|---|---|---|
| SPARC, an Indian NGO, has built 300 community-managed toilet blocks | 550 smart traffic signals installed to regulate vehicle flow | Investment in electric buses and Chalo app to make payments quicker and easier | Clean Up Mumbai campaign (Municipal Corporation of Greater Mumbai) to remove litter and inform residents about the benefits of reducing waste |

# RETRIEVAL

Learn the answers to the questions below, then cover the answers column with a piece of paper and write down as many as you can. Check and repeat.

## Questions

## Answers

| | Questions | | Answers |
|---|---|---|---|
| 1 | Why did London's population decline after the Second World War? | Put paper here | Decrease in employment opportunities (dock closure and deindustrialisation); counter-urbanisation |
| 2 | Give two criticisms of the London Docklands development. | | Lack of affordable housing; lack of appropriate employment opportunities |
| 3 | Give an example of a sustainability-led development strategy in London. | | BedZED |
| 4 | In 2021, what percentage of London's population was born outside the UK? | Put paper here | 40% |
| 5 | Give two reasons why Mumbai is at risk of flooding. | | Two from: monsoon climate / cyclone and tropical storm risks / river drainage patterns / tsunami risk |
| 6 | Name four urban issues faced by Mumbai. | Put paper here | Four from: lack of effective sanitation / overcrowding / water shortages / traffic congestion / air pollution / waste disposal / pressure on medical services |
| 7 | What did the NGO SPARC do to improve sanitation in Mumbai? | | Built 300 community-managed toilet blocks |

## Previous questions

Now go back and use these questions to check your knowledge of previous topics.

## Questions

## Answers

| | Questions | | Answers |
|---|---|---|---|
| 1 | Which category of urban waste has the largest share? | Put paper here | Food and green waste (44%) |
| 2 | What are Singapore's incinerators called? | | Waste to energy plants |
| 3 | Name three gases that contribute to atmospheric pollution. | | Three from: carbon monoxide / carbon dioxide / sulphur dioxide / nitrogen dioxide |
| 4 | What is planning blight? | Put paper here | When delays in plans for redevelopment leave buildings empty and derelict over time |
| 5 | What is the ecological footprint of an urban area? | | The environmental resources necessary to produce its goods and services and manage its waste |
| 6 | What are the four dimensions of urban sustainability? | | Natural, physical, social, economic |

# PRACTICE

## Exam-style questions

**1**   Outline how deindustrialisation has changed urban areas. **[4 marks]**

**2**   Outline issues associated with social segregation in urban areas. **[4 marks]**

**3**   Explain the impact of different surfaces that are characteristic of towns and cities on the urban water cycle. **[4 marks]**

**4**   Outline why dereliction can be an environmental problem in urban areas. **[4 marks]**

**5**   Outline the relationship between waste streams and economic characteristics of populations, their lifestyles and attitudes. **[4 marks]**

**6**   Analyse the data shown in **Figure 1**. **[6 marks]**

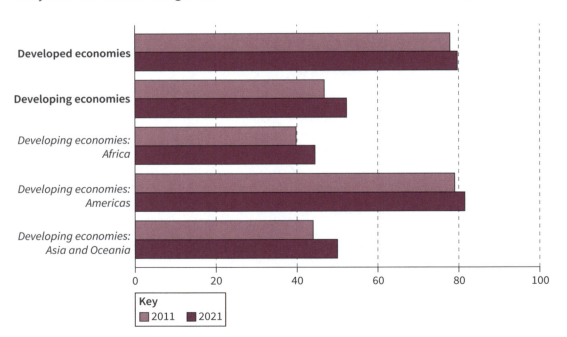

▲ **Figure 1** Urban population by group of economies (Percentage of total population)

7  Evaluate the usefulness of **Figure 2** in depicting data about the emergence of megacities around the world.  **[6 marks]**

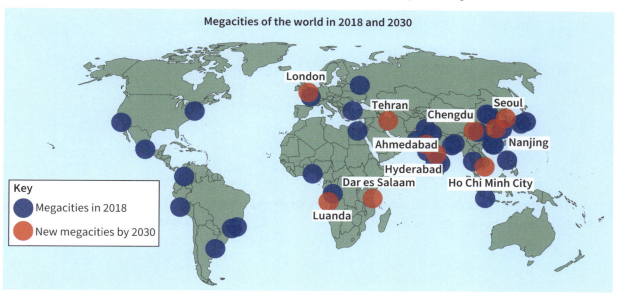

**Megacities of the world in 2018 and 2030**

Key
●  Megacities in 2018
●  New megacities by 2030

▲ **Figure 2** *Megacities of the world 2018 and 2030*

8  Analyse the data in **Figure 3**.  **[6 marks]**

| 0–5 Meets WHO guideline | 5.1–10 Exceeds by 1 to 2 times | 10.1–15 Exceeds by 2 to 3 times | 15.1–25 Exceeds by 3 to 5 times | 25.1–35 Exceeds by 5 to 7 times | 35.1–50 Exceeds by 7 to 10 times | >50.1 Exceeds by over 10 times |
|---|---|---|---|---|---|---|

| Rank | City | 2022 | Jan | Feb | Mar | Apr | May | Jun | Jul | Aug | Sep | Oct | Nov | Dec | 2021 | 2020 |
|---|---|---|---|---|---|---|---|---|---|---|---|---|---|---|---|---|
| 1 | Lahore, Pakistan | 97.4 | 133 | 102.5 | 85.6 | 69.3 | 60.9 | 52.1 | 47.8 | 46.2 | 64.2 | 123.2 | 190.5 | 192.9 | 86.5 | 79.2 |
| 2 | Hotan, China | 94.3 | 61.7 | 91.6 | 132.7 | 106.2 | 120.5 | 69.5 | 172.5 | 75 | 65.1 | 75 | 50.3 | 120 | 101.5 | 110.2 |
| 3 | Bhiwadi, India | 92.7 | 110.6 | 98 | 116.2 | 149.5 | 123.8 | 102.8 | 38.8 | 36.5 | 59.9 | 85.4 | 111.4 | 86.9 | 106.2 | 95.5 |
| 4 | Delhi (NCT), India | 92.8 | 141 | 100.9 | 91 | 98 | 73.2 | 56.2 | 34.3 | 31.1 | 38.3 | 99.7 | 176.8 | 171.9 | 96.4 | 84.1 |
| 5 | Peshawar, Pakistan | 91.8 | 110.2 | 103.5 | 78.3 | 68.5 | 53.5 | 56.3 | 51.8 | 57.8 | 79 | 100 | 132 | 212.1 | 89.6 | ... |

▲ **Figure 3** *Top five air polluted cities in 2022, yearly average, and per month, and 2021 and 2020 yearly averages, based on PM2.5 (particules less than 2.5 micrometres in diameter) readings*

9  Assess the role of demographic processes in the growth of megacities.  **[9 marks]**

> **EXAM TIP**
>
> These 9-mark questions have 4 marks for AO1 (knowledge and understanding of the concepts in the figure) and 5 marks for AO2 (application of knowledge and evaluation).

10  Assess the extent to which new urban landscapes, such as town centre mixed developments, cultural and heritage quarters and fortress developments, lead to cultural diversity.  **[9 marks]**

**11** Assess the impact of economic inequality and social segregation on 'liveability' in urban areas. **[9 marks]**

**12** With reference to contrasting urban areas, compare strategies used to develop their environmental sustainability. **[9 marks]**

**13** 'Dealing with environmental issues is more challenging than tackling social problems when managing urban areas.'

How far do you agree with this statement? **[20 marks]**

> **EXAM TIP**
>
> 20-mark questions have 10 marks for AO1 (knowledge and understanding of the concepts in the question) and 10 marks for AO2 (application of knowledge to provide a detailed analysis and evaluation in response to the question).

**14** Evaluate the impact of economic characteristics, lifestyles and attitudes on waste generation and disposal. **[20 marks]**

 # KNOWLEDGE

## 10 Population and the environment

### 10.1 The relationship between physical geography and population

## Physical geography and demography

There is a relationship between physical geography and demography (characteristics of the population). Aspects of the physical environment, climate, soils, and resources distributions (food, water, energy) all affect human population characteristics and change.

## Global population patterns

In mid-November 2022, the global population reached 8 billion. It is expected to increase by another 2 billion in the next 30 years. Population is still growing but the rate of growth is slowing down.

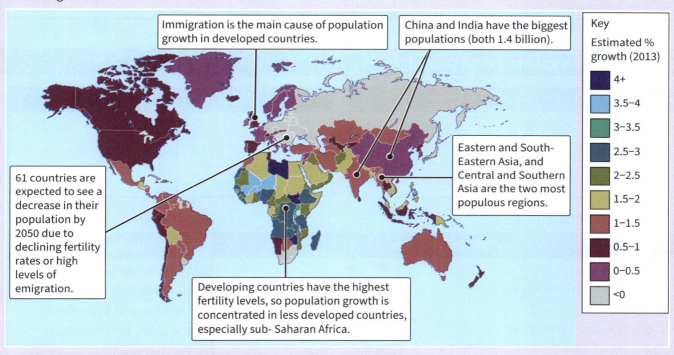

Immigration is the main cause of population growth in developed countries.

China and India have the biggest populations (both 1.4 billion).

61 countries are expected to see a decrease in their population by 2050 due to declining fertility rates or high levels of emigration.

Eastern and South-Eastern Asia, and Central and Southern Asia are the two most populous regions.

Developing countries have the highest fertility levels, so population growth is concentrated in less developed countries, especially sub- Saharan Africa.

Key
Estimated % growth (2013)

- 4+
- 3.5–4
- 3–3.5
- 2.5–3
- 2–2.5
- 1.5–2
- 1–1.5
- 0.5–1
- 0–0.5
- <0

▲ **Figure 1** *The global pattern of estimated population growth (2013)*

## The role of development

- The 46 least developed economies in the world are some of the world's fastest-growing in terms of population.
- More than half of the estimated future population growth until 2050 will be concentrated in just eight countries: the DR Congo, Egypt, Ethiopia, India, Nigeria, Pakistan, the Philippines and Tanzania (between 2–3% population growth).

- Lack of development causes high **fertility** rates, the main driver of rapid population growth.
- The concentration of population growth in less developed countries may result in them struggling to develop and to eradicate poverty, as there will be additional pressure on resources.

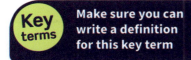 **Key terms** Make sure you can write a definition for this key term

fertility

 **KNOWLEDGE**

## 10 Population and the environment

### 10.2 Global food production and consumption

## Global and regional patterns

### Food production

- Agricultural production is highest in Asia, Europe and the Americas and lowest in Sub-Saharan and North Africa.
- The global production of crops, livestock and fish is expected to increase by 1% a year over the next 10 years.
- Increased food production in Asia will drive future growth.

### Food consumption

- Consumption is increasing globally, with North America consuming most per head and Africa least.
- Rate of undernourishment is highest in Sub-Saharan Africa. In 2020, 53% of the population in Somalia were undernourished. Rates in North America and Europe are below 2.5%.
- Rates of overconsumption are increasing due to increased intake of fatty and sugary foods, particularly in LDE and EME countries.

## Agricultural systems and productivity

### Inputs

- Labour
- Relief/slope
- Drainage
- Seeds
- Livestock
- Chemicals
- Climate
- Soil
- Machinery
- Capital
- Fodder

### Processes

- Planting
- Growing
- Harvesting crops
- Rearing livestock

### Outputs

- Animals
- Crops
- Waste

### Factors affecting productivity

#### Physical factors

- Temperature
- Precipitation and water supply
- Altitude
- Gradient, which affects soil characteristics and the use of machinery
- Aspect
- Soil depth
- Soil type – water-retention capacity, structure, pH, **leaching**, mineral content

#### Human factors

- Land ownership
- Farm size
- Transport to markets
- Government subsidies
- Technology

## Different types of agriculture

- Arable farming – the growing of crops on flat land and high-quality soil
- Pastoral farming – raising animals where the land is less suitable for arable farming (mixed farming is both arable and pastoral)
- Shifting cultivation/nomadic herding – farmers move with seasonal rainfall, common in areas of poor soil (sedentary farmers remain in the same location)
- Subsistence farming – growing food and raising animals to feed the farmer and a small group
- Commercial farming – growing food and raising animals for profit
- Extensive farming – amount of labour and capital are small in relation to the area being farmed; yields are low
- Intensive farming – amounts of labour and capital are high in relation to the area being farmed; yields are high

 **Key terms** Make sure you can write a definition for this key term

leaching

 **256**

# ⇄ RETRIEVAL

Learn the answers to the questions below, then cover the answers column with a piece of paper and write down as many as you can. Check and repeat.

| | Questions | | Answers |
|---|---|---|---|
| 1 | Which continents consume the most and least food per head? | Put paper here | North America (most) and Africa (least) |
| 2 | Name five inputs to the agricultural system. | | Five from: labour / relief or slope / drainage / seeds / livestock / climate / soil / machinery / capital / fodder |
| 3 | What processes take place in an agricultural system? | | Planting, growing, harvesting crops, rearing livestock |
| 4 | Name four physical factors affecting agricultural productivity. | Put paper here | Four from: temperature / precipitation and water supply / altitude / gradient of slope / aspect / soil depth / soil type |
| 5 | What is extensive farming? | | Farming where amounts of labour and capital are small in relation to the area being farmed, and yields are low |
| 6 | What is the difference between nomadic farmers and sedentary farmers? | Put paper here | Nomadic – move with the seasonal rainfall; sedentary – stay in one location |
| 7 | What is mixed farming? | | Farming that involves both arable (crop growing) and pastoral (animals) farming |
| 8 | What is subsistence farming? | Put paper here | Growing food and raising animals to feed the farmer and a small group |
| 9 | What is arable farming? | | The growing of crops on flat land and high-quality soil |
| 10 | What is pastoral farming? | | Raising animals where the land is less suitable for arable farming |
| 11 | What is commercial farming? | Put paper here | Growing food and raising animals for profit |
| 12 | What is intensive farming? | | Where amounts of labour and capital are high in relation to the area being farmed, and yields are high |

# 10 Population and the environment

## 10.3 Climate

## Two major climactic types

| Tropical monsoon, e.g. central Asia |
|---|

**Summer**
- Insolation increases over northern India, Pakistan, and central Asia
- Heat rises creating an area of low pressure
- Air is drawn in from the Indian Ocean leading to substantial precipitation

**Winter**
- Cooler temperatures
- A high-pressure system develops leading to dry air and little rainfall
- Monsoon floods deposit alluvium to enrich the soil for farming
- Rice is planted when monsoon rains flood the paddy fields and is harvested in October during the dry season
- Wheat, barley and peas are grown during the dry season
- More than 60% of the global population live in areas with a monsoonal climate

| Semi-arid, e.g. Canadian prairies |
|---|

- Long, sunny days and low precipitation
- Frost and snows in winter
- Strong, warm and dry 'Chinook' wind in spring melts the snow and enables wheat to be sown
- Short growing season
- Tornadoes in summer can lead to crop damage
- The main industries are agriculture, services and oil production
- Relief is gentle and undulating so farming is extensive, commercial and highly mechanised
- Chemicals are used to improve yields
- The Canadian prairie provinces are home to nearly 6 million people

## Climate change as it affects agriculture

- Water-intensive crops, such as rice, sugar, cotton and wheat, may be impacted by unpredictable variations in precipitation that could threaten their cultivation.
- Warmer temperatures may increase the length of growing seasons, e.g. in the UK.
- Extreme weather events, such as heatwaves and drought, may lead to crop failure and heat stress in livestock.

- More intense storms may lead to crop destruction in areas that experience tropical cyclones.
- Current rainfall patterns are expected to change and warmer climates may see diseases and pests in new areas.

◀ **Figure 1** *Wheat crops impacted by drought in the Netherlands*

# 10.4 Soils

## Two key zonal soils

- Soils could be viewed as the world's most important natural resource.
- They are irreplaceable if lost.
- There are two key zonal soils.

### Tropical latosols

- Found under tropical rainforests
- High iron and aluminium oxide content
- Red or yellow-red in colour
- No distinct **horizons**
- Deep – often 20–30 m
- Thin but fertile top layer due to rapid nutrient cycling
- Infertile second layer due to rapid leaching
- Shifting cultivation takes place in areas of cleared rainforest as soil rapidly loses its fertility without the nutrient cycling from the forest

### Podsol

- Found under coniferous or boreal forests, heathland in southern Australia and British moorlands
- Formed by **podsolisation** where rain percolates through the soil and washes nutrients out of the upper horizon, leaving an ash-grey bleached layer and a reddish-brown layer where the materials have accumulated
- Iron deposits create an iron pan between the two horizons
- Poor soils for agriculture due to low levels of nutrients and a low pH
- Areas of podsols are best used for pastoral farming

## Soil problems for agriculture

```
                        Soil
                      problems
```

### Soil erosion

Top layers of soil are eroded by wind, intensive farming, deforestation, overcropping or prolonged and heavy rainfall. The rate of erosion is determined by climate, topography, soil type and vegetation cover.

### Waterlogging

Over-irrigation, heavy rainfall and low evaporation, where surplus water is not sufficiently drained away, can lead to waterlogged soil where roots suffer from a lack of oxygen.

### Salination

Waterlogged soil can increase the water table and bring dissolved salts into the top soil affecting the roots of salt-intolerant plants.

### Structural deterioration

Compaction of soil particles leads to less drainage and waterlogging.

**Key terms** Make sure you can write a definition for these key terms

horizon    podsolisation    salination
structural deterioration    waterlogging

## 10 Population and the environment

### 10.4 Soils

## Managing soil problems

There are several ways to manage soil problems to improve agriculture:

- Crop rotation to balance nutrient use
- Circles of stones (bunds) to prevent overland flow
- Organic farming – reducing the use of chemicals to improve fertility
- Tilling and aerating the soil to add moisture and air to improve conditions for seedlings and control weeds
- Cover crops to manage soil erosion by offering protection from heavy rainfall.

## Strategies to ensure food security

In 2023, domestic food price inflation affected countries at all stages of development, leading to rising food insecurity. In 2022, 11.7% of the global population faced food insecurity at severe levels, and progress towards the Sustainable Development Goal (SDG) 2, of 'Zero Hunger by 2030', is at risk.

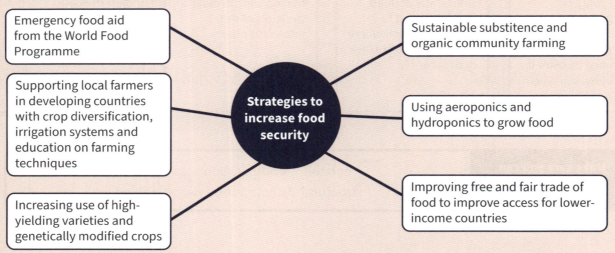

# RETRIEVAL

Learn the answers to the questions below, then cover the answers column with a piece of paper and write down as many as you can. Check and repeat.

## Questions

## Answers

| | Questions | Answers |
|---|---|---|
| 1 | How does the tropical monsoon climate influence rice cultivation? | Rice is planted when monsoon rains flood the paddy fields and is harvested in October during the dry season |
| 2 | What can cause damage to crops on the Canadian prairies? | Tornadoes in summer |
| 3 | Name four ways in which climate change may affect agriculture. | Warmer temperatures may increase the length of growing seasons; more extreme weather events may lead to crop failure and heat stress in livestock; more intense storms may lead to crop destruction; rainfall patterns may change and warmer climates may see diseases and pests in new areas |
| 4 | Where is podsol soil found? | Under coniferous or boreal forests, heathland in southern Australia and British moorlands |
| 5 | What does 'horizons' mean in terms of soils? | Layers in the soil which have different characteristics |
| 6 | What soil problems affect agriculture? | Soil erosion, waterlogging, salinisation and structural deterioration |
| 7 | What does structural deterioration of soil mean? | Compaction of soil particles leads to less drainage and waterlogging |
| 8 | How can soil problems be managed? | Crop rotation, cover crops, tilling, organic farming, stone bunds |

*Put paper here*

## Previous questions

Now go back and use these questions to check your knowledge of previous topics.

## Questions

## Answers

| | Questions | Answers |
|---|---|---|
| 1 | Name five inputs to the agricultural system. | Five from: labour / relief or slope / drainage / seeds / livestock / climate / soil / machinery / capital / fodder |
| 2 | What processes take place in an agricultural system? | Planting, growing, harvesting crops, rearing livestock |
| 3 | Name four physical factors affecting agricultural productivity. | Four from: temperature / precipitation and water supply / altitude / gradient of slope / aspect / soil depth / soil type |
| 4 | What is the difference between nomadic farmers and sedentary farmers? | Nomadic – move with the seasonal rainfall; sedentary – stay in one location |
| 5 | What is subsistence farming? | Growing food and raising animals to feed the farmer and a small group |

*Put paper here*

# 10 Population and the environment

## 10.5 The environment and health

### Global patterns of health, mortality and morbidity

- In 2019, heart disease was the leading cause of mortality (deaths) worldwide, followed by stroke and chronic obstructive pulmonary disease (COPD).
- The rates of morbidity (illness) and mortality (deaths) for different diseases varies between countries.
- Global patterns of health are affected by a number of different human and environmental factors, such as the income of the country, the availability of accessible, high quality healthcare, the climate, access to sanitation, lifestyle factors and air and water quality.

### Epidemiological transition

**Epidemiological transition** refers to how the types of disease affecting the population of a country change as the country develops economically and socially.

| Low income countries | Middle income countries | High income countries |
|---|---|---|
| • Causes of death mainly from biologically transmitted (communicable) diseases e.g. malaria and tuberculosis<br>• High birth/death/infant mortality rates<br>• Low life expectancy<br>• Healthcare is poor and less accessible<br>• Poor access to clean water and sanitation | • Falling death rate from biologically transmitted diseases due to improved healthcare and sanitation<br>• Improving life expectancy<br>• Increasing prevalence of non-communicable disease as rising incomes increases lifestyle factors such as drinking alcohol, smoking and unbalanced diets<br>• Increased risk of respiratory disease and cancers due to air pollution | • Causes of death mainly from non-communicable diseases e.g. cancer and dementia<br>• Onset of degenerative diseases can be delayed with advanced healthcare and medicine<br>• Low birth/death rates<br>• High life expectancy<br>• High quality healthcare and easily accessible |

**Increasing economic and social development** →

**Key terms** — Make sure you can write a definition for these key terms

disease vector    epidemiological transition
non-communicable disease    particulate matter    pathogen
vector-borne disease    water-borne disease

# The relationship between environment variables and disease

**Environmental variables**

**Climatic factors**
(e.g. temperature, rainfall and average daily sunshine hours)

**Topographical factors**
(the natural features of the land, e.g. drainage and relief)

**Vector-borne diseases**
- Many **disease vectors**, e.g. mosquitos or parasitic worms, thrive in warm temperatures with abundant rainfall.
- Heavy rainfall creates stagnant water, providing habitats and breeding grounds for vectors.
- Areas of the world with a warm and wet climate, e.g. between the Tropics, will have a higher incidence of diseases spread by vectors, such as malaria and bilharzia.

**Non-communicable diseases**
- A lack of sun exposure can lead to a deficiency in vitamin D and rickets.
- Depletion of the ozone layer increases the amount of UV radiation reaching the Earth, which can lead to an increased risk of skin cancer and cataracts.

- These factors link with climatic factors.
- In areas of good drainage, stagnant pools are less likely to accumulate.
- At higher altitudes, temperatures drop, so there isn't a suitable habitat for disease vectors.

## Pollution and health

- Air pollution and water pollution are important risk factors in mortality and morbidity worldwide.

- Human activity can lead to pollutants in the air and water, which can lead to a decrease in air and water quality and an increased incidence of some diseases.

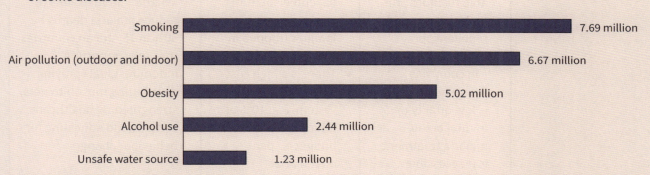

| | |
|---|---|
| Smoking | 7.69 million |
| Air pollution (outdoor and indoor) | 6.67 million |
| Obesity | 5.02 million |
| Alcohol use | 2.44 million |
| Unsafe water source | 1.23 million |

▲ **Figure 1** *Number of global deaths by risk factor, 2019*

### Air quality

- In 2019, 99% of the world's population were exposed to air that exceeds the WHO's safe limits for human health.

- Air pollution is a contributing factor in over 7 million deaths a year (WHO).

- An increase in harmful **particulate matter** in the air can affect cardiovascular, cerebrovascular and respiratory health, increasing cases of cancers, stroke, heart disease and asthma.

- Rapid development (industrialisation, car ownership, energy production) in EME countries has led to an increase in air pollution.

Air pollution in India

- In 2019, New Delhi recorded its highest concentration of PM2.5 (the most hazardous type of particulate matter) to date, with approximately 54,000 premature deaths in the city due to the air pollution in 2020. The pollution is caused by industry, motor vehicles, domestic cooking and power generation and is leading to an increase of cases of lung and oesophageal cancer. The incidence of these cancers is higher in the North of India - the region with the poorest air quality.

### Water quality

- Pollutants in water results in consumption of unsafe water.

- **Pathogens** enter the water supply through contact with untreated sewage and human/animal waste.

- This increases the incidence of **water-borne diseases**, such as diarrhoea, typhoid, cholera, polio and bilharzia.

- Deaths attributed to consuming unsafe water are more common in low-income countries that have a lack of basic sanitation and clean water sources.

## 10 Population and the environment

### 10.6 Biologically transmitted and non-communicable diseases

#### A biologically transmitted disease: malaria

| Global prevalence, distribution, and seasonal incidence | Factors affecting the incidence | Impacts | Management and mitigation strategies |
|---|---|---|---|
| • Malaria transmission occurs in countries between the Tropics and is higher during rainy seasons.<br>• The African continent has 95% of global malaria cases.<br>• In 2021, there were 247 million cases. | • Mosquitos thrive in warm, wet climates where stagnant pools of water allow them to breed.<br>• Locations at higher altitudes are malaria-free due to lower temperatures.<br>• Income – lower-income households are less likely to be able to afford anti-malarial nets and repellent sprays.<br>• Access to healthcare – in countries where healthcare is low quality or in rural areas, malaria treatment is inaccessible. | • In 2021 there were 619,000 malaria deaths<br>• In many countries, peak malaria transmission coincides with harvest season, threatening the agricultural economy.<br>• Malaria illness increases absenteeism from work and keeps families trapped in the poverty cycle.<br>• There is a high cost to the healthcare system, e.g. 10% of Ethiopia's healthcare budget is spent on malaria. | • The WHO's Global Malaria Programme coordinates the efforts to eliminate malaria, monitors national programmes, and sets the standards for malaria mitigation.<br>• Sustainable Development Goal 3.3 aims to end epidemics of malaria by 2030.<br>• The USA's President's Malaria Initiative aims to reduce malaria deaths by 50% in 15 African countries through supporting indoor residual spraying, distributing mosquito nets, strengthening healthcare systems and educating for behaviour change. |

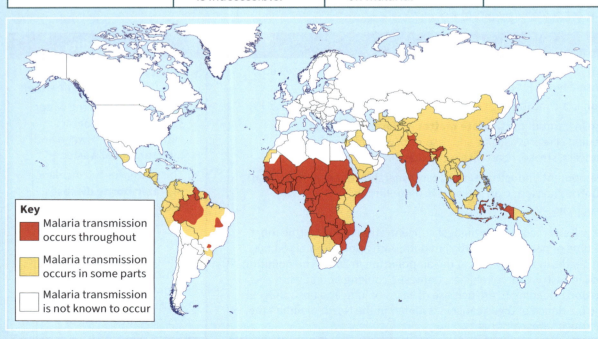

**Key**
- Malaria transmission occurs throughout
- Malaria transmission occurs in some parts
- Malaria transmission is not known to occur

▲ *Figure 1 The global distribution of malaria*

# A non-communicable disease: cancer

| Global prevalence and distribution | Factors affecting the incidence | Impacts | Management and mitigation strategies |
|---|---|---|---|
| • In 2020, there were over 18 million new global cancer cases.<br>• Asia accounted for nearly 50% of new cases.<br>• Cancer incidence is increasing more significantly in LDE and EME countries, which also have lower survival rates | • Many cancers are caused by modifiable lifestyle risk factors, e.g. smoking, alcohol consumption, sunbed use and diet.<br>• Air pollution was recognised as **carcinogenic** in 2013.<br>• Poverty – within high income countries, cancer survival rates are lower in areas of deprivation.<br>• Increasing wealth – as personal incomes rise, so does standard of living and lifestyle risk factors.<br>• Age – incidence of cancer increases with age as cellular repair is poorer and risk factors have built-up. | • In 2020 there were over 10 million deaths.<br>• People are unable to work.<br>• Cancer spending is one of the top three uses of the UK's NHS budget.<br>• Individual cancer patients experience loss of income, high cost of medical appointments, treatment, as well as social and psychological costs. | • Direct strategies involve investments in medical technology for diagnosis and treatment.<br>• Indirect strategies involve education campaigns to reduce engagement in lifestyle risk factors, e.g. warnings on cigarette packaging.<br>• The WHO's Global Action Plan aims to promote national cancer plans and set standards for early diagnosis, treatment and monitoring. |

# Promoting health and combating disease at the global scale

A range of organisations are involved in promoting global health and combating disease, including:

• The WHO's goal is 'to promote health, keep the world safe, and serve the vulnerable' (WHO, 2018). They respond to global health emergencies, strengthen healthcare systems and research and monitor health situations. They are aiming to offer, by 2023, one billion more people universal health coverage, better protection from health emergencies and improved health and wellbeing.

• UNICEF is responsible for children's humanitarian aid, focusing on reducing deaths from childhood diseases and maternal mortality. It provides vaccines, assists with risk factors for disease, e.g. sanitation, and provides emergency healthcare.

• Médecins Sans Frontières (MSF) is an NGO providing medical aid to countries in conflict zones, vaccinations in epidemic outbreaks, and medical assistance for refugees during hazard events.

**Key terms** Make sure you can write a definition for these key terms

biologically transmitted disease    carcinogenic

# RETRIEVAL

Learn the answers to the questions below, then cover the answers column with a piece of paper and write down as many as you can. Check and repeat.

## Questions | Answers

| # | Questions | | Answers |
|---|---|---|---|
| 1 | Which type of diseases are more prevalent in high-income countries? | Put paper here | Non-communicable, e.g. cancer and dementia |
| 2 | Name two climatic factors that affect the incidence of disease. | | Two from: Temperature / rainfall / average daily sunshine hours |
| 3 | Name two topographic factors that affect the incidence of disease. | | Drainage, relief (altitude) |
| 4 | What are disease vectors? | Put paper here | Carriers of an infectious disease |
| 5 | What are the main causes of poor air quality? | | Air pollution from industrialisation, private car ownership and energy production |
| 6 | What percentage of the world's population are exposed to air that exceeds the WHO's safe limits? | Put paper here | 99% |
| 7 | Give an example of a water-borne disease. | | One from: diarrhoea / typhoid / cholera / polio / bilharzia |
| 8 | Give two examples of international agencies and NGOs who work to promote global health. | | Two from: WHO / UNICEF / Médecins Sans Frontières |

## Previous questions

Now go back and use these questions to check your knowledge of previous topics.

## Questions | Answers

| # | Questions | | Answers |
|---|---|---|---|
| 1 | What is extensive farming? | Put paper here | Farming where amounts of labour and capital are small in relation to the area being farmed and yields per hectare are low |
| 2 | What can cause damage to crops on the Canadian prairies? | | Tornadoes in summer |
| 3 | How does the tropical monsoon climate influence rice cultivation? | | Rice is planted when monsoon rains flood the paddy fields and is harvested in October during the dry season |
| 4 | What soil problems affect agriculture? | Put paper here | Soil erosion, waterlogging, salinisation and structural deterioration |
| 5 | How can soil problems be managed? | | Crop rotation, cover crops, tilling, organic farming, stone bunds |

# 10 Population and the environment

## 10.7 Natural population change

### Factors in population change

**Natural population change** occurs when birth and death rates differ within a place.

- If birth rates are higher than death rates there will be a natural increase in population.
- If death rates are higher than birth rates then there will be a natural decrease in population

Population also changes when people migrate into and out of an area.

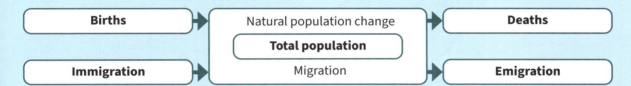

| Births | Natural population change | Deaths |
|--------|---------------------------|--------|
| | **Total population** | |
| Immigration | Migration | Emigration |

### Key vital rates

- Population change is also influenced by key vital rates, which can help us to understand national population change.
- Key vital rates show the most important aspects of the population change, and the speed at which they change.

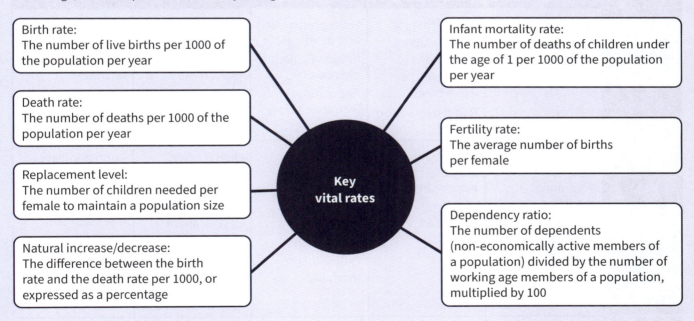

Birth rate:
The number of live births per 1000 of the population per year

Death rate:
The number of deaths per 1000 of the population per year

Replacement level:
The number of children needed per female to maintain a population size

Natural increase/decrease:
The difference between the birth rate and the death rate per 1000, or expressed as a percentage

**Key vital rates**

Infant mortality rate:
The number of deaths of children under the age of 1 per 1000 of the population per year

Fertility rate:
The average number of births per female

Dependency ratio:
The number of dependents (non-economically active members of a population) divided by the number of working age members of a population, multiplied by 100

# 10 Population and the environment

## 10.7 Natural population change

### The demographic transition model

- The **demographic transition model** describes how a country's population changes over time. Countries move through the stages at different rates.
- The model is simplistic and does not account for the different cultural and historical situations of countries, nor interventions to address demographic issues.

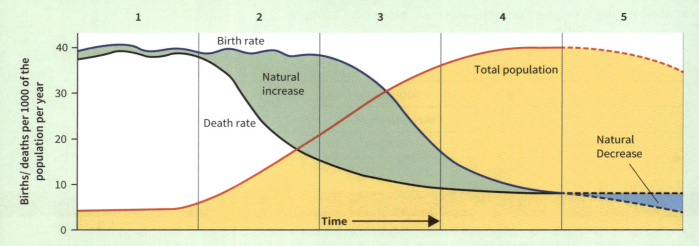

| Stage | 1 High fluctuating | 2 Early expanding | 3 Late expanding | 4 Low fluctuating | 5 Natural decrease |
|---|---|---|---|---|---|
| **Vital rates** | High birth rate<br>High death rate<br>Low life expectancy | High birth rate<br>Rapidly falling death rate<br>Increasing life expectancy<br>Rapid population growth | Falling birth rate and death rate<br>Increasing life expectancy<br>Population increases slowly | Low birth rate and death rate<br>High life expectancy<br>Population growth plateaus | More deaths than births<br>Population decreases |
| **Reasons** | Lack of healthcare<br>Traditional medicine<br>No birth control<br>High infant mortality | Poor quality, inaccessible healthcare<br>Improved sanitation | Wider use of birth control<br>Improved healthcare<br>Child vaccinations<br>Falling infant mortality | High quality, accessible healthcare<br>High levels of development<br>Greater gender equality leads to more people having children later or not at all | Population ages and deaths from non-communicable diseases rise, births are below replacement level |
| **Countries in this stage today** | Remote indigenous communities | Least developed economies, e.g. Eritrea, Bangladesh | Industrialising, EME countries, e.g. India, Mexico | HDEs, e.g. UK, Canada | Japan<br>China for the first time in 2022 |

▲ *Figure 1* The demographic transition model

# Population pyramids

Population pyramids show the age–sex structure of a population for a given year. Changes to a population over time can be tracked by looking at the changing shape of pyramids.

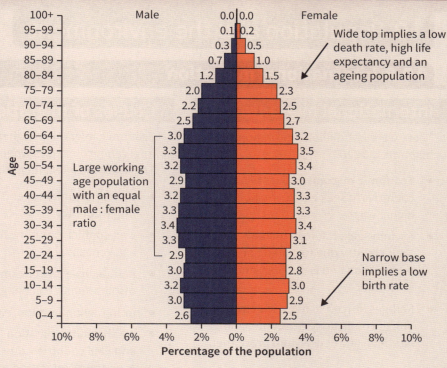

▲ **Figure 2** *A population pyramid for the UK (2023)*

# Cultural controls on population

## China's one-child policy

- In 1979, after severe famine and predicted rapid population growth, China's government implemented a policy where most families were only permitted to have one child.
- Incentives included free medical care, education, and salary benefits.
- If families were to have a second child, benefits were removed and fines were imposed.

- There were reports of forced abortions, sterilisations and female infanticide (Chinese society favoured males for their greater earning potential).
- It has resulted in a gender imbalance, an increasing dependency ratio and the threat of a too small working age population by 2050.
- In 2016, when the fertility rate was 1.62, the policy was relaxed to a two-child policy for all families, and three children in 2021.
- In 2022, it was reported that China's population shrank for the first time since the 1960s.

# The demographic dividend

A decline in fertility rates can lead to economic growth due to the productivity of the working age population and the low number of young dependents. Money can be invested in social development and a country can see improvements. This time is limited, as a prolonged fall in fertility rates will lead to those who were economically-active becoming elderly dependents alongside a shrinking workforce that will reduce economic growth.

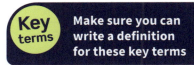

**Key terms** Make sure you can write a definition for these key terms

demographic dividend   demographic transition model
natural population change

 **KNOWLEDGE**

## 10 Population and the environment

### 10.8 International migration

## International migrants

There were 281 million international migrants in 2020.

- **Economic migrant:** moves between countries to seek employment and higher wages.
- Asylum seeker: forced to flee their country due to conflict or persecution and is seeking sanctuary in another.
- Refugee: forced to flee their country due to conflict or persecution and has been granted asylum in another.

## Causes of migration

Economic migrants weigh up push and pull factors and assess intervening obstacles when deciding whether to migrate. These factors affect migration flows:

- the opportunity to send their wages to their family in the origin country
- war, conflict, or political instability in the origin country, e.g. Venezuela
- natural disasters or climate-related events causing displacement, e.g. Bangladesh
- retirement migration, e.g. UK to Spain
- south–south migration, where economic migrants move to LDE and EME countries, due to the restrictive nature of advanced economies' migration policies, growing opportunities in rapidly developing industrialising countries and the ease of travelling shorter distances.

## Implications of migration

**Economic:**
- Migrant **remittances** benefit origin countries on an individual, local and national scale
- Immigrants increase the tax base in destination countries and act as consumers themselves
- Migrant remittances are lost from the host country's economy
- Migrants fill labour shortages and skills gaps

**Demographic:**
- Immigration can balance out an ageing population and improve the dependency ratio

**Environmental:**
- Increased pressure on resources in areas of high migrant populations, e.g. water supply in California

**Implications of migration**

**Political:**
- Governments implement migration policies to meet the social and economic needs of their country
- Political decisions around migration can be decisive and contested

**Health:**
- Healthcare systems in the host country may benefit from the influx of health workers, or may become strained with increased population

**Social:**
- There is a risk of human trafficking and forced labour
- Origin country can experience 'brain drain'
- Segregation and tensions between migrants and the host population can lead to discrimination and unrest

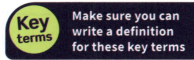 **Make sure you can write a definition for these key terms**

economic migrant    remittance

# RETRIEVAL

Learn the answers to the questions below, then cover the answers column with
a piece of paper and write down as many as you can. Check and repeat.

## Questions | Answers

| # | Questions | | Answers |
|---|-----------|---|---------|
| 1 | Name two factors affecting the total population of a country. | | Natural population change, migration |
| 2 | Define the term 'birth rate'. | Put paper here | The number of live births per 1000 of the population per year |
| 3 | What is replacement level? | | The number of children needed per female to maintain a population size |
| 4 | What type of countries are in the late expanding stage of demographic transition? | | Industrialising, EME countries, e.g. India, Mexico |
| 5 | What does a narrow base of a population pyramid imply? | Put paper here | A low birth rate / fertility rate |
| 6 | Why did China implement the one-child policy? | | As a result of severe famine and predicted rapid population growth |
| 7 | What is the demographic dividend? | | When falling fertility rates offer the opportunity for economic growth |
| 8 | State three causes of migration. | Put paper here | Three from: to earn money to send back to their country of origin / war, conflict, political instability / natural disasters or climate change-related events / retirement / economic migration to LDE and EME countries due to the restrictive nature of advanced economies' migration policies and shorter distance to travel |
| 9 | What is a refugee? | Put paper here | Someone who has been forced to flee their country due to conflict or persecution and has been granted asylum in another |
| 10 | What are the environmental implications of migration? | | Increased pressure on resources in areas of high migrant populations |

## Previous questions

Now go back and use these questions to check your knowledge of previous topics.

## Questions | Answers

| # | Questions | | Answers |
|---|-----------|---|---------|
| 1 | Which continents consume most and least food? | Put paper here | North America (most) and Africa (least) |
| 2 | What does structural deterioration of soil mean? | | Compaction of soil particles leads to less drainage and waterlogging |
| 3 | Which type of diseases are more prevalent in high-income countries? | | Non-communicable, e.g. cancer and dementia |
| 4 | What are disease vectors? | Put paper here | Carriers of an infectious disease |
| 5 | Give an example of a water-borne disease. | | One from: diarrhoea / typhoid / cholera / polio / bilharzia |

# 10 Population and the environment

## 10.9 Population ecology

### Population growth dynamics

| Underpopulation | Optimum population | Overpopulation |
|---|---|---|
| Too few people relative to the resources in the area to realise the area's economic potential. | The population size is in harmony with the available resources to produce the highest possible economic return per capita and an associated high standard of living. | Too many people in an area relative to the available resources to maintain an adequate standard of living. |

### The balance between population and resources

The relationship between population and resources varies between countries at different stages of economic development. The population size and per capita consumption of resources must be considered in order to understand this balance. These concepts help to assess this balance:

- **Carrying capacity** – the maximum population that a particular area can support sustainably. It varies in terms of lifestyle aspirations and consumption levels of an area.

- **Ecological footprint** – a measure of human impact on the environment, expressed as the number of hectares of land required to sustain a population's resource use.

### Population, resources and pollution model

Daniel Chiras' model illustrates the relationship between humans and the environment. It shows that when managing environmental degradation, population growth will be a determining factor.

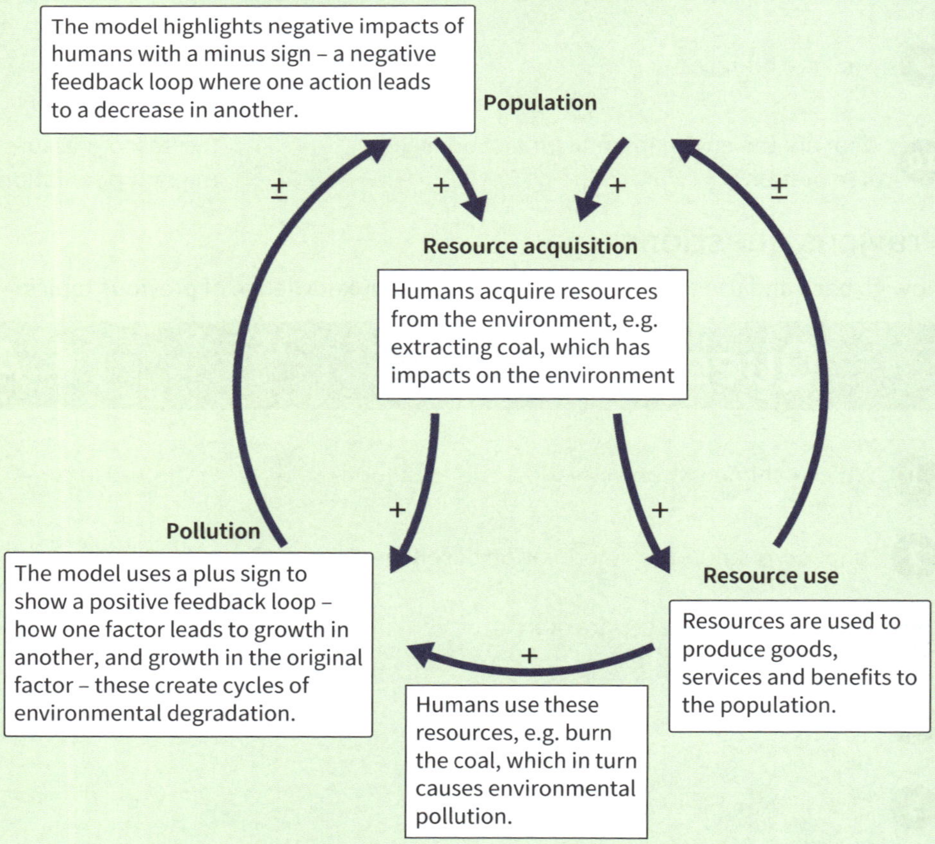

The model highlights negative impacts of humans with a minus sign – a negative feedback loop where one action leads to a decrease in another.

**Population**

**Resource acquisition**

Humans acquire resources from the environment, e.g. extracting coal, which has impacts on the environment

**Pollution**

The model uses a plus sign to show a positive feedback loop – how one factor leads to growth in another, and growth in the original factor – these create cycles of environmental degradation.

Humans use these resources, e.g. burn the coal, which in turn causes environmental pollution.

**Resource use**

Resources are used to produce goods, services and benefits to the population.

▶ **Figure 1** *Daniel Chiras' population, resources and pollution model*

## Contrasting perspectives on population growth

There are different theories relating to the relationship between population and resources.

| Pessimistic theory – Thomas Malthus (1798) | Optimistic theory – Ester Boserup (1965) |
|---|---|
| • Population grows geometrically or exponentially, but food supply grows arithmetically.<br>• Once global carrying capacity is reached, negative or positive checks would affect population growth.<br>• Negative checks: factors that decrease the birth rate.<br>• Positive checks: a reduction in population size due to famine, war and disease.<br>• Neo-Malthusian views widen the link to consider the limits of all natural resources and the use of artificial birth control as a negative population check.<br>• In 1972, the Club of Rome computer modelled the negative economic impacts that would occur if population growth, industrialisation and resource depletion were to continue. | • Population growth will instigate technological innovation in food production to meet the needs of the population.<br>• In 1981, Julian Simon agreed that human innovation will challenge resource depletion by finding alternatives, recycling resources and determining new strategies for their management. |

## Health impacts of global environmental change

Increasing health concerns caused by environmental change

**Climate change**
The change in temperatures and weather patterns around the world, recently due to human activity

**Ozone depletion**
The thinning of the ozone layer, caused by human-produced chlorofluorocarbons (CFCs), increasing the amount of UV radiation that reaches Earth

**Agricultural productivity and nutritional standards** Changing weather patterns, sea-level rise and extreme weather events will lead to crop destruction and disrupt agricultural systems leading to increasing health impacts of undernutrition and malnutrition

**Emergent and changing distribution of vector borne diseases** Changing weather patterns will result in more areas becoming habitable for vectors, such as mosquitos, leading to the increasing geographical spread of disease such as malaria and West Nile virus

**Thermal stress** Rising temperatures may result in increased cases of heat stroke, exhaustion, cramps and rashes, which is greater in the elderly or those with comorbidities

**Skin cancer** UV radiation is a carcinogen that increases the risk of someone developing two types of skin cancer

**Cataracts** Exposure to UV B radiation increases the risk of cortical cataracts developing in the eyes

**Key terms** Make sure you can write a definition for this key term

carrying capacity    ecological footprint

## 10 Population and the environment

### 10.10 Global population futures

## Projections for future growth

In 2022, the UN estimated that the population will rise to over 10 billion by 2059. There are different scenarios to make projections for future growth based on different fertility and migration levels.

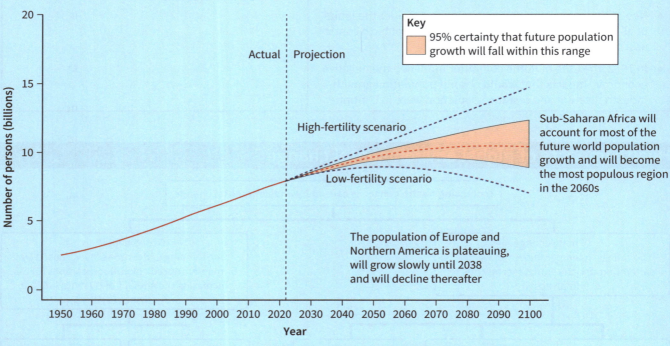

▲ **Figure 1** *Global population estimates, 1950–2022, and future high- and low-fertility scenarios, 2022–2100 (UN, 2022)*

## Critical appraisal of future population–environment relationships

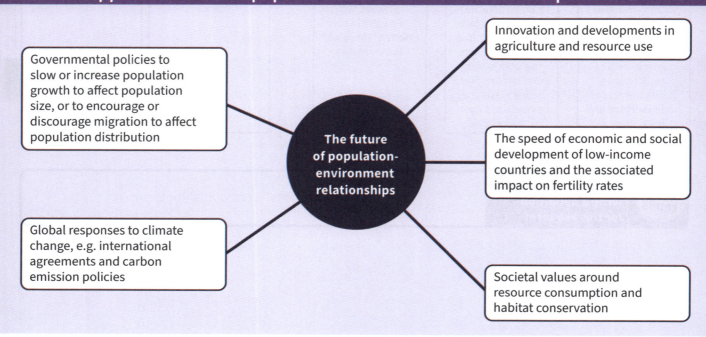

Learn the answers to the questions below, then cover the answers column with a piece of paper and write down as many as you can. Check and repeat.

## Questions | Answers

| # | Questions | Answers |
|---|-----------|---------|
| 1 | What is 'underpopulation'? | When there are too few people relative to the resources in the area to realise the area's economic potential |
| 2 | What is 'carrying capacity'? | The maximum population that a particular area can support sustainably |
| 3 | What is 'ecological footprint'? | A measure of human impact on the environment |
| 4 | What is a negative feedback loop? | Where one action leads to a decrease in another |
| 5 | What was Boserup's theory of the relationship between population and resources? | That population growth will instigate technological innovation in food production to meet the needs of the population |
| 6 | What are Malthusian positive checks on population? | A reduction in population size due to famine, war and disease |
| 7 | What are two ways that ozone depletion can affect health? | Increase in skin cancer and cataracts |
| 8 | What is thermal stress? | Where rising temperatures may result in increased cases of heat stroke, exhaustion, cramps and rashes |
| 9 | By 2059, what does the UN estimate the global population size to be? | Over 10 billion |
| 10 | How might governments affect the future population–environment relationship? | Through policies to slow or increase population growth to affect population size, or to encourage or discourage migration to affect population distribution |

*Put paper here*

## Previous questions

Now go back and use these questions to check your knowledge of previous topics.

## Questions | Answers

| # | Questions | Answers |
|---|-----------|---------|
| 1 | What is mixed farming? | Farming that involves both arable (crop growing) and pastoral (animals) farming |
| 2 | What does a narrow base of a population pyramid imply? | A low birth rate / fertility rate |
| 3 | What is replacement level? | The number of children needed per female to maintain a population size |
| 4 | What is the demographic dividend? | When falling fertility rates offer the opportunity for economic growth |
| 5 | What are the environmental implications of migration? | Increased pressure on resources in areas of high migrant populations |

*Put paper here*

## 10 Population and the environment

### 10.11 Case study: Population change

 **Case Study: China**

In 2022, China's population shrank for the first time in 60 years. In the next 25 years, the World Bank projects China's population will fall by 80 million people. In 2023, India overtook China as the most populated country in the world.

## Factors affecting change

- The legacy of the one-child policy lowering the fertility rate to 1.2 per woman, which is below replacement level.
- Changing societal attitudes towards marriage and family size.
- Women dissatisfied with traditional gender roles such as the expectation for them to take the burden of household work and childcare.
- Increased cost of living and education in China's cities.

▲ **Figure 1** China's fertility rate 1960–2021 (UN)

## Implications for Chinese society

- Ageing population increases the dependency ratio, from 37% in 2010 to 45% in 2021.
- Decline in economic growth due to a shrinking workforce – it no longer has a large, cheap labour force to drive industrialisation and growth.
- Risk to China's status as an emerging superpower threatening to compete with the USA.
- Risk to social security as there is less money to put into pensions and healthcare.
- Less environmental degradation due to a smaller population and a shift from the polluting industrial sector to cleaner technologies and service sector jobs.
- Some provinces have introduced paternity leave and expanded healthcare services and housing subsidies for couples with multiple children.
- President Xi Jinping announced a policy system to boost birth rates and reduce the costs of bringing up children.

# 10.12 Case study: Place and health

▲ **Figure 1** *Birmingham has experienced deindustrialisation and growth in the service sector*

Birmingham is the UK's second-largest city with nearly 1.2 million people. Experiencing rapid expansion during the industrial revolution, its recent history has been characterised by de-industrialisation and growth in the service sector.

- It is ethnically diverse and young – 40% of the population are under 25 years old.

- There are high levels of **deprivation** – ten per cent of the most deprived areas in England are in Birmingham, and 43% of the city's population live in these areas. Deprivation is highest in areas immediately surrounding the city centre.

- Deaths due to cardiovascular disease in Birmingham were 57.3 per 100 000 compared to 43.4 for England.

- Life expectancy has increased, but male life expectancy in Birmingham is 77.2, compared to 79.5 in England.

## Physical environment

- There is poor air quality around the CBD and inner city – air pollution causes 900 deaths a year. The council introduced an ultra-low emission zone in 2021 to reduce the number of polluting cars within the CBD.

- The proportion of mortality attributable to particulate matter air pollution is 5.8% in Birmingham. It is 5.1% for England.

## Socio-economic character

Social, economic and environmental factors contribute to health inequalities experienced in Birmingham:

- There are wide variations in life expectancy across the city.

- Infant mortality rate is 7.9 per 1000 in Birmingham compared with 3.9 per 1000 in England – 28.1% of children in Birmingham live in low-income families, compared with 17% in England.

- Prevalence of childhood obesity is higher in children from LDE neighbourhoods.

## 10 Population and the environment

Figure 2 shows the life expectancy at birth of males (blue) and females (green) at various places in Birmingham related to the rail map. There is a nine year difference for males (83–74) between affluent and less affluent areas and an eight year difference for females (87–79).

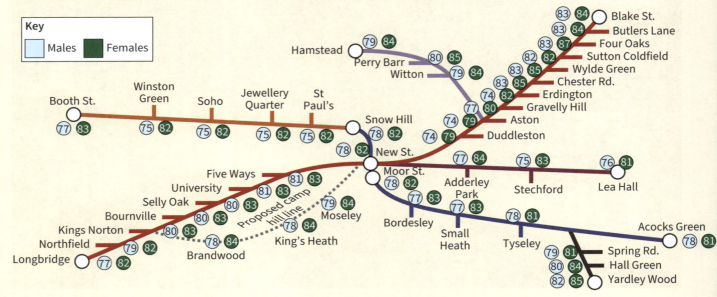

▲ *Figure 2* Life expectancy at birth by Birmingham railway stations (2017–19)

## Experience and attitudes of the population

- The percentage of adults regularly eating five portions of fruit and vegetables a day is 52.6% in Birmingham compared to 55.40% in England.
- The percentage of adults who are physically inactive is 28.9% in Birmingham compared to 22.9% in England.
- Deaths due to smoking are 274.8 per 100,000 in Birmingham, which is higher than 250.2 for England.

To improve health, the city has a Birmingham City Health and Wellbeing Strategy 2022–2030, which consists of five core themes:

1. 'Healthy and Affordable Food' – reduce the prevalence of obesity by 10%; increase the uptake of Healthy Start vouchers in eligible families.
2. 'Mental Wellness and Balance' – creating a mentally healthy city, e.g. reduce the prevalence of depression and anxiety in adults by 12%.
3. 'Active at Every Age and Ability' – increase the percentage of adults using active travel (walking or cycling) at least three days a week by at least 25%.
4. 'Contributing to a Green and Sustainable Future' – reduce the percentage of mortality attributable to air pollution to less than 4.5%.
5. 'Protect and Detect' – achieve national immunisation and screening targets.

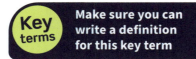 **Key terms** | Make sure you can write a definition for this key term | deprivation

# RETRIEVAL

Learn the answers to the questions below, then cover the answers column with a piece of paper and write down as many as you can. Check and repeat.

## Questions | Answers

| | Questions | | Answers |
|---|---|---|---|
| 1 | What is China's fertility rate? | | 1.2 children per woman |
| 2 | In the next 25 years, by how many people does the UN predict China's population will fall? | | 80 million people |
| 3 | What economic factor has led to the population decline? | | The increased cost of living and education in China's cities |
| 4 | What has happened to China's dependency ratio? | Put paper here | Increased, from 37% in 2010 to 45% in 2021 |
| 5 | What might the environmental impacts of China's population change be? | | Less environmental degradation due to a smaller population and a shift from the polluting industrial sector to cleaner technologies and service sector jobs |
| 6 | Where in Birmingham is deprivation highest? | | Around the city centre |
| 7 | How many deaths in Birmingham are attributable to air pollution? | | 900 a year |
| 8 | What is male life expectancy in Birmingham? | | 77.2 years |
| 9 | How does infant mortality in Birmingham compare with national data? | | Infant mortality rate is 7.9 per 1000 in Birmingham compared with 3.9 per 1000 in England |
| 10 | How is Birmingham trying to address health challenges related to food? | | Reduce the prevalence of obesity by 10%, increase the uptake of Healthy Start vouchers in eligible families |

## Previous questions

Now go back and use these questions to check your knowledge of previous topics.

## Questions | Answers

| | Questions | | Answers |
|---|---|---|---|
| 1 | Where is podsol soil found? | | Under coniferous or boreal forests, heathland in southern Australia and British moorlands |
| 2 | Name two climatic factors that affect the incidence of disease. | | Two from: temperature / rainfall / average daily sunshine hours |
| 3 | What is a refugee? | Put paper here | Someone who has been forced to flee their country due to conflict or persecution and has been granted asylum in another |
| 4 | What is 'carrying capacity'? | | The maximum population that a particular area can support sustainably |
| 5 | What are Malthusian positive checks on population? | | A reduction in population size due to famine, war and disease |

## Exam-style questions

1  Outline the relationship between climate and incidence of transmission vectors of disease. **[4 marks]**

2  Outline global and regional patterns of food consumption. **[4 marks]**

3  Outline how soil erosion can create problems for agriculture. **[4 marks]**

4  Outline the concept of the demographic dividend. **[4 marks]**

5  Outline the socio-economic causes of migration. **[4 marks]**

6  **Figure 1a** shows the share of the population with access to drinking water in different areas of the world in 2020.

**Figure 1b** shows the estimated annual number of deaths attributed to unsafe water, per 100,000 people, in 2019.

Analyse the data shown in **Figures 1a** and **1b**. **[6 marks]**

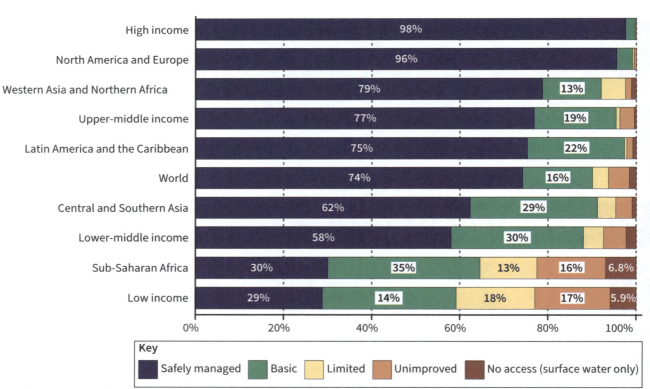

▲ *Figure 1a The share of the population with access to drinking water in different areas of the world, 2020*

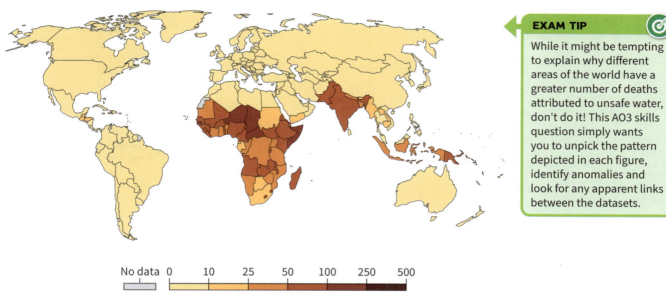

**EXAM TIP**

While it might be tempting to explain why different areas of the world have a greater number of deaths attributed to unsafe water, don't do it! This AO3 skills question simply wants you to unpick the pattern depicted in each figure, identify anomalies and look for any apparent links between the datasets.

| No data | 0 | 10 | 25 | 50 | 100 | 250 | 500 |

▲ **Figure 1b** *The estimated annual number of deaths attributed to unsafe water, per 100,000 people, 2019*

7    **Figure 2a** shows the leading causes of death in low-income countries in 2000 and 2019.

**Figure 2b** shows the leading causes of death in high-income countries in 2000 and 2019.

Evaluate the usefulness of these sources in depicting causes of death in low-income and high-income countries.          **[6 marks]**

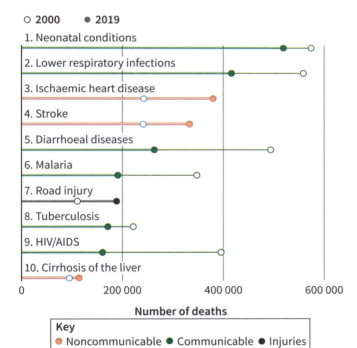

○ **2000**    ● **2019**

1. Neonatal conditions
2. Lower respiratory infections
3. Ischaemic heart disease
4. Stroke
5. Diarrhoeal diseases
6. Malaria
7. Road injury
8. Tuberculosis
9. HIV/AIDS
10. Cirrhosis of the liver

**Number of deaths**

**Key**
● Noncommunicable ● Communicable ● Injuries

◄ **Figure 2a** *The leading causes of death in low-income countries in 2000 and 2019 according to the WHO*

# Exam-style questions

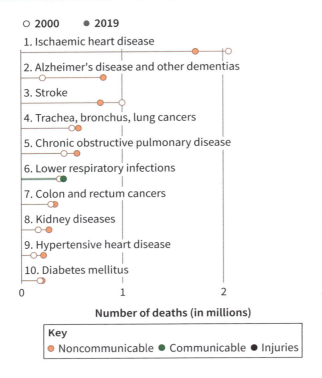

○ 2000  ● 2019

1. Ischaemic heart disease
2. Alzheimer's disease and other dementias
3. Stroke
4. Trachea, bronchus, lung cancers
5. Chronic obstructive pulmonary disease
6. Lower respiratory infections
7. Colon and rectum cancers
8. Kidney diseases
9. Hypertensive heart disease
10. Diabetes mellitus

Number of deaths (in millions)

**Key**
● Noncommunicable ● Communicable ● Injuries

◄ **Figure 2b** *The leading causes of death in high-income countries in 2000 and 2019*

**EXAM TIP**

Use evidence, e.g. numbers and observations, from the figures to support your analysis. Try to make connections between both of the figures in your answer.

**8** For a country/society you have studied, assess the relative impact of the socio-economic and environmental implications of population change. **[9 marks]**

**9** Using **Figure 3** and your own knowledge, assess the importance of environmental change on health. **[9 marks]**

**EXAM TIP**

These 9-mark questions have 4 marks for AO1 (knowledge and understanding of the concepts in the figure) and 5 marks for AO2 (application of knowledge and evaluation).

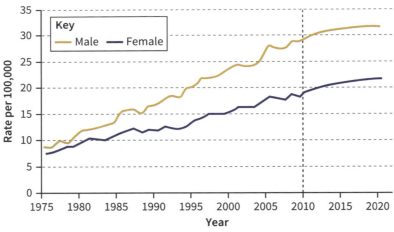

▲ **Figure 3** *Age-adjusted melanoma incidence rates (actual and projected) by sex, National Cancer Institute 2010*

**10** 'Global governance of health depends on the work of NGOs rather than international agencies.'

To what extent to you agree with this statement? **[9 marks]**

**11** 'Current population growth dynamics mean that the future population-environment relationship is hard to predict.'

How far do you agree with this statement? **[9 marks]**

**12** For a biologically transmitted disease you have studied, assess the relationship between environmental variables and the incidence of the disease. **[20 marks]**

**EXAM TIP**

20-mark questions have 10 marks for AO1 (knowledge and understanding of the concepts in the question) and 10 marks for AO2 (application of knowledge to provide a detailed analysis and evaluation in response to the question).

**13** 'The key issues facing agricultural productivity are soil problems.'

To what extent do you agree with this statement? **[20 marks]**

## 11 Resource security

### 11.1 Resource development

## Key concepts for resource development

| Concept | Explanation |
|---------|-------------|
| Natural resource | Any material, substance or organism found within the environment that humans perceive as useful for economic production or consumption. |
| **Stock resource** | A non-renewable, finite and therefore exhaustible resource. They are created at rates considerably slower than their use. |
| **Flow resource** | A renewable, ongoing resource that is either immediately available (e.g. geothermal power) or created at comparable rates to their consumption (e.g. trees for fuelwood). |
| **Measured reserves** | Measured reserves are when geological conditions including grade of deposit can be confirmed to allow detailed planning for extraction. |
| **Indicated reserves** | An indicated reserve is where the quality and quantity of a resource can be estimated with a sufficient level of confidence to allow further evaluation of the economic viability, with conversion to a possible reserve. |
| **Inferred resources** | An inferred resource is where quality and quantity can only be assessed through limited information – it suggests there might be a resource but it is not certain. |
| Possible resources | A possible resource is when it can be reasonably expected that the majority of inferred resources could be upgraded to indicated mineral resources with continued exploration. |
| **Resource exploration** | Resource exploration is the initial stage in the resource development process, involving the search for valuable natural resources through geological surveys, aerial photography, mapping and drilling. The goal is to identify the presence and extent of economically viable potential resources. |
| **Resource exploitation** | The phase following the discovery of a resource where it is extracted for economic gain. This involves the development of mines, drilling of wells, or other extraction methods, as well as the infrastructure to support such activities like roads, processing facilities and transportation networks. |
| Resource development | Encompasses the full process of bringing a resource from discovery to market. This includes not only exploration and exploitation but also the assessment of feasibility, securing funding and permissions, building necessary infrastructure and actual production. |
| **Resource frontier** | An area on the periphery of a country, where resources are produced for the first time. They are frequently found in locations that are difficult to exploit. |
| **Resource peak** | The phase of maximum production of a resource before depletion occurs. |
| **Sustainable resource development** | The practice of managing natural resources to fulfil current economic demands without jeopardising future prospects. It integrates extraction with conservation, ensuring renewability, efficiency and minimal ecological impact. This concept calls for the equitable distribution of resources, reduced pollution and the restoration of exploited ecosystems. |
| **Environmental Impact Assessment (EIA)** | The assessment of the environmental consequences, positive and negative, usually completed for a planned resource development project. The EIA helps in the decision over whether the project should go ahead, so as not to cause long-term negative environmental consequences. |

# EIAs and resource development projects

An EIA provides the basis for an overall decision on whether a project should go ahead, considers the scale of impacts and mitigation strategies and considers what will happen after the project.

| ENVIRONMENTAL INVENTORY | | | 1 | 2 | 3 | 4 | 5 | 6 | 7 | 8 | 9 | 10 | 11 | 12 | 13 | 14 | 15 | 16 | 17 | 18 | 19 | 20 |
|---|---|---|---|---|---|---|---|---|---|---|---|---|---|---|---|---|---|---|---|---|---|---|
| | | **ACTIVITIES THAT MIGHT HAVE IMPACT** | | | | | | | | | | | | | | | | | | | | |
| | | Landscape | -3/3 | | | | | | | | | | | | | | | | | | | |
| | | Flora | -4/2 | | | | | | -2/2 | | | | | -3/1 | | | | | | | | |
| | POPU | Demographics | -1/8 | -2/8 | -6/8 | -4/1 | -1/1 | | | | | | | -1/1 | -3/2 | | | | -5/5 | | | |
| | POPU | Economic activities | -1/6 | -1/6 | -1/6 | -1/4 | -1/4 | | | | | | | -1/1 | | | | | | | | |
| | FAUNA | Sea Fauna | | | | | | | -8/7 | -7/7 | -4/1 | -3/2 | -7/7 | | | -2/1 | -7/7 | | | | | |
| | FAUNA | Birds | -2/1 | | | | | | | | | | | | | | | | | -2/2 | | |

| | | | | | | |
|---|---|---|---|---|---|---|
| 1 | Energy source | 8 | Artificial reef effect / contamination by heavy metal and salts influences | 14 | Turtle nesting alteration |
| 2 | Freshwater production | | | 15 | Brine discharge |
| 3 | Production of cooling systems | 9 | Organism drag and compression influences | 16 | Cause sociocultural impacts |
| 4 | Mineral production | 10 | Redistribution of ocean water bodies influences | 17 | Significant public controversy |
| 5 | Lithium production | | | 18 | Migration routes interruption |
| 6 | CO$_2$ emissions | 11 | Impact by organic antifouling chemicals influences | 19 | Waste |
| 7 | Dragging nutrients to the surface influences | 12 | Noise | 20 | Sanitary discharges from the station |
| | | 13 | Lighting | | |

▲ **Figure 1** *A Leopold Matrix records the likely impacts of resource exploitation using a series of scores for various activities*

# Hubbert's Curve, 1956

- Hubbert's calculations were accurate for when the peak was achieved, but the predicted decline, although initially accurate, was not.

- This was because of the development of new techniques that allowed oil and gas to be extracted from unconventional sources.

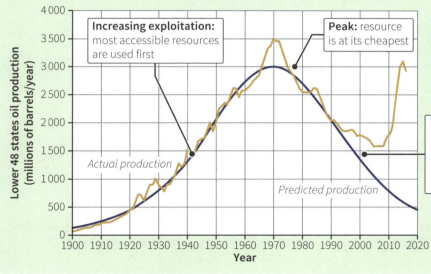

**Increasing exploitation:** most accessible resources are used first

**Peak:** resource is at its cheapest

*Actual production*

*Predicted production*

**Depletion:** prices rise with depletion as extraction and processing is more challenging

◀ **Figure 2** *Hubbert's predictions about the resource peaks of US crude oil and natural gas productions*

**Key terms**

**Make sure you can write a definition for these key terms**

EIA   flow resource   indicated reserve   inferred resource
measured reserve   resource exploitation   resource exploration
resource frontier   resource peak   stock resource
sustainable resource development

Learn the answers to the questions below, then cover the answers column with a piece of paper and write down as many as you can. Check and repeat.

## Questions | Answers

| # | Question | Answer |
|---|----------|--------|
| 1 | Define a stock resource. | Stock resources are non-renewable – they are finite and therefore exhaustible |
| 2 | Define a flow resource. | Flow resources are renewable – they are ongoing |
| 3 | Define an indicated reserve. | An indicated reserve is where the quality and quantity of a resource can be estimated with a sufficient level of confidence to allow further evaluation of the economic viability |
| 4 | What usually follows identification of an indicated reserve? | A feasibility study to look at economic viability |
| 5 | Define a possible reserve. | When it can be reasonably expected that the majority of inferred resources could be upgraded to indicated mineral resources with continued exploration |
| 6 | What is the initial stage in the resource development process? | Resource exploration |
| 7 | What phase in the resource development process follows the discovery of a resource? | Resource exploitation |
| 8 | Define resource development. | Resource development encompasses the full process of bringing a resource from discovery to market |
| 9 | What is a resource frontier? | An area on the periphery of a country, where resources are produced for the first time |
| 10 | What is the resource peak? | The phase of maximum production of a resource before depletion occurs |
| 11 | M King Hubbert proposed that most finite resources will follow Hubbert's Curve, which is what shape of curve? | A bell-shaped curve |
| 12 | Why do resource prices often rise after the resource peak (generally where they are cheapest)? | Because of more challenging extraction and processing as the most accessible resources have been used first |
| 13 | What does EIA stand for? | Environmental Impact Assessment |
| 14 | EIA assessments are recorded using which matrix? | The Leopold Matrix |
| 15 | Hubbert's calculations for the resource peaks of US crude and gas were accurate – why did his predictions for decline prove inaccurate? | Because of the development of new techniques that allowed oil and gas to be extracted from unconventional sources |

*Put paper here*

## 11 Resource security

### 11.2 Natural resource issues

## Global patterns of energy

### Energy production

- Fossil fuels: the majority of global energy production is still dominated by stock energy resources such as coal, oil and natural gas. For example, the USA, Russia and Saudi Arabia are key crude oil producers.

- Renewable energy: there's a growing shift towards flow resources such as solar, wind, hydroelectric, and geothermal power. China leads in total renewable energy production, while European countries like Ireland, Germany, UK, and Spain excel in wind power.

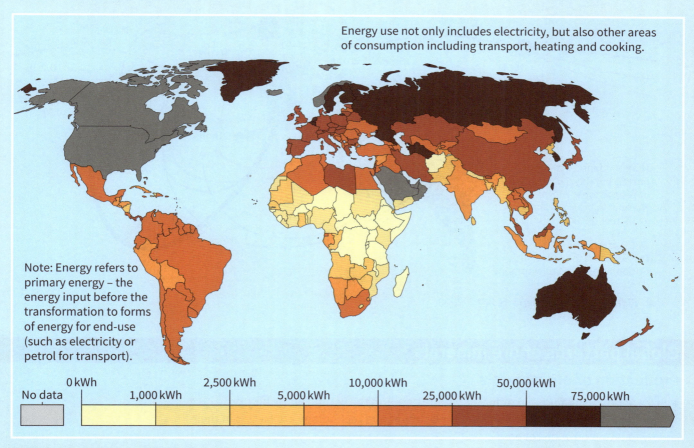

Energy use not only includes electricity, but also other areas of consumption including transport, heating and cooking.

Note: Energy refers to primary energy – the energy input before the transformation to forms of energy for end-use (such as electricity or petrol for transport).

| No data | 0 kWh | 1,000 kWh | 2,500 kWh | 5,000 kWh | 10,000 kWh | 25,000 kWh | 50,000 kWh | 75,000 kWh |

▲ *Figure 1 Energy use per capita 2021*

# 11 Resource security

## 11.2 Natural resource issues

### Energy trade and movement

- The unequal distribution of energy sources means there is a global trade in energy.
- Flow resources cannot be physically transported, but the stock resources of fossil fuels can be, with higher costs for heavy, bulky coal than for oil or gas.
- Trade and movement of energy is affected by **geopolitics**.

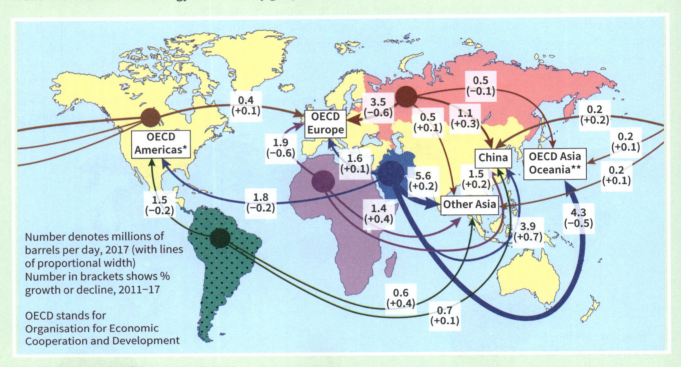

Number denotes millions of barrels per day, 2017 (with lines of proportional width)
Number in brackets shows % growth or decline, 2011–17

OECD stands for Organisation for Economic Cooperation and Development

▲ **Figure 2** *Crude oil exports 2017–2020*

## Global patterns of ore minerals

### Production of ore minerals

- Ore deposits are found in different patterns around the world and result from geological processes. Development, investment and technology are factors in their exploitation once discovered.
- Iron ore is the most important in terms of mass of production, with 2.6 billion metric tons (MT) produced per year worldwide, followed by potash (40 million MT) and copper (22 million MT) (2022).

| Iron ore | Potash | Copper | Uranium |
|----------|--------|--------|---------|
| 1 Australia | 1 Canada | 1 Chile | 1 Kazakhstan |
| 2 Brazil | 2 Russia | 2 Peru | 2 Canada |
| 3 China | 3 Belarus | 3 China | 3 Namibia |
| 4 India | 4 China | 4 DRC | 4 Australia |
| 5 Russia | 5 Germany | 5 USA | 5 Uzbekistan |

# Consumption and movement of ore minerals

- The factors affecting consumption of mineral ores are primarily industrial and manufacturing diversity, and financial strength. The largest importers tend to be countries with extensive agricultural and industrial bases.

- China is the world's biggest importer of iron ore, as well as being its third largest producer. It is the world's leading stainless steel producer, but has insufficient iron ore to meet its domestic demands.

- Consumption often takes place in countries without access to their own deposits of the ore minerals, especially in advanced economies with smaller territories, for example in the EU and East Asia. This creates geopolitical issues as trade in the ore minerals has to be negotiated.

- The most significant change in consumption patterns has been in the growing consumption of ore minerals in Asia. New or increased demands have also developed. For example, copper is a key component in electricity-related technologies, including electric vehicles (EVs). There are concerns that copper's resource peak has already been reached.

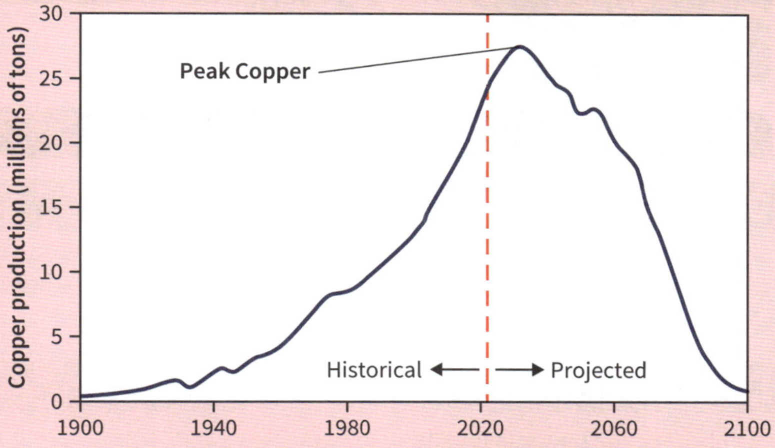

▲ **Figure 3** Copper production, historic and projected

# Global patterns of water availability and demand

Water availability refers to the quantity of water that is physically accessible for use, and the quality of that water – whether it is safe to use.

**Factors affecting water availability**

Rainfall:
High annual rainfall increases the quantity of water available.

Temperature:
High temperatures can cause high evaporation rates and can encourage bacterial and algal growth, reducing water quality.

Climate change:
Reduces water availability through warmer temperatures and reduction in rainfall. The disappearance of glaciers is likely to cause water stress in some areas.

Geology:
Permeable rocks encourage groundwater storage (e.g. chalk); water is less likely to be contaminated underground. Many large cities depend on aquifers for their water supply.

Water demand:
Increased consumption and pollution of water by humans is reducing water availability; many aquifers are being depleted faster than they are being recharged. Global demand may outstrip supply by 2030.

## 11 Resource security

### 11.2 Natural resource issues

## Geopolitics of natural energy resources

Geopolitics is the study of the ways in which political decisions and processes affect the use of resources (and the space through which they may be moved).

### Energy

In February 2022, Russia invaded Ukraine, creating an energy crisis as sanctions were imposed against Russia.

- A decision was made to reduce Europe's high dependence on Russian oil and gas (increasing energy security) as oil and gas prices rocketed.

- This meant new trade routes, new infrastructure (new liquid gas terminals), a push towards renewables, and a renewed use of high-emission coal.

- The share of Russian oil imported fell from 31% of EU's imports in January 2022 to 3% in March 2023.

- Oil production increased by 5% globally, and there was a 65% increase in exports to the EU among alternative suppliers. The USA became the EU's main supplier, followed by Norway and Saudi Arabia. The EU abandoned previous concerns about environmental consequences.

- At first, Russia responded by increasing supplies to India and China, which obtained 50% discounts. It also developed new customers including Brazil and Ghana.

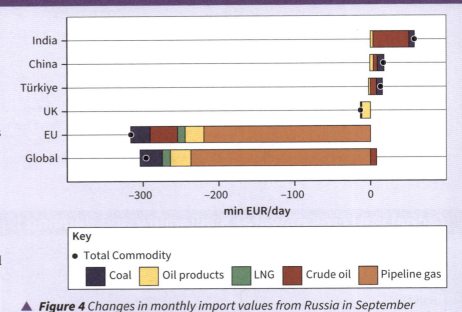

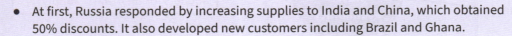

▲ **Figure 4** *Changes in monthly import values from Russia in September 2022 compared to February 2022*

### Ore minerals

TNCs are often in control of ore mineral production, leasing land from national governments. Governments that earn a lot from mineral exports are then highly dependent on TNCs, giving TNCs influence over governments.

When countries control supply of ore minerals that are in high demand, this has geopolitical implications. For example, 70% of the world's known rare earth metals (key materials for much modern tech) are in China's Bayan Obo mine. Any restriction in exports can lead to panic buying outside China, and intensified exploration efforts.

### Water resource distribution

The geopolitics of water resource distributions often relate to situations where river basins cover more than one country or region of a country. Decisions about who gets access to how much water can lead to geopolitical disputes and conflicts.

An example is the Indus River, which is crucial for Pakistan's farming. The partition of India in 1946 created conflict between the newly formed states over the waters of the Indus basin. In 1960, international mediation was required to divide the Indus and its tributaries between the two countries.

 **Key terms** Make sure you can write a definition for these key terms

*geopolitics*

# ⇄ RETRIEVAL

Learn the answers to the questions below, then cover the answers column with a piece of paper and write down as many as you can. Check and repeat.

## Questions | Answers

| | Questions | Answers |
|---|---|---|
| 1 | Name one of the top three producers of crude oil (a country). | One from: USA / Russia / Saudi Arabia |
| 2 | Which country leads renewable energy production? | China |
| 3 | Name two events that both produced a dip in global energy consumption this century (so far). | Global Financial Crisis and COVID-19 pandemic |
| 4 | Which country leads production of iron ore? | Australia |
| 5 | What are the two factors primarily affecting consumption of mineral ores? | Industrial and manufacturing diversity and financial strength; extensive agricultural and industrial bases |
| 6 | There are concerns about which ore mineral – a key ore mineral in all electricity-related tech – reaching its resource peak? | Copper |
| 7 | What two factors does the concept of water availability refer to? | The quantity of water that is physically accessible for use, and the quality of that water – whether it is safe to use |
| 8 | Define geopolitics. | Geopolitics is the study of the ways in which political decisions and processes affect the use of resources |
| 9 | From 31% of the EU's imports the year before, what did the share of Russian oil imported into the EU fall to in March 2023? | 3% |
| 10 | Which countries became Russia's most important customers after its invasion of Ukraine? | India and China |

*Put paper here*

## Previous questions

Now go back and use these questions to check your knowledge of previous topics.

## Questions | Answers

| | Questions | Answers |
|---|---|---|
| 1 | Define a stock resource. | Stock resources are non-renewable – they are finite and therefore exhaustible |
| 2 | Define a flow resource. | Flow resources are renewable – they are ongoing |
| 3 | What phase in the resource development process follows the discovery of a resource? | Resource exploitation |
| 4 | What is the resource peak? | The phase of maximum production of a resource before depletion occurs |
| 5 | EIA assessments are recorded using which matrix? | The Leopold Matrix |

*Put paper here*

# 11 Resource security

## 11.3 Water security

## Water sources and demand

Freshwater makes up just 2.5% of water on Earth, and of that, only around a third is economically accessible – surface and groundwater.

The main components of demand for water are agricultural, industrial and domestic. On average, around 70% of global freshwater withdrawals are to meet agricultural demand, 19% go to industry and 11% go to meet domestic (including municipal) demand.

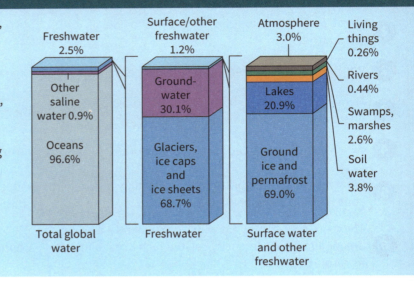

▶ **Figure 1** *Where is the Earth's water?*

## Water stress

**Water stress** occurs when demand for safe, usable water in a given area exceeds the supply.

Even if there are significant natural reasons for **water scarcity** (low rainfall, high temperatures), it is usually human factors that tip regions into water stress. These human factors make up **economic water scarcity** – inadequate water infrastructure. Insufficient investment to bring clean water to people, or infrastructure being cut off, for example by conflict, are the principal causes of water stress. Climate change is also a factor in water stress.

## Water supply and physical geography

The volume and quality of water supply is strongly related to key aspects of physical geography – climate, geology and drainage.

| Climate | Geology | Drainage |
|---|---|---|
| • Precipitation | • Rock type | • River systems |
| • Rainfall intensity (and infiltration) | • Permeability and porosity | • Watershed / catchment areas |
| • Snowfall and glacial melt | • Synclines and artesian basins – aquifer potential | • Drainage basin types |
| • Seasonal variations in precipitation (e.g. monsoon rains) | • Soil composition | • Groundwater flow |
| • Temperature | • Water retention ability | • Aquifer recharge |
| • Evapotranspiration rates | • Filtration and purification | • Springs, oases |
| • Low and high pressure systems | • Tectonic processes | • Surface runoff |
| • Droughts | • Uplands and rain shadows | • Urban drainage |
| • Floods | • Basin formation | • Agricultural runoff |
| | | • Dams, barrages and reservoirs |
| | | • Irrigation systems |

# Strategies to increase water supply

| Strategy | Advantages | Disadvantages |
| --- | --- | --- |
| **Catchment**<br>e.g. The Murray-Darling Basin Plan in Australia, which manages water resources sustainably across four states. | • Enhances water quality by managing pollutants and land use.<br>• Can increase water quantity through improved land and waterway management practices.<br>• Supports biodiversity and ecosystem services. | • Complex to manage due to the range of stakeholders involved.<br>• May require significant changes to agricultural practices.<br>• Potential for conflict between environmental objectives and existing land uses. |
| **Diversion**<br>e.g. The California Aqueduct in the USA diverts water from mountains in northern California to the drier south: watering 5.7 million acres of farmland. | • Transports water from areas of relative abundance to areas of scarcity.<br>• Supports agriculture and enables agricultural development.<br>• Enables hydropower generation, providing a renewable energy source. | • Large-scale schemes disrupt ecosystems at both source and destination.<br>• Can lead to legal and geopolitical disputes.<br>• High construction, maintenance and operation costs. |
| Storage<br>e.g. The Three Gorges Dam in China, which has created a massive reservoir for water supply and power generation. | • Ensures a steady supply of water during dry/drier periods.<br>• Can be used for flood control and irrigation.<br>• Large dams are used for the generation of HEP – clean, cheap electricity. | • Large dams lead to displacement of people and loss of land.<br>• High evaporation losses in some climates.<br>• Siltation is a major problem that limits the useful life of water storage systems. |
| **Water transfer**<br>e.g. The South–North Water Transfer Project in China redirects water from the Yangtze River to the drier northern regions. | • Alleviates water shortages in water-scarce areas.<br>• Supports economic development in recipient regions.<br>• Can be integrated with other water management strategies. | • High economic costs and significant energy requirements.<br>• Risk of spreading pollution and invasive species.<br>• The recipient area may simply continue to increase consumption. |
| **Desalination**<br>e.g. The Sorek Desalination Plant in Israel is one of the largest desalination facilities in the world. | • Provides a reliable, climate-independent supply of water.<br>• Enhances water security for coastal cities or regions with saline water sources.<br>• Technological advancements are reducing costs and energy consumption. | • High energy requirements lead to significant costs and environmental impacts.<br>• Brine discharge can harm marine ecosystems.<br>• Requires access to seas or saline groundwater, limiting inland applicability. |

# 11 Resource security

## 11.3 Water security

### Environmental impacts of the South–North Water Transfer Project

**Background**

The South–North Water Transfer Project (SNWTP), due for completion in 2050 at a cost of US$62 billion, aims to channel 45 billion m³ of fresh water annually from the Yangtze River in southern China to the arid and semi-arid regions of northern China, including Beijing and Tianjin where groundwater is seriously depleted and facing water stress.

The project consists of the completed eastern (2013) and central (2014) routes and the planned western route. By 2021, the central route was bringing 9.5 billion m³ of water annually to 191 rural and urban areas, benefiting about 79 million people.

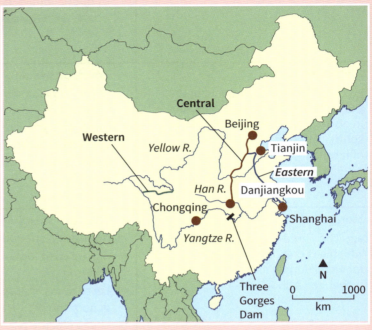

▲ **Figure 2** China's South–North Water Transfer Project

### Ecosystem disruption

The project involves massive infrastructure construction, including canals, tunnels and aqueducts, across diverse ecological zones. This construction has led to significant ecological changes, including habitat loss and alterations in local hydrology. Wetlands are particularly at risk. There are plans for industrialisation to take place along the route of the major canals, further increasing the risk of pollution.

### Water quality and sediment transport

Water diversion affects the quality and quantity of water downstream, impacting on agricultural lands, fisheries and natural ecosystems. In 2013, fish farmers in Shandong complained that the polluted Yangtze River water entering a lake was killing their fish. Algal blooms have become more common. The reduction of river flow is decreasing fish biodiversity (though species that prefer slow-flowing river water are increasing). Sediment transport dynamics are also altered, which can affect river morphology and ecosystems both upstream and downstream of the diversion points.

> **REVISION TIP**
>
> You need to know the environmental impacts of a major water supply scheme. Revise the scheme you have studied.

### Biodiversity and endangered species

The SNWTP has the potential to impact local biodiversity, especially aquatic species that depend on stable river ecosystems. There is a concern for species in the Yangtze River basin, which is already stressed from other large-scale projects like the Three Gorges Dam. There is also the risk of diseases being spread from the south to the north, including snail-borne schistosomiasis.

### Greenhouse gas emissions

The SNWTP's construction and operation involve significant energy use and associated greenhouse gas emissions. The impact of such emissions is an essential factor in evaluating the project's environmental footprint.

### Water waste

The scheme has been criticised for encouraging continued wasteful use of water in the northern cities, instead of the overuse of water there being a stimulus for using water more efficiently.

## Strategies to manage water consumption

- Technology can be used to reduce consumption. In the UK, about a third of water use is in toilet flushing, so dual-flush toilets, which cut the amount of water used for flushing (reducing demand) can be effective at the domestic scale.

- Water meters can also be fitted domestically, which encourage householders to use water more carefully because only the water used is paid for.

- Leakage from water pipes in the UK is estimated at 3 billion litres every day. This is related to the UK's ageing infrastructure. Water companies have committed to reducing leakage by 50% by 2050, at a cost of around £96 billion.

- Governments use a mix of rewards and punishments to encourage water saving. For example, in Bengaluru water supply is cut off if rainwater-collection tanks are not installed. In Australia, rebates are offered for installing water-saving devices.

- Agriculture dominates water use, so strategies to reduce the amount of water that is lost from fields due to evaporation (e.g. drip-feed irrigation, using mulches) or through runoff (reducing ploughing, using bunds or zai pits) are crucial.

## Sustainability issues

Sustainable water management requires a shift in thinking about water among those societies that have always previously taken it for granted.

### Conservation

- Strategies to conserve water include the implementation of water-efficient domestic appliances, rainwater-harvesting systems, sustainable farming practices, and the development of drought-resistant crop varieties.

- Effectiveness can be impacted by negative public views because of increased costs, regulation and inconvenience. Where farmers are subsidised for water, for example, there is little incentive to introduce conservation measures.

### Virtual water trading

- **Virtual water** is the volume of fresh water used to produce a product, measured at the place where the product was actually made.

- A trade in virtual water therefore allows water-scarce regions to import products that require a lot of water to produce, so conserving water.

- There are impacts on exporting regions; for example, avocados require around 70 litres of water to grow. High demand for avocados in countries like the UK (water surplus) is causing severe water shortages in countries like Mexico and Chile.

### Recycling

Water recycling involves treating wastewater to remove solids and impurities, making it suitable for reuse. Recycled water reduces the demand on freshwater supplies and is a reliable source, unlike rainwater.

- **Greywater** is relatively clean water directly reused from baths, sinks, washing machines and other kitchen appliances. It can be used for irrigation, flushing toilets and watering gardens.

- **Blackwater** is wastewater from toilets and waste disposal units. It can be filtered and decontaminated to be reused, even as drinking water, though this is expensive to achieve.

- People can have negative reactions to the concept of recycled water, especially in domestic use. The costs of maintaining the systems for recycling water means that integrated systems to recycle water tend to be found in hotels and larger residential complexes.

### Groundwater management

- Over-extraction of groundwater can lead to a lower water table, land subsidence and reduced water quality.

- Pollution can contaminate groundwater, making it unsafe or unusable.

- Sustainable groundwater management includes aquifer recharge (redirected surface water to replenish aquifers) and regulation to ensure water withdrawal does not exceed the natural recharge rate.

## 11 Resource security

### 11.3 Water security

## Water conflicts

As water becomes scarce or water supplies become threatened, conflicts over shared water resources (e.g. river basins, aquifers) can occur at local, national and international scales.

Different factors can contribute to conflicts over shared water resources, for example:

- drought, increasing frequency of drought (climate change, El Niño events)
- population increase
- pollution and contamination events
- large water transfer or storage projects (often a factor in international conflicts)
- attempts at mitigation to resolve conflicts, for example regulation.

### Local scale:
### the Cochabamba Water War, Bolivia

- Background: in the late 1990s, the Bolivian government privatised the water supply in Cochabamba, Bolivia's third-largest city.
- Conflict: drastic increases in water prices sparked widespread protests and civil unrest. Legal measures even restricted the collection of rainwater.
- Resolution: the protests led to a reversal of the privatisation policy, and the water utility was returned to public control.

### National scale:
### the Kaveri Water Dispute, India

- Background: the Kaveri River flows through the Indian states of Karnataka and Tamil Nadu.
- Conflict: both states have agricultural lands that depend on the Kaveri river. The dispute goes back to 1892 over the sharing of water resources and has often led to violence.
- Resolution: the Supreme Court of India has issued multiple orders to resolve the dispute, including setting up the Kaveri Water Management Authority to ensure fair distribution. For example, in August 2023 it directed a release of water by Karnataka for 15 days.

### International scale:
### the Tigris–Euphrates Basin

- Background: the Tigris and Euphrates rivers are crucial for Iraq, Syria and Türkiye. By the end of the century, the flows of the Euphrates and Tigris are forecast to decrease by 30 and 60% respectively.
- Conflict: Türkiye's south-eastern Anatolia Project (GAP) involves building a series of dams that reduces the flow of water to Syria and Iraq, impacting agriculture and leading to water shortages.
- Resolution: the conflict is ongoing with no comprehensive agreement in place. There have been temporary agreements, but tensions persist, especially during periods of drought.

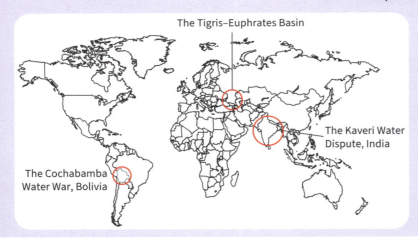

▲ **Figure 3** *The location of these water conflicts*

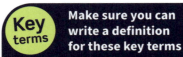

Make sure you can write a definition for these key terms

blackwater   catchment   desalination   diversion
economic water scarcity   greywater   virtual water   water scarcity
water stress   water transfer

# RETRIEVAL

Learn the answers to the questions below, then cover the answers column with a piece of paper and write down as many as you can. Check and repeat.

## Questions | Answers

| # | Question | Answer |
|---|----------|--------|
| 1 | What percentage of global freshwater withdrawals are to meet agricultural demand? | 70% |
| 2 | What is the definition of water stress? | When demand for safe, usable water in a given area exceeds the supply |
| 3 | The volume and quality of water supply is strongly related to which three key aspects of physical geography? | Climate, geology and drainage |
| 4 | A water transfer scheme is one strategy for increasing water supply: name two others. | Two from: catchment (management) / diversion / storage / desalination |
| 5 | Suggest one way in which farmers can reduce water loss from evaporation. | One from: drip-feed irrigation / use of mulches |
| 6 | Rainwater harvesting systems are an example of what category of sustainable water management? | Conservation |
| 7 | What is the definition of virtual water? | The volume of fresh water used to produce a product, measured at the place where the product was actually made |
| 8 | What term is given to water from toilets and waste disposal units, which is heavily contaminated? | Blackwater |
| 9 | Aquifer recharge is an example of a sustainable groundwater management strategy. What does it involve? | Excess surface water is directed to aquifers to replenish them |
| 10 | What triggered the Cochabamba Water War? | The privatisation of the city's water supply and drastic price increases for water |

*Put paper here* (repeated in answer column dividers)

## Previous questions

Now go back and use these questions to check your knowledge of previous topics.

## Questions | Answers

| # | Question | Answer |
|---|----------|--------|
| 1 | Define a possible reserve. | When it can be reasonably expected that the majority of inferred resources could be upgraded to indicated mineral resources with continued exploration |
| 2 | Define resource development. | Resource development encompasses the full process of bringing a resource from discovery to market |
| 3 | Name two events that both produced a dip in global energy consumption this century (so far). | Global Financial Crisis and COVID-19 pandemic |
| 4 | What two factors does the concept of water availability refer to? | The quantity of water that is physically accessible for use, and the quality of that water – whether it is safe to use |
| 5 | Which countries became Russia's most important customers after its invasion of Ukraine? | India and China |

*Put paper here* (repeated in answer column dividers)

## 11 Resource security

### 11.4 Energy security

## Sources of energy

### Primary sources of energy

These are forms of energy found in nature that have not been subjected to any human-engineered conversion process. They are used in the form they are found or are converted to secondary sources.

- Fossil fuels: They are used directly for heating or to generate electricity; for example, coal, natural gas, and oil.
- Nuclear energy: derived from the nuclear reactions of elements such as uranium and thorium.
- Renewable energy: for example, solar radiation, wind energy, hydro-energy from water, geothermal energy from the heat inside the Earth, and biomass from organic materials.

### Secondary sources of energy

These are derived from the transformation of primary sources.

- The most common secondary source is electricity, which can be generated from the conversion of various primary sources like coal, natural gas, wind or sunlight.
- Hydrogen can be produced from water through electrolysis.
- Ethanol and biodiesel are created from biomass and used in vehicles as a cleaner alternative to fossil fuels.

## Components of demand

**Population size:**
Larger populations generally require more energy for residential heating, cooling, transportation and industry.

**Economic development:**
Developed countries with more industrialisation and technological advancements tend to use more energy.

**Lifestyle:**
Lifestyles that include high levels of consumption, larger homes and more vehicles contribute to higher per capita energy demand.

**Factors affecting demand**

**Climate:**
Places with very hot or cold seasons or climates may require more energy to cool or heat buildings.

**Government policies:**
Policies promoting energy efficiency or conservation can reduce demand, while subsidies for energy use can increase it.

**Energy pricing:**
High prices can suppress demand, while low prices may encourage increased consumption.

# Energy mix

The **energy mix** refers to the combination of different energy sources used by a country or region. The mix changes according to different factors including time.

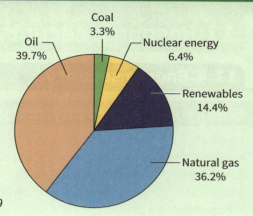

▶ **Figure 1** *UK energy mix 2019*

## The energy mix in developed countries

Developed economies:

- have a diverse mix, typically mixing fossil fuels, nuclear power and a growing proportion of renewable sources
- are actively trying to shift their energy mix towards more renewable sources to reduce carbon emissions. This is made more complex and expensive by the existing infrastructure for energy.

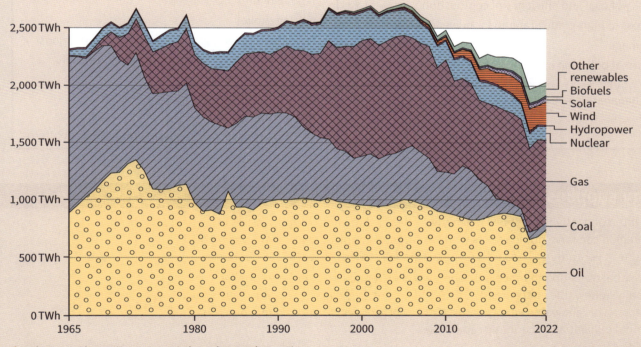

▲ **Figure 2** *The UK's energy mix has changed over time*

## The energy mix in developing economies

Developing economies:

- may rely heavily on one type of energy source, often fossil fuels, due to their availability and lower initial costs
- often leapfrog to renewable technologies, bypassing the traditional energy infrastructure development of more developed nations
- face challenges in energy infrastructure, which can affect the efficiency and reliability of energy supply

- typically have a more reliable energy supply and a greater mix of energy sources in urban areas than rural areas, including more access to electricity and cleaner fuels
- may rely more on traditional biomass in rural areas for heating and cooking, with less access to electricity or modern energy services.

# ⚙ KNOWLEDGE

## 11 Resource security

### 11.4 Energy security

## Energy supply and physical geography

Physical geography (climate, geology and drainage) can determine or heavily influence a country's energy mix and the volume and quality of energy supply. For example, the energy mix in Nigeria is still dominated by fuelwood and oil. This is due to regional geological factors and the presence of large, local oil reserves, while lack of rainfall means low levels of HEP.

| Factors affecting the quality of energy supply | Factors affecting the volume of energy supply |
|---|---|
| <ul><li>Climate</li><li>Consistency of solar radiation in deserts vs. variable cloud cover</li><li>Seasonal variability of wind speeds and its impact on wind energy</li><li>Geology</li><li>Purity of coal and efficiency of combustion</li><li>Access to high-grade uranium for nuclear power</li><li>Drainage</li><li>Thermal pollution in aquatic ecosystems from power plants</li><li>Contamination risks from fracking and oil drilling</li></ul> | <ul><li>Climate</li><li>Solar energy potential in tropical versus temperate climates</li><li>Wind energy availability in coastal versus inland areas</li><li>Hydropower potential in mountainous regions with high rainfall</li><li>Geology/relief</li><li>Fossil fuel deposits in sedimentary rock formations</li><li>Geothermal energy in volcanic and tectonically active regions</li><li>High relief for HEP and wind energy</li><li>Drainage patterns</li><li>Dams for hydropower need specific drainage patterns</li></ul> |

## Energy supplies in a globalising world

- The interconnectivity of globalised markets means that energy supply issues in one region can have ripple effects worldwide. For example, the Russian invasion of Ukraine in February 2022 caused oil prices globally to rocket to US$125 per barrel.
- The global transition to renewable energy is also impacting on energy supplies. Globally, renewable energy made up 75% of new energy capacity added in 2019. This expansion of renewable energy varies significantly by region and is dependent on national policies and economic capabilities.
- The competition for energy resources can lead to geopolitical conflicts. Nations may prioritise energy security over cooperation, leading to tensions. For example, Arctic nations are competing to gain access to fossil fuel reserves made exploitable by climate change.
- Nations can also cooperate over energy supplies. For example, the EU's energy policy aims to ensure energy security for all member states by diversifying energy sources and creating a common energy market.

## TNCs and energy supplies

- TNCs are big players in the energy market. Some of the world's biggest companies are energy companies, for example Shell and Exxon Mobil. National governments compete to attract investment from these TNCs.

- TNCs can use their influence to obtain concessions, tax breaks and deregulation from governments. This can sometimes go against the national interest, such as maintaining energy affordability for the population and ensuring environmental protection.

- There has been a shift in recent years towards TNCs promoting sustainable energy; for example, BP and Shell are investing in renewable energy. Others are working with national governments to promote sustainable energy, or setting ambitious targets for their own operations. For example, Walmart aims to run its operations on 100% renewable energy by 2035.

## Environmental impacts of the Marcellus shale gas development, USA

### Background

The Marcellus Shale formation extends across the Appalachian Basin and is a significant source of natural gas. Its exploitation has utilised hydraulic fracturing ('fracking') techniques, which have been the subject of environmental concerns.

### Location and scale

The Marcellus Shale region spans approximately 95,000 square miles across Pennsylvania, West Virginia, Ohio, and New York. The development of this resource has been rapid since the mid-2000s, with thousands of wells drilled.

**REVISION TIP**

You need to know the environmental impacts of a major energy resource development such as an oil, coal or gas field. Revise the example you have studied.

## Environmental impacts

- Fracking involves the use of highly pressurised water that contains chemicals. Spills can lead to contamination of surface and groundwater. The disposal of fracking wastewater also poses risks of contaminating water supplies if not managed properly.

- Gas extraction and distribution networks are sources of methane leaks, contributing to greenhouse gas emissions.

- The extraction process can release Volatile Organic Compounds (VOCs), contributing to air pollution and potential health risks.

- The expansion of infrastructure (drilling sites, roads and pipelines) leads to habitat fragmentation and changes in land use patterns.

- Drilling activities disrupt forested areas and can lead to soil erosion and compaction.

- The injection of wastewater into deep wells has been linked to a rise in minor earthquakes in regions not typically prone to seismic activity.

Regulations and industry best practices aim to improve the design and integrity of wells, reduce spills and manage wastewater.

Monitoring and controlling emissions from drilling operations help to reduce air quality impacts.

**Mitigation measures**

Post-drilling land restoration aims to minimise long-term habitat disruption.

## 11 Resource security

### 11.4 Energy security

## Strategies to increase energy supply

Not only is global demand for energy increasing, but geopolitical issues have underlined the importance of energy security for many nations, while the climate crisis also urges a shift towards renewable energy.

## Oil and gas exploration

Oil and gas remain primary energy sources worldwide. Advances in technology have enabled deeper and more complex exploration, expanding the potential for discovery.

- Arctic exploration: countries like Norway and Russia are exploring Arctic regions for oil and gas reserves made more accessible by technological advances and climate warming.
- Shale gas: the USA has significantly increased its energy independence through shale gas extraction. 70% of natural gas production in the USA is now from fracking (2022).
- Deepwater exploration: TNCs like Gazprom, Shell and Exxon Mobil are making big investments in offshore drilling technology that enables oil and gas to be extracted from deeper water as onshore and continental shelf fields are now running out.

Challenges include:

- the increased cost of extraction, which reduces profits
- the geopolitical tensions involved as countries stake claims to new fields
- the environmental concerns about extracting more fossil fuels.

## Nuclear energy

Advances in nuclear energy technology have made production of low-carbon energy more efficient and safer, although still very expensive.

- Small modular reactors (SMRs) are safer alternatives to traditional large reactors that can be built more cheaply and more quickly. The UK government has invested £200 million with Rolls-Royce on an SMR design which could be ready for the 2030s.
- **Nuclear fusion**: nuclear fusion has the potential to produce the same near zero-carbon energy as **nuclear fission** but without the dangerous radioactive waste. However, that potential may not be realised until the 2050s at the earliest.

Challenges include:

- the 2011 Fukushima disaster in Japan led to increased safety protocols worldwide and a reduction in public and government trust
- long-term waste disposal is a major challenge
- cost is significant; the UK's Hinkley Point C nuclear power station is predicted to cost £33 billion.

## The development of renewable sources

Renewables are the fastest-growing energy sector, driven by technological advancements and environmental concerns.

| Solar energy | Wind energy | Hydroelectric power |
|---|---|---|
| • Innovations in photovoltaic (PV) cell technology have reduced costs, making solar energy much more affordable.<br>• Large-scale solar farms are being developed worldwide, e.g. China's Tengger Desert Solar Park and Morocco's Ouarzazate Solar Complex. | • Offshore wind farms are expanding; the UK's Hornsea Project is the world's largest.<br>• Advancements in turbine technology have increased efficiency and reduced environmental impact.<br>• Costs of producing wind power have dropped by 90% since the 1980s. | • There is a shift towards smaller, less intrusive hydroelectric projects.<br>• Pumped storage hydroelectricity (e.g. Bath County Pumped Storage Station, USA) offers energy storage solutions in the 'largest battery in the world'. |

Challenges include:

- high initial investment costs
- intermittency (solar and wind do not provide a constant energy supply)
- integration (existing energy infrastructure is often in the wrong place for renewable energy)
- public acceptance and changes in government policies and subsidies for renewable energy production.

## Managing energy consumption

Strategies to manage energy consumption can be achieved by maximising energy efficiency and reducing demand at a variety of scales.

For example:

The Paris Agreement on Climate Change: 196 countries have committed to reducing their greenhouse gas emissions in order to limit global temperature increases to 1.5 °C.

⬇

National commitments, known as Nationally Determined Contributions (NDCs), set out how countries plan to do this. In 2020, the UK committed to reducing greenhouse gas emissions by at least 68% by 2030 (compared to 1990 levels).

⬇

National government agrees policies to meet its legally binding targets. For the UK, these originally included a phase-out of new petrol and diesel cars in 2030, a phase-out of gas boilers in 2035 (replaced in part by low-carbon heat-pump systems) and a requirement for landlords to improve the energy efficiency of their properties. Coal-fired power plants are to be phased out by 2030, with offshore wind power increased to 40 GW by 2030.

⬇

Local government shares responsibilities to meet the Paris Agreement NDCs too: for example, local authorities can require minimum energy efficiency standards for all new housing and work to decarbonise public transport in their region, as well as installing charging points for electric vehicles.

⬇

Government policies, subsidies and standards incentivise companies to develop their products and run their businesses in ways that reduce energy consumption. Reducing costs is also an incentive: many companies invest in insulation, energy-efficient lighting and HVAC (heating, ventilation and air con) and solar panels for this reason.

## 11 Resource security

## 11.4 Energy security

### Sustainability issues

#### Acid rain

Acid rain is a form of precipitation with high levels of sulphuric and nitric acids, primarily caused by the release of sulphur dioxide ($SO_2$) and nitrogen oxides (NOx) from burning fossil fuels. It can cause severe damage to forests, freshwater bodies, soil, and aquatic life. It also erodes buildings and monuments and affects human health.

In the UK, the use of 'flue gas desulphurisation' in power stations has significantly reduced $SO_2$ emissions.

#### Enhanced greenhouse effect

This effect refers to the additional warming caused by increased levels of greenhouse gases (GHGs) such as $CO_2$ and methane due to human activities, primarily fossil fuel combustion. It leads to global warming and climate change, causing extreme weather events, rising sea levels and ecological disruptions.

The Paris Agreement aims to limit global warming below 1.5°C.

To achieve this, greenhouse gas emissions should have peaked by 2025 at the latest and declined 43% by 2030.

#### Nuclear waste

Nuclear waste, the by-product of nuclear reactors, remains radioactive and hazardous for thousands of years, requiring safe disposal.

The UK's Nuclear Decommissioning Authority manages nuclear waste through interim storage, followed by plans for a Geological Disposal Facility. This involves storing the waste deep underground behind multiple barriers that will keep it completely safe for hundreds of thousands of years. Currently no site for the GDF has been selected.

#### Energy conservation

Reducing energy consumption is crucial for sustainability, decreasing environmental impacts and enhancing energy security. Governments and international organisations can legislate to encourage energy conservation. For example:

- The EU has an energy labelling system for electrical goods, which rates their efficiency on a scale from A (most efficient) to G (least efficient). This helps consumers make informed decisions based on a product's energy efficiency.

- Some countries have introduced Minimum Energy Performance Standards (MEPS) which products must meet to be sold.

- The Energy Performance of Buildings Directive sets out minimum standards for buildings. As of 2030, all new buildings in the EU must be zero-emission (2027 for all new public buildings). Buildings that are in low energy-efficiency classes will have to be renovated to become more energy efficient.

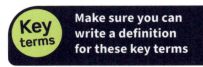

**Key terms** — Make sure you can write a definition for these key terms

energy mix    nuclear fusion    nuclear fission
primary source of energy    secondary source of energy

# RETRIEVAL

Learn the answers to the questions below, then cover the answers column with a piece of paper and write down as many as you can. Check and repeat.

## Questions | Answers

| | Questions | | Answers |
|---|---|---|---|
| 1 | Define primary sources of energy. | | Forms of energy found in nature that have not been subjected to any human-engineered conversion process |
| 2 | Give an example of a secondary source of energy. | Put paper here | One from: electricity generated from conversion of primary sources (e.g. sunlight) / hydrogen from water through electrolysis / ethanol or biodiesel from biomass |
| 3 | What is the energy mix of a country? | | The combination of different energy sources used by that country |
| 4 | How is the EU aiming to ensure energy security for all member states? | Put paper here | By diversifying energy sources and creating a common energy market |
| 5 | Name two environmental impacts caused by fracking. | | Two from: contamination of water supplies / methane emissions / release of VOCs / habitat fragmentation / soil erosion and compaction / earthquakes |
| 6 | Name one challenge associated with increasing energy supplies by oil and gas exploration. | Put paper here | One from: increased cost of extraction / geopolitical tensions from competing claims / the environmental concerns and pressures about extracting more fossil fuels |
| 7 | What is the advantage of nuclear fusion over nuclear fission, if the technology to produce energy from nuclear fusion were to be successfully developed? | | Fusion produces the same near zero-carbon energy as nuclear fission but with no dangerous radioactive waste |
| 8 | How have technological developments affected the costs of producing wind power since the 1980s? | | They have reduced the cost of producing wind power by 90% since the 1980s |

## Previous questions

Now go back and use these questions to check your knowledge of previous topics.

## Questions | Answers

| | Questions | | Answers |
|---|---|---|---|
| 1 | What is a resource frontier? | | An area on the periphery of a country, where resources are produced for the first time |
| 2 | What are the two factors primarily affecting consumption of mineral ores? | Put paper here | Industrial and manufacturing diversity and financial strength; extensive agricultural and industrial bases |
| 3 | What is the definition of virtual water? | | The volume of fresh water used to produce a product, measured at the place where the product was actually made |
| 4 | What is the definition of water stress? | Put paper here | When demand for safe, usable water in a given area exceeds the supply |
| 5 | Aquifer recharge is an example of a sustainable groundwater management strategy: what does it involve? | | Excess surface water is directed to aquifers to replenish them |

# KNOWLEDGE

## 11 Resource security

### 11.5 Mineral security

### Sources of copper

- Copper is sourced from mining and recycling to supply a rising global consumption that reached 25.1 million MT in 2022.
- Copper ore deposits are highly concentrated spatially (unlike iron ore), which is because the ore is only found in the veins and cavities of igneous rocks.
- Porphyry copper deposits (formed in magma chambers beneath volcanoes) are the biggest source and can be bulk-mined, but the metal content of the rock is very low, typically 0.5% copper. The most economic way of mining these deposits is by open-pit mining and then primary processing on site to extract the copper metal.
- Copper can be recycled repeatedly without loss of quality and requires less than 15% of the energy needed for primary copper production. Recycled copper makes up approximately a third of all global copper consumption.

### Distribution of copper ore reserves

More than half of global copper ore deposits are in South America. Chile has the largest copper reserves, with significant mines such as Chuquicamata and Escondida, and dominates production. Peru is a major copper producer. Outside of South America, China, Australia, USA, Russia, Zambia and the DRC are big producers.

### Physical geography and location of copper deposits

- Mountain ranges: copper deposits are often found in mountainous regions, particularly in areas with a history of volcanic activity: the Andes in South America and the Rockies in the USA, for example.
- Desert regions: some major copper deposits are located in arid, desert regions like the Atacama Desert in Chile. These areas can pose challenges for mining due to water scarcity.
- Proximity to plate boundaries: because of the geological conditions that produce copper, copper ore deposits are often found near to plate margins.
- Accessibility: physical geography can impact the feasibility and cost of mining operations. Remote or inaccessible areas increase the challenges and costs of mining.

### End uses of copper

**Electrical applications**

Due to its excellent conductivity, copper is widely used in wires and cables

**Construction**

Copper is used in plumbing, roofing and cladding due to its durability and resistance to corrosion

**Renewable energy**

Essential in wind turbines, solar panels and other renewable energy infrastructure

**Electronics**

Found in mobile phones, computers and other electronic devices

**Alloys**

Copper forms alloys like bronze and brass, used in various applications from musical instruments to naval hardware

# Components of demand for copper

- While production is concentrated in South America, more than 70% of global consumption is in Asia.

- China consumes 55% of copper, with a quarter of that being used in construction.

- Global copper consumption has more than doubled since 1990, much of this because of rapid urbanisation in Asia. Demand is predicted to rise by a further 20% by 2035. This is driven by the importance of copper in a wide range of industries.

- Renewable energy: copper is a key component in solar panels, wind turbines and the infrastructure required for transmitting and distributing renewable energy.

- EVs: the move towards electric vehicles is a major driver of copper demand. EVs require more copper than traditional engines, particularly for batteries and electric motors.

- Consumer electronics: copper is a key material for smartphones, tablets and computers.

- Energy efficiency: copper plays a key role in energy-efficient systems and equipment due to its excellent thermal and electrical conductivity.

- Public health and medical uses: copper has antimicrobial properties and it is increasingly used for surfaces and equipment to improve hygiene.

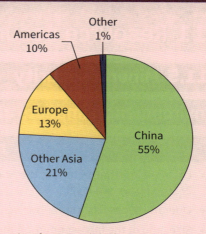

▲ **Figure 1** *Global consumption of refined copper by region, 2022*

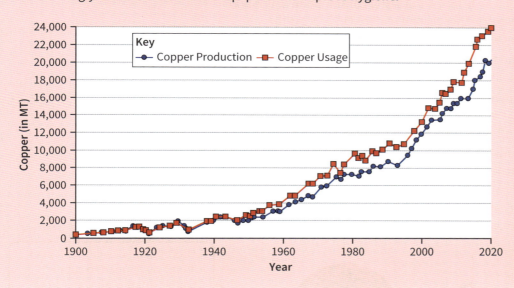

◀ **Figure 2** *Copper production and copper consumption from 1900 to 2020*

# Sustainability issues

Despite the high recyclability and therefore sustainability of copper, which means that a third of all global copper consumed is recycled copper, there are serious sustainability issues with copper production, trade and consumption.

- Land degradation: open-pit mining activities can lead to deforestation, soil erosion and habitat loss.

- Water pollution: copper mining and processing require substantial amounts of water, which can strain local water resources, especially in arid regions. Additionally, chemicals used in mining and metal compounds can contaminate water resources.

- Air pollution: the processing of copper ore can release dust, sulphur dioxide and other pollutants into the air, contributing to air quality issues.

- Energy consumption: copper extraction and processing are energy-intensive, contributing to greenhouse gas emissions.

- Waste management: tailings, the materials left over after separating the copper from the rock, can be a major source of pollution if not properly managed.

- Sea contamination: Chile, the world's biggest producer, exports 89% of its copper production by sea, risking sea water contamination by heavy metal compounds and hydrocarbons from the sea tankers.

# KNOWLEDGE

## 11 Resource security

### 11.5 Mineral security

## Environmental impacts of the Chuquicamata copper mine, Chile

### Background

- Location: northern Chile, in the Atacama Desert; 1650 km from the capital, Santiago; 2850 m above sea level
- Operator: Codelco, a state-owned Chilean mining company
- Size: one of the largest open-pit mines globally, 5 km long, 3 km wide and 1 km deep
- Production: slumped from 1.7MT in 2020 to 1.4MT in 2022, due to declining quality of ore grades and also a mine collapse

**REVISION TIP**

You need to know the environmental impacts of a mineral resource extraction scheme. Revise the example you have studied.

### Environmental impacts

- Land degradation: the sheer size of the mine has led to extensive alteration of the local landscape.
- Water usage: Mining is responsible for 70% of water consumption in this arid region of northern Chile, with 2 m³ per second required to process the ore. Much of the water has been extracted from groundwater.
- Processing: China's demand for high quality metal means more energy-intensive processing that produces toxic chemical waste material.
- Air pollution: unsafe emissions of arsenic and sulphur dioxide in 2008 caused Codelco to relocate its 7000 workers to a town 17 km away. Children under six and pregnant people are forbidden to visit because of pollution.
- Waste management: the unwanted materials (tailings) including toxic metals such as arsenic and mercury, are kept in pools as a slurry. There are many concerns about the damage caused to health from tailings leaks and pollution.
- Hundreds of large trucks, each carrying between 330 and 400MT, transport waste and ore from the mine each day. Carbon emissions are therefore significant.

Facing declining yields, underground mining began in 2016 that will secure the mine's life for another 40 years.

Codelco has announced goals to cut carbon emissions by 70%, reduce inland water consumption by 60% and to be recycling 65% of its industrial waste by 2030.

**Future developments**

A US$1 billion desalination plant is being developed to supply Chuquicamata with water from the sea instead of from inland groundwater sources.

An activated sludge biological treatment plant has been built to treat the wastewater from copper processing: this allows the water to be reused for processing.

The underground mine will operate using electric power and all vehicle tyres will be recycled.

# ⇄ RETRIEVAL

Learn the answers to the questions below, then cover the answers column with a piece of paper and write down as many as you can. Check and repeat

| Questions | Answers |
|---|---|
| **1** Under what geological conditions are porphyry copper deposits formed? | In the magma chambers beneath volcanoes |
| **2** What percentage of the energy used to refine primary copper production is required to recycle copper? | 15% |
| **3** Approximately what proportion of copper consumption globally is recycled copper? | A third |
| **4** More than half of all copper ore deposits are in which continent? | South America |
| **5** Which country consumes 55% of global copper? | China |
| **6** Name two industrial sectors that are driving rising demand for copper. | Two from: renewable energy technology / EVs / consumer electronics / energy efficiency technology / public health and medical uses |
| **7** Name two environmental issues associated with copper production. | Two from: water pollution / air pollution / energy consumption / waste management / sea contamination |
| **8** In which desert is the Chuquicamata copper mine located? | The Atacama Desert |
| **9** What is the name of the state-owned company that runs Chuquicamata? | Codelco |
| **10** What forced the closure of the mining settlement at Chuquicamata in 2008? | Air pollution due to emissions of arsenic and sulphur dioxide from the processing plant |

*Put paper here*

## Previous questions

Now go back and use these questions to check your knowledge of previous topics.

| Questions | Answers |
|---|---|
| **1** What usually follows identification of an indicated reserve? | A feasibility study to look at economic viability |
| **2** The volume and quality of water supply is strongly related to which three key aspects of physical geography? | Climate, geology and drainage |
| **3** Give an example of a secondary source of energy. | One from: electricity generated from conversion of primary sources (e.g. sunlight) / hydrogen from water through electrolysis / ethanol or biodiesel from biomass |
| **4** Name two environmental impacts caused by fracking. | Two from: contamination of water supplies / methane emissions / release of VOCs / habitat fragmentation / soil erosion and compaction / earthquakes |
| **5** How have technological developments affected the costs of producing wind power since the 1980s? | They have reduced the cost of producing wind power by 90% since the 1980s |

*Put paper here*

## 11 Resource security

### 11.6 Resource futures

## Future changes in supply and demand

Resources will be affected by growing populations, levels of business activity, urbanisation patterns and climate change but, future changes in resource supply and demand are hard to predict precisely.

| Water supply and demand | Energy supply and demand | Mineral supply and demand |
|---|---|---|
| **Technological developments** | | |
| • GM crops are being designed so they need less water to grow.<br>• Improved desalination plants increase freshwater supply and are becoming cheaper to run. | • New energy reserves become exploitable with application of new technologies (e.g. fracking in USA in 2000s).<br>• Battery tech and smart grids may manage intermittency of renewable energy. | • New processing technologies may reduce demand for rare earth minerals.<br>• Innovations in mining technologies will lead to more efficient extraction processes. |
| **Economic developments** | | |
| • Increasing costs of storage and transfer might force end users to reduce their water footprint.<br>• Significant investments in water infrastructure will continue to be necessary to ensure sustainable water management. | • The cost of renewable energy should continue to fall, making it increasingly competitive with fossil fuels.<br>• Previously uneconomically recoverable resources may become viable (e.g. oil in the Arctic Ocean). | • Global economic growth, particularly in emerging economies, will drive up demand for mineral ores.<br>• Increased demand may mean that even lower grades of ore are mined, resulting in increased tailings. |
| **Environmental developments** | | |
| • Changes in precipitation patterns and more extreme weather will affect water availability.<br>• Improvements in EIAs may mean fewer negative impacts from large schemes and more reversal of old damage. | • As reserves become exhausted, protected and reserved areas may be opened for exploitation.<br>• Delayed effects of climate change might force reduction in use of fossil fuels. | • Increasing demand for non-renewable ore minerals may mean increased extraction in environmentally sensitive areas.<br>• Impacts of ore extraction may be better managed. |
| **Political developments** | | |
| • Fewer large dams will be built if environmental and social opinions influence policy.<br>• Political decisions around water rights and distribution will continue to lead to conflicts or may spark cooperation between regions and countries. | • Climate change agreements should reduce demand on fossil fuels.<br>• Nuclear energy remains politically contentious, but governments are increasingly worried over meeting future energy gaps. | • Political protection for wilderness areas may be extended or come under threat.<br>• Political decisions regarding the control and export of mineral ores may impact supply. |

# ⚙ KNOWLEDGE

## 11 Resource security

### 11.7 Case study: Resource issues

## 📖 Case Study: Energy issues in Slovakia

### Slovakia and the invasion of Ukraine

Before the Russian invasion of Ukraine in February 2022, Slovakia was almost 100% dependent on Russian imports of natural gas, oil and nuclear fuel. It then faced the challenge of finding alternative energy sources for its developed, high-income economy. Slovakia is moving away from using its own coal – state subsidies for electricity generation from coal ended in 2023.

> **REVISION TIP**
>
> You should revise the case study you have studied, as you will have covered it in depth.

### Slovakia's energy mix

- The main domestic-produced sources of energy are renewables (mainly biomass and HEP) and brown coal, although coal is now being phased out.

- Oil is vital for Slovakia's industry sector and for transportation, while gas is used for over 90% of all heating in Slovakia.

- Nuclear power is used to generate most of Slovakia's electricity and also to supply heat to neighbouring towns.

- The coal-mining region of the Upper Nitra is receiving EU funding to enable it to transition to a post-coal economy. The two remaining coal-powered power stations are being converted to biomass fuel.

- Biomass (wood chips) and HEP make up most of Slovakia's renewable energy. There are only five wind turbines in the whole of the country, and solar and geothermal are also undeveloped.

### Diversification since 2022

- Gas pipelines now connect Slovakia to all its neighbours, including to a Polish LNG terminal (liquid natural gas).

- This new infrastructure means that Slovakia is now supplied with gas by Norway and discussions with Italy, Germany, Qatar, Asia and the USA are ongoing.

- Slovakia is one of seven EU countries that have not reduced their gas consumption – its industrial sector relies heavily on gas.

- It remains dependent on Russia for 60% of its natural gas supply.

- It still relies on Russian oil because of difficulties in adapting the refinery to different oil types and the risk of damage to the economy from a sudden cut-off.

- While the process of certifying new suppliers for nuclear fuel is under way, it is a long process and Slovakia is also still reliant on Russia for nuclear fuel.

- Its largest industries are investing in solar and wind power. Most heating is from central municipal sources, which makes it relatively easy to switch to renewable energy such as biomass and geothermal.

### Impacts of the invasion of Ukraine

- A surge in global energy prices occurred, and as Slovakia is heavily reliant on imports for its energy needs, this led to higher prices for energy and fuel directly impacting on consumers and industries.

- The higher energy prices affected the whole economy by contributing to inflation, increased production costs for businesses, especially those reliant on energy-intensive processes, and the job market and consumer spending.

- The rise in energy costs and the broader economic impact had a social impact on the cost of living in Slovakia.

- Elections in October 2023 led to a coalition government headed by a pro-Russian politician. This reflected dissatisfaction among many Slovaks with the impact of the Russian–Ukrainian war on their cost of living.

# ⚙ KNOWLEDGE

## 11 Resource security

### 11.8 Case study: How physical environment affects resources

 **Case Study: Water management in Mexico City**

## Physical environment

- Mexico City, one of the world's largest cities, is situated in a high-altitude valley in central Mexico. The city has a population of over 21 million people in its metropolitan area, making it one of the most populous urban centres in the world.

- The city relies heavily on groundwater from aquifers, which are being depleted faster than they can be replenished. This causes subsidence – in the oldest residential areas, the ground has sunk by 9 m.

- Most of the rainwater from the May to October rainy season is not captured, the impermeable urban surfaces prevent infiltration and much of the rainwater then mixes with sewage and becomes unusable.

- Climate change is making rainfall more unpredictable, exacerbating water scarcity. In 2023, an El Niño year, Mexico as a whole had 25% less rainfall than average.

- The water system loses approximately 40% of its water to leaks due to ageing infrastructure. As a result, one in five people in Mexico City are restricted to a few hours of running water per day.

- To the north, the city's untreated wastewater (blackwater) is used to irrigate the Mezquital Valley – which used to be desert. In the 1990s, farmers there were forbidden from growing crops for human consumption, but animal feed can be grown there. An aquifer is now forming beneath the valley.

> **REVISION TIP**
> You should revise the case study you have studied, as you will have covered it in depth.

Water scarcity disproportionately affects lower-income neighbourhoods, where residents often rely on water truck deliveries.

**Socio-economic impacts**

The financial cost of water delivery and infrastructure maintenance is significant, often straining the city's budget.

Inadequate water supply leads to sanitation issues, impacting on public health.

## Responses

- The water supply to parts of Mexico City has been reduced still further.

- To combat the continued drought, Mexico's air force releases silver iodine into the atmosphere to promote condensation and rainfall – cloudseeding.

- There is investment in repairing and upgrading water distribution systems to reduce leakage, but the scale of the infrastructure challenge is enormous, and the cost is prohibitive.

At the same time, sustainable water management solutions are also being implemented:

- Since 2016, 500 rainwater harvesting systems have been installed in houses and public buildings, reducing the pressure on aquifer use. Each system provides 40,000 litres of drinking water per year, which saves a family US $200 in water bills.

- Parks and public spaces have been built that contain large tanks for capturing rainwater and also use local volcanic gravel to create permeable surfaces that allow rainwater to filter through and recharge the aquifer.

- A series of artificial wetland areas that naturally filter and clean wastewater have also been constructed within the city.

# RETRIEVAL

Learn the answers to the questions below, then cover the answers column with a piece of paper and write down as many as you can. Check and repeat.

## Questions

## Answers

| | Questions | | Answers |
|---|---|---|---|
| 1 | Suggest one way in which technology may affect resource futures for water supply or demand. | Put paper here | One from: GM crops are being designed so they need less water to grow / improved desalination plants increase freshwater supply and are becoming cheaper to run |
| 2 | What are the main domestically produced sources of energy in Slovakia? | | Renewables (mainly biomass and HEP) and brown coal |
| 3 | What type of energy is used to generate most of Slovakia's electricity? | Put paper here | Nuclear energy |
| 4 | Explain how Slovakia has been able to diversify its supply of natural gas since February 2022. | | Slovakia is now connected by gas pipelines to all its neighbours, which means it can receive gas supplies from them and from other suppliers through them |
| 5 | A year after the Russian invasion of Ukraine, how far had Slovakia been able to diversify its gas supply (as a percentage)? | Put paper here | By 40% (it still relied on Russia for 60% of its gas) |
| 6 | What percentage of Mexico City's water is wasted through leaking pipes? | | Approximately 40% |
| 7 | When is the rainy season in Mexico City? | Put paper here | From May to October |
| 8 | Explain one reason why the rainy season doesn't recharge Mexico City's aquifers. | | One from: the huge demand for water from a population of over 21 million people / impermeable urban surfaces that reduce infiltration |

## Previous questions

Now go back and use these questions to check your knowledge of previous topics.

## Questions

## Answers

| | Questions | | Answers |
|---|---|---|---|
| 1 | Why do resource prices often rise after the resource peak (generally where they are cheapest)? | Put paper here | Because of more challenging extraction and processing as the most accessible resources have been used first |
| 2 | Define geopolitics. | | Geopolitics is the study of the ways in which political decisions and processes affect the use of resources |
| 3 | Rainwater harvesting systems are an example of what category of sustainable water management? | | Conservation |
| 4 | Name two challenges associated with increasing energy supplies through developing renewable energy. | Put paper here | Two from: high initial investment costs / problems with intermittency / integration problems / public acceptance / changes in government policies and subsidies |
| 5 | Name two environmental issues associated with copper production. | | Two from: water pollution / air pollution / energy consumption / waste management / sea contamination |

# PRACTICE

## Exam-style questions

1   In stock resource evaluation, what is meant by the term possible resource?                    **[4 marks]**

2   Outline the difference between primary and secondary sources of energy.                       **[4 marks]**

3   Outline the stages of stock resource evaluation.             **[4 marks]**

4   Outline the concept of virtual water trade.                   **[4 marks]**

5   Analyse the data shown in **Figures 1a** and **1b**.          **[6 marks]**

**EXAM TIP**

6-mark analyse questions all have 6 marks for AO3: identify trends in the data ('the big picture'), anomalies and any connections within the figures and/or between them.

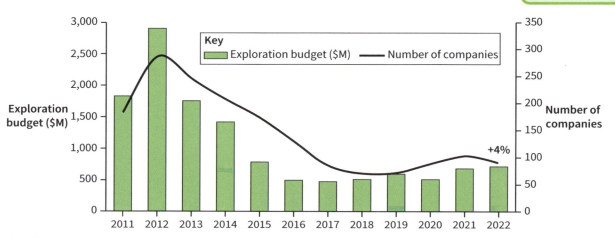

▲ **Figure 1a** *Iron ore exploration budgets and number of companies involved, 2011–2022*

| Mine production (thousand metric tons) | | | | |
|---|---|---|---|---|
| | Usable ore | | Iron content | |
| | 2021 | 2022 | 2021 | 2022 |
| Australia | 912,000 | 880,000 | 565,000 | 540,000 |
| Brazil | 431,000 | 410,000 | 273,000 | 260,000 |
| Canada | 57,500 | 58,000 | 34,500 | 35,000 |
| Chile | 17,700 | 16,000 | 11,200 | 10,000 |
| China | 394,000 | 380,000 | 246,000 | 240,000 |
| India | 273,000 | 290,000 | 169,000 | 180,000 |
| Iran | 72,900 | 75,000 | 47,900 | 49,000 |
| Kazakhstan | 64,100 | 66,000 | 13,100 | 14,000 |
| Russia | 96,000 | 90,000 | 66,700 | 63,000 |
| South Africa | 73,100 | 76,000 | 46,500 | 48,000 |
| Ukraine | 83,800 | 76,000 | 52,400 | 47,000 |
| United States | 47,500 | 46,000 | 30,100 | 29,000 |

▲ **Figure 1b** *Major producers of iron ore*

6   Using **Figures 2a, 2b** and your own knowledge, to what extent do you agree
that strategies to increase water supply in this region require high levels of
economic development?                                          **[9 marks]**

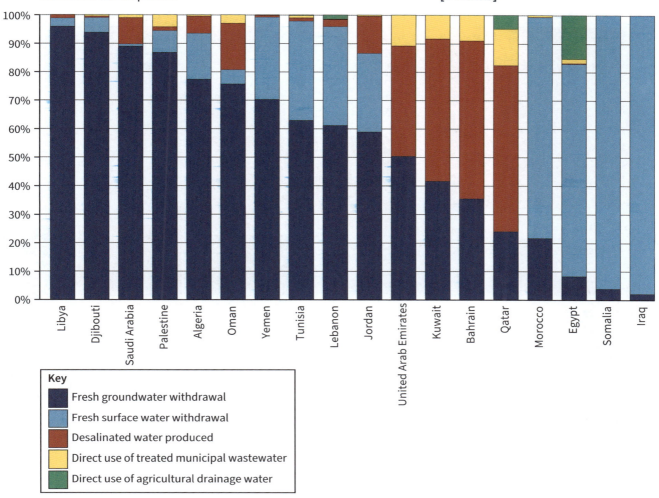

▲ *Figure 2a* Origin of water withdrawal in selected Arab states by source (2020)

| Country | GDP per capita (US$) | Country | GDP per capita (US$) |
|---|---|---|---|
| Libya | 6,357 | Jordan | 10,007 |
| Djibouti | 3,150 | United Arab Emirates | 78,255 |
| Saudi Arabia | 51,600 | Kuwait | 67,891 |
| Palestine | 5,795 | Bahrain | 52,129 |
| Algeria | 3,690 | Qatar | 112,789 |
| Oman | 35,286 | Morocco | 3,795 |
| Yemen | 2,078 | Egypt | 14,226 |
| Tunisia | 3,807 | Somalia | 447 |
| Lebanon | 11,561 | Iraq | 10,175 |

▲ *Figure 2b* GDP per capita of selected Arab states (2021)

7    With reference to **Figures 3a** and **3b** and your own knowledge, assess
     the security of energy supplies in a globalising world.        **[9 marks]**

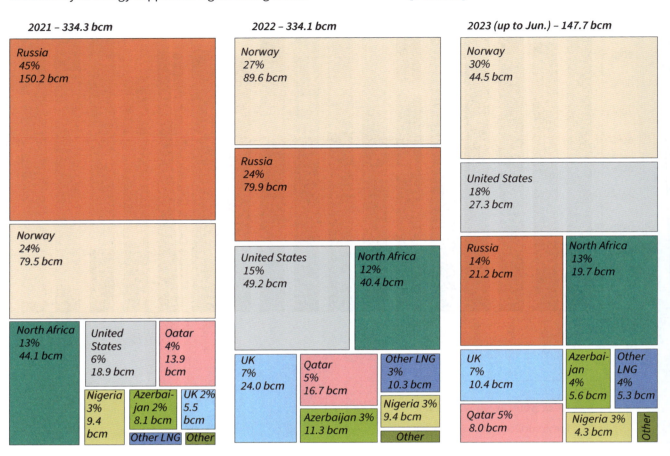

▲ **Figure 3a** *The composition (make-up) of natural gas imports into the EU for the period 2021–2023. Bcm stands for billion cubic metres of natural gas*

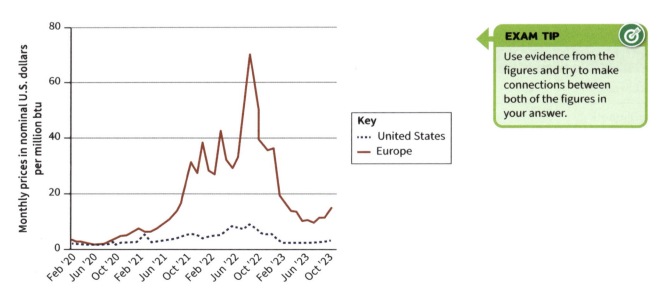

**Key**
···· United States
— Europe

**EXAM TIP**

Use evidence from the figures and try to make connections between both of the figures in your answer.

▲ **Figure 3b** *Monthly prices for natural gas in Europe compared to the USA, 2020–2023. (btu stands for British thermal units. 1 million btu is equivalent to 28 cubic metres of gas)*

8    For a specified place you have studied, assess the relative importance of physical and human factors in determining how either water, energy or a mineral ore is used.    **[9 marks]**

9    Assess the degree to which the strategies to increase energy supply have proved more effective than managing energy consumption at a global scale.    **[20 marks]**

10    For a mineral ore you have studied, how far do you agree that its geopolitics has been shaped by the development of technologies linked to globalisation?    **[20 marks]**

11    For a place you have studied, evaluate the influence of exogenous and endogenous factors on its resource security with regard to water **or** energy **or** a mineral ore.    **[20 marks]**

12    For either a water **or** energy **or** mineral ore resource, evaluate the extent to which addressing associated sustainability issues will always be less important than human welfare.    **[20 marks]**

**EXAM TIP**

20-mark questions have 10 marks available for AO1 and 10 marks for AO2 (your argument). Some degree of balance in your answer is needed. Could you present a counterargument in your penultimate paragraph, ahead of your conclusion?

# KNOWLEDGE

## 12 Fieldwork and skills

### 12.1 Fieldwork

## Preparation for fieldwork

Before fieldwork is conducted, preparation involves:

- background reading into the geographical theories and concepts to ensure that the enquiry is connected to existing theories and that there is literature to support the analysis of findings
- drawing up aims and **objectives** for the enquiry, including an overall aim and sub-aims, which may take the form of a **hypothesis** to be tested
- planning the collection of **primary data** and the use of sources of **secondary data**
- planning the **sampling** techniques that will be used for primary data collection to eliminate bias.

> **REVISION TIP**
>
> Ensure that you can justify the use of different sampling techniques, as well as being able to discuss their advantages and drawbacks.

### Ethical considerations

Ethical considerations involve:

- minimising environmental damage
- ensuring anonymity and confidentiality when working with participants
- ensuring consent is obtained for data collection at the field site
- ensuring that no harm is caused when interacting with participants.

### Sampling strategy

A sampling strategy should be devised to ensure the whole population is accurately represented in the data collection. Sampling strategies can be:

- systematic (every 5 m)
- stratified (representative of the proportions of sub-groups in a data set)
- random (selection through random number generators)
- opportunistic (asking passers-by to complete a questionnaire)
- as well as sampling across a transect
- an area, or using point sampling.

### Risk assessment

A risk assessment carried out should identify health and safety issues in the field, e.g. working close to the sea, and put steps in place to mitigate the risks, e.g. choosing study sites away from the coastline and knowing the tide times.

### Planning primary data collection

An OS map can be used to plan primary data collection as the whole sample area can be seen, access can be determined and physical features, such as relief and water sources, can be seen.

## Collection of primary data

Primary data can be **quantitative** (numerical) or **qualitative** (words). The methods used to collect the primary data should be appropriate for the aims of the investigation. Secondary data can be used to help reach the aims of geographical enquiry, e.g. house prices, crime data, weather data, census data.

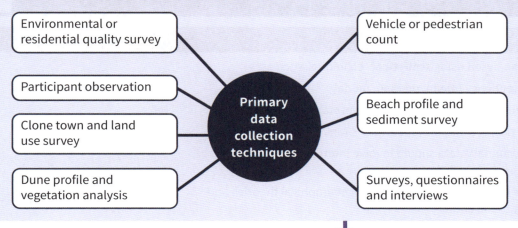

## Processing and presenting data

Once data has been collected it is processed and presented using relevant graphical and cartographical techniques.

- Graphical and cartographical techniques are used to present quantitative data and analyse patterns, trends and relationships.

- **Geolocated** data is data which is presented on a map, such as a proportional symbols displaying vehicle or footfall counts on a base map of a central urban area.

- GIS can be used to present data, as different types of data collected can be layered on a base map to reveal and better analyse patterns.

- Other techniques such as annotated photographs, field sketches, quotes and word clouds can be used to present qualitative data.

**REVISION TIP**

Ensure that you can choose appropriate presentation techniques by understanding how effective each one is at presenting data.

**REVISION TIP**

Look at the Skills section to see examples of graphs and maps that can be used for data presentation.

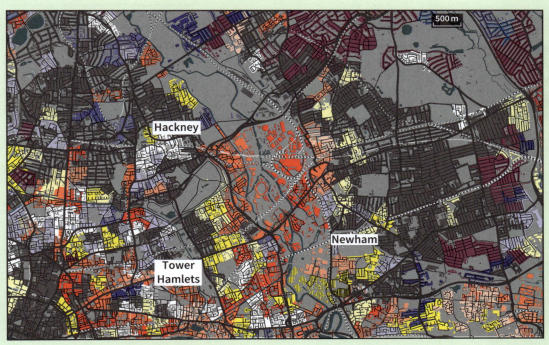

▲ **Figure 1** *An example of GIS data plotted on a base map; the age of housing mapped across the Queen Elizabeth Olympic Park and surrounding areas in East London*

## 12 Fieldwork and skills

### 12.1 Fieldwork

## Analysing data

- Quantitative data can be analysed using **statistical** techniques such as:
  - measures of central tendency (mean, median, mode)
  - measures of dispersion (range, inter-quartile range, standard deviation)
  - inferential and statistical techniques (Spearman's rank and Chi-squared).
- Analysis of primary and secondary data should involve interpreting the patterns and trends shown on the presented data.
- Analysis should link to the background reading on the geographical theories and concepts that are relevant to the enquiry.

## Drawing conclusions

- Conclusions should relate back to the original aims and objectives of the enquiry.
- Essentially, sub-questions should be answers and comments made as to whether the hypothesis has been proved or disproved.

## Reviewing the enquiry

- All stages of a fieldwork-based enquiry should be **evaluated**, e.g. were appropriate methods chosen? Were appropriate sampling methods used?
- Limitations of the enquiry should be addressed and comments on validity, reliability and accuracy can be made.
- Consideration should also be given as to how the enquiry could be further developed.

> **REVISION TIP**
>
> Ensure that you can evaluate the planned data collection methods of an unfamiliar fieldwork-based enquiry.

 **Key terms** Make sure you can write a definition for these key terms

| evaluated | geolocated | hypothesis | objectives | primary data |
| qualitative | quantitative | sampling | secondary data | statistical |

# RETRIEVAL

Learn the answers to the questions below, then cover the answers column with a piece of paper and write down as many as you can. Check and repeat.

| | Questions | | Answers |
|---|---|---|---|
| 1 | Why is background reading important in preparing for fieldwork? | Put paper here | To ensure that the enquiry is connected to existing theories and that there is literature to support the analysis of findings |
| 2 | Why do sampling techniques need to be planned? | | To eliminate bias in primary data collection |
| 3 | What is quantitative data? | | Numerical data |
| 4 | What is qualitative data? | Put paper here | Non-numerical data, e.g. words |
| 5 | Give two examples of secondary data. | | House prices, crime data, weather data, Census data |
| 6 | Give two examples of primary data collection techniques for physical geography enquiry. | Put paper here | Beach profile, sediment survey, dune profile, vegetation analysis |
| 7 | What is geolocated data? | | Data which is presented on a map |
| 8 | How can qualitative data be presented? | | Annotated photographs, field sketches, quotes and word clouds |
| 9 | What are three measures of central tendency? | Put paper here | Mean, median, mode |
| 10 | What are three measures of dispersion? | | Range, inter-quartile range, standard deviation |
| 11 | What should conclusions of a geographical enquiry relate back to? | | The original aims and objectives of the enquiry |

### 12.2 Cartographic skills

## Weather maps

A **synoptic chart** is a type of weather map that summarises the atmospheric conditions at a particular time over a certain area. Some weather maps include a range of symbols that indicate specific cloud, wind, temperature, pressure and weather conditions.

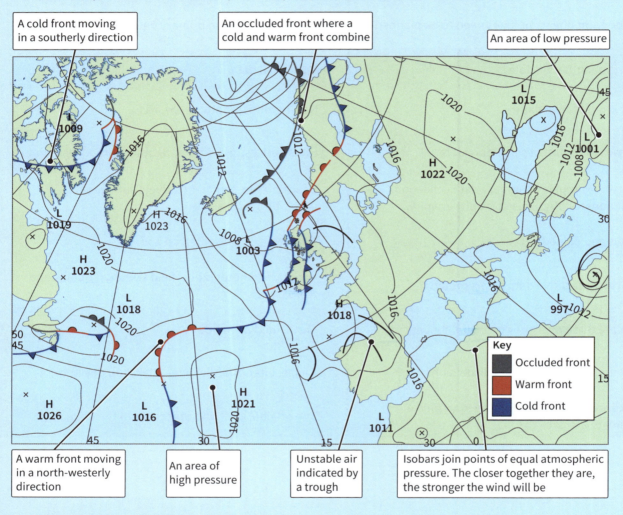

▲ **Figure 1** A synoptic chart from September 2023

## Proportional symbol maps

- **Proportional symbol maps** use symbols, most commonly circles, that are proportionate to the value they are representing.
- Symbols are placed on a base map at the location of the data collected.
- They clearly show the spatial variation of values across the map area.
- However, it can be difficult to decipher exact values and symbols may obscure each other, and the base map.

## Maps showing movement

- **Flow line**, **desire line**, or **trip line** maps use arrows on a base map to show a movement of people or goods from one place to another.
- The lines may be drawn with a thickness proportional to the value they are representing.
- Often lines shown a general direction of movement rather than an exact path taken, and do not always point to the exact location of departure and arrival.

**REVISION TIP**

Look at an example of a flow map on page 288.

## Dot maps

**Dot maps** use small dots that each represent the same value (e.g. 7500 people) to show the distribution of data within an area.

2020 POPULATION DISTRIBUTION IN THE UNITED STATES AND PUERTO RICO

The absence of a dot may not always imply that there is no data there – the threshold for a dot (e.g. 7500 people) may not have been reached

The spatial variation of values across the map area is shown by interpreting clusters and absences of dots

One dot = 7,500 people

United States Census 2020

U.S. Department of Commerce  U.S. Census Bureau

▲ **Figure 2** *USA population distribution, dot map 2020*

## Choropleth maps

- **Choropleth maps** use a scale of graduated colours to show spatial variations of data across different areal units, e.g. countries, states, regions.
- It is easy to see how the data varies and to compare the colour of an areal unit to the value, or range of values, it represents.
- A choropleth cannot show the spatial variation of data within an areal unit and, when a range is used in the key, it cannot give exact values.

**REVISION TIP**

Look at an example of a choropleth map on page 27 or page 255.

## 12 Fieldwork and skills

### 12.2 Cartographic skills

## Isoline maps

An **isoline map** joins points of equal value to show how these values vary spatially.

| Spaces between isolines can be shaded to make interpretation of the pattern easier | Isolines can only be drawn with data that changes gradually and for which there are many data points | The relationship between atmospheric pollution and traffic (thicker lines represent ring roads) can be analysed |
|---|---|---|

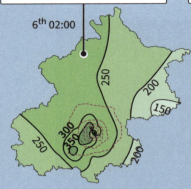

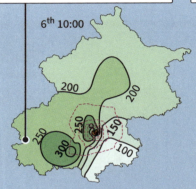

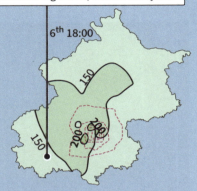

▲ **Figure 3** An isoline map showing particulate matter that are 2.5 microns or less (PM$_{2.5}$) in Beijing on 6 October 2013

 **Key terms**  Make sure you can write a definition for these key terms

choropleth map    desire line    dot map
flow line    isoline map    proportional symbol map
synoptic chart    trip line

# 12.3 Graphical skills

## Line graphs

- **Line graphs** are used most commonly to show temporal variations, e.g. climate graphs showing change over a year.

- Comparative line graphs show several different data sets on one graph for comparison.

- As data points are joined by a straight line, a continuous change between data points, e.g. years, is implied, which may be inaccurate.

- A compound line graph shows subdivisions of the area beneath the line graph to show different proportions of the total.

**REVISION TIP**

Look at an example of a line graph on page 276.

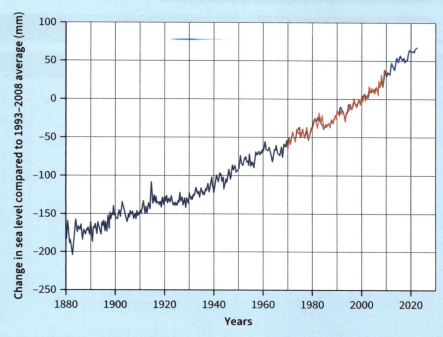

▲ **Figure 1** This divergent line graph shows how values deviate, positively and negatively, from a mean or average

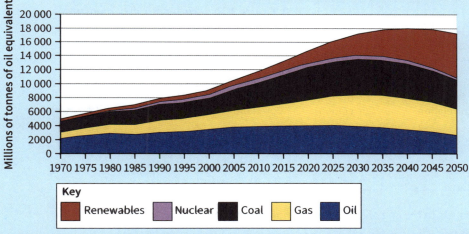

▲ **Figure 2** This compound line graph shows the predicted global energy demand, divided into five components that show their contribution to the total amount

## Bar graphs

- A **bar graph** shows the frequency or amount of discrete data in a number of different categories.

- It is clear to see the value represented on a bar chart by looking at the length of the bar.

- A comparative or compound bar chart shows a single bar divided into sub categories, to show their contribution to the total value represented by the bar.

- It is harder to see the values represented on both compound line graphs and bar charts, as the beginning and end of each section needs to be

determined and then the difference between them shows the total value.

- A divergent bar chart shows positive and negative values on the same chart.

**REVISION TIP**

Look at an example of a simple bar graph on page 263 and an example of a percentage compound bar chart on page 280.

## 12 Fieldwork and skills

### 12.3 Graphical skills

## Scatter graphs

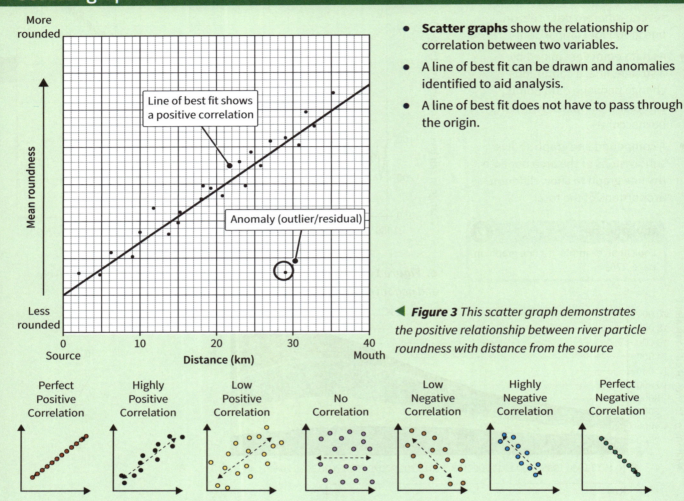

- **Scatter graphs** show the relationship or correlation between two variables.
- A line of best fit can be drawn and anomalies identified to aid analysis.
- A line of best fit does not have to pass through the origin.

◀ **Figure 3** This scatter graph demonstrates the positive relationship between river particle roundness with distance from the source

▲ **Figure 4** How close the plotted data sits to the line of best fit shows how strong the relationship between the two variables is

# Pie charts

- **Pie charts** use segments to show proportions of a total.
- They provide a clear visual representation of larger and smaller sections within the data set.
- Data is converted into percentages, so the values represented by the percentages are not often known.
- Pie charts often contain an 'other' category, where the features of this category are not often known.

**REVISION TIP**

Look at an example of a pie chart on page 299.

# Triangular graphs

- **Triangular graphs** show data that has three variables.
- Data is manipulated into a percentage and plotted against three axes.

The data point plotted here represents
A = 60%
B = 37%
C = 5%

Triangular graphs (Figure 8) are limited to data which can be divided into three parts, e.g. the percentage of people working in primary, secondary and tertiary job sectors within a country

**REVISION TIP** 

Make sure you know how to read the data off the axes of a triangular graph.

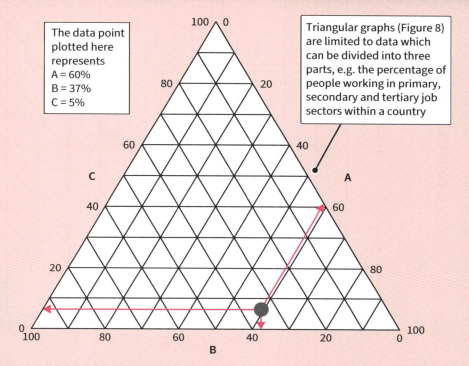

▲ **Figure 5** *A triangular graph*

## 10 Population and the environment

### Logarithmic scales

- Graphs can use **logarithmic scales** when the data to be presented has a large range.
- The scale increases on a cycle of 10, e.g. 10 to 100, 100 to 1000, 1000 to 10,000.
- Greater space is given to smaller values, while less space is given for larger values.

▲ **Figure 6** *Semi-log scales have one normal axis and one logarithmic axis*

### Dispersion diagrams

- **Dispersion** refers to how a data set is distributed and can be shown on a dispersion diagram.
- The range and any outliers can be easily understood, and the distribution of more than two data sets can be compared.

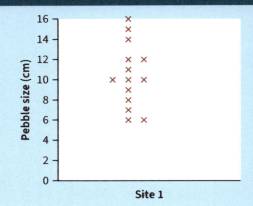

▶ **Figure 7** *A dispersion diagram*

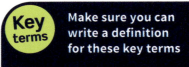 **Key terms** — Make sure you can write a definition for these key terms

| bar chart | dispersion | line graph |
| logarithmic scale | pie chart | scatter graph |
| | triangular graph | |

## 12.4 Statistical skills

### Measures of central tendency

Measures of **central tendency** are used to describe a data set:

- the mean is used to show the average of a data set
- the median is used to show the mid-point of a data set
- the mode is used to show the most common value in a data set
- the mean is very commonly used, but can be influenced by extreme outliers and does not give any information about how the data is spread around the mean.

### Measures of dispersion

Measures of dispersion are used to show how data is distributed:

- the range is calculated by looking at the difference between the highest and lowest values in a data set
- the inter-quartile range omits extreme outliers and gives the range of the middle 50% of the data in the data set – the mid-point of the bottom 25% of the data (Q1) is subtracted from the mid-point of the top 25% of the data (Q3) to find the inter-quartile range
- the standard deviation can be calculated to show how far a data set is spread around the mean.

> **REVISION TIP**
>
> Make sure you can calculate and interpret a value for the standard deviation.

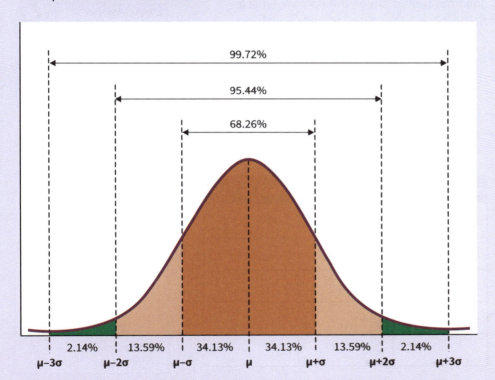

▲ **Figure 1** *The normal distribution curve tells us that 68.26% of the data within a data set will lie within 1 standard deviation (σ) either side of the mean*

## Inferential and relational statistical techniques

- These statistical techniques are used to make inferences about a data set as a sample of a larger population.

- **Spearman's rank** correlation gives a value, or correlation coefficient (Rs), to the relationship between two variables. An Rs value closer to –1 indicates a negative relationship and an Rs value closer to 1 indicates a positive relationship. The closer the value is to –1 or 1 indicates the strength of the relationship.

- The **Chi-squared** test can be used to see if there is a significant difference between two data sets.

  - Data must be in the form of frequencies, grouped or categorised.

  - The total number of observations must exceed 20.

  - The expected frequency for any one group must exceed 5.

  - The categories should not have a directly causal link.

**REVISION TIP**

Make sure you can calculate and interpret a value for Spearman's rank correlation and Chi-squared.

## Application of significance tests

- The results of Spearman's rank and Chi-squared can be compared against a critical values table to examine how likely it is that the observed result has occurred by chance.

- If the calculated value is higher than the number given in the critical value table at both the 95% and 99% significance level, then it can be said that the result is highly significant.

- For Spearman's rank, the higher the value of Rs, the more significant the result is. The larger the value of Chi-squared, the smaller the possibility that there is no significant difference between the two data sets.

**REVISION TIP**

When testing the Chi-squared value for significance, remember to use N-1 in the critical values table.

> N is the number of values in the data set

> The 99% or 0.01 significance level is a 1 in 100 probability that the result has occurred by chance

> If the calculated value is higher than the number given in the critical value table at both the 95% and 99% significance level, then it can be said that the result is highly significant

| N | 95% (0.05) | 99% (0.01) |
|---|---|---|
| 5 | 0.90 | 1.00 |
| 6 | 0.83 | 0.94 |
| 7 | 0.71 | 0.89 |
| 8 | 0.64 | 0.83 |
| 9 | 0.60 | 0.78 |

▲ **Figure 2** *The critical values for Spearman's rank for a data set of between 5 and 9 values*

**Key terms** Make sure you can write a definition for these key terms

central tendency     Chi-squared     Spearman's rank

# RETRIEVAL

Learn the answers to the questions below, then cover the answers column with a piece of paper and write down as many as you can. Check and repeat.

## Questions

## Answers

| | Questions | Answers |
|---|---|---|
| 1 | What does a synoptic chart show? | The atmospheric conditions at a particular time over a certain area |
| 2 | Which maps use a scale of graduated colours to show spatial variations of data across different areal units? | Choropleth maps |
| 3 | Which maps joins points of equal value? | Isoline maps |
| 4 | What does a divergent line graph show? | How values deviate, positively and negatively, from a mean or average |
| 5 | What does a scatter graph show? | The relationship or correlation between two variables |
| 6 | What type of graph uses segments to show proportions of a total? | Pie chart |
| 7 | How many variables does a triangular graph present? | Three |
| 8 | How does the scale on a logarithmic graph increase? | On a cycle of 10, e.g. 10 to 100, 100 to 1000, 1000 to 10,000 |
| 9 | What does dispersion mean? | How a data set is distributed |
| 10 | What are three examples of measures of central tendency? | Mean, median and mode |
| 11 | What is the range? | The difference between the highest and lowest values in a data set |
| 12 | Which measure of dispersion gives the range of the middle 50% of the data set? | Inter-quartile range |
| 13 | When is a Chi-squared test used? | To see if there is a significant difference between two data sets |

Put paper here

# OS MAPS SYMBOLS

## Symbols on Ordnance Survey maps (1:50 000 and 1:25 000)

### ROADS AND PATHS

| | |
|---|---|
| M1 or A6(M) | Motorway |
| A35 | Dual carriageway |
| A31(T) or A35 | Trunk or main road |
| B3074 | Secondary road |
| | Narrow road with passing places |
| | Road under construction |
| | Road generally more than 4 m wide |
| | Road generally less than 4 m wide |
| | Other road, drive or track, fenced and unfenced |
| >> | Gradient: steeper than 1 in 5; 1 in 7 to 1 in 5 |
| Ferry | Ferry: Ferry P – passenger only |
| | Path |

### PUBLIC RIGHTS OF WAY

(Not applicable to Scotland)

| 1:25 000 | 1:50 000 | |
|---|---|---|
| – – – – – | ·········· | Footpath |
| – · – · – · | – · – · – · | Road used as a public footpath |
| + + + + + + | – – – – – | Bridleway |
| –+–+–+– | +–+–+–+ | Byway open to all traffic |

### RAILWAYS

| | |
|---|---|
| | Multiple track |
| | Single track |
| | Narrow gauge/Light rapid transit system |
| | Road over; road under; level crossing |
| | Cutting; tunnel; embankment |
| | Station, open to passengers; siding |

### BOUNDARIES

| | |
|---|---|
| –+–––+– | National |
| –+–+–+– | District |
| –·––·––· | County, Unitary Authority, Metropolitan District or London Borough |
| | National Park |

### HEIGHTS/ROCK FEATURES

| | |
|---|---|
| —50— | Contour lines |
| ·144 | Spot height to the nearest metre above sea level |

### ABBREVIATIONS

| | | | |
|---|---|---|---|
| P | Post office | PC | Public convenience (rural areas) |
| PH | Public house | TH | Town Hall, Guildhall or equivalent |
| MS | Milestone | Sch | School |
| MP | Milepost | Coll | College |
| CH | Clubhouse | Mus | Museum |
| CG | Coastguard | Cemy | Cemetery |
| Fm | Farm | | |

### ANTIQUITIES

| | | | |
|---|---|---|---|
| VILLA | Roman | ✗ | Battlefield (with date) |
| Castle | Non-Roman | * | Tumulus/tumuli (mound over burial place) |

### LAND FEATURES

| | |
|---|---|
| | Buildings |
| | Public building |
| | Bus or coach station |
| | Place of worship with tower |
| | Place of worship with spire, minaret or dome |
| | Place of worship without such additions |
| ∘ | Chimney or tower |
| | Glass structure |
| Ⓗ | Heliport |
| △ | Triangulation pillar |
| | Mast |
| | Wind pump; wind generator |
| | Windmill |
| + | Graticule intersection |
| | Cutting; embankment |
| | Quarry |
| | Spoil heap, refuse tip or dump |

Coniferous wood
Non-coniferous wood
Mixed wood

Orchard
Park or ornamental ground

Forestry Commission access land

National Trust – always open

National Trust, limited access, observe local signs

National Trust for Scotland

### TOURIST INFORMATION

| | |
|---|---|
| P | Parking |
| P&R | Park & Ride |
| V | Visitor centre |
| i | Information centre |
| ✆ | Telephone |
| | Camp site/ Caravan site |
| | Golf course or links |
| | Viewpoint |
| PC | Public convenience |
| ✗ | Picnic site |
| | Pub/s |
| | Museum |
| | Castle/fort |
| | Building of historic interest |
| | Steam railway |
| | English Heritage |
| | Garden |
| | Nature reserve |
| | Water activities |
| | Fishing |
| ☆ | Other tourist feature |
| | Moorings (free) |
| | Electric boat charging point |
| | Recreation/leisure/ sports centre |

### WATER FEATURES

# OXFORD
## UNIVERSITY PRESS

Great Clarendon Street, Oxford, OX2 6DP, United Kingdom

Oxford University Press is a department of the University of Oxford. It furthers the University's objective of excellence in research, scholarship, and education by publishing worldwide. Oxford is a registered trade mark of Oxford University Press in the UK and in certain other countries.

Written by Rob Bircher, Alice Griffiths, Rebecca Priest and Lucy Scovell

British Library Cataloguing in Publication Data

Data available

978-1-382-05242-9

978-1-382-05243-6 (ebook)

10 9 8 7 6 5 4 3 2 1

The manufacturing process conforms to the environmental regulations of the country of origin.

Printed in the UK by Bell and Bain Ltd, Glasgow

## Acknowledgements

The publisher and authors would like to thank the following for permission to use photographs and other copyright material:

**Photos: p4**: george green / Shutterstock; **p5**: Gardawind / Shutterstock; **p18**: colin grice / Shutterstock; **p21**: Covey, K., et al. (2021); **p24(t), 196(l)**: Clare Louise Jackson / Shutterstock; **p24(b)**: NOAA; **p28**: Stephen Firmender / Shutterstock; **p32(l)**: Karl W. Wegmann / Stockimo / Alamy Stock Photo; **p32(r), 35** HYPERLINK "http://www.sandatlas.org"www.sandatlas.org / Shutterstock; **p47(t)**: Moehring / Shutterstock; **p47(m)**: Christian SAPPA / Getty Images; **p47(b)**: Corbin17 / Alamy Stock Photo; **p48**: Hoberman Collection / Getty Images; **p66**: Michael Brian Shannon / Shutterstock; **p69**: Church et al. (2013), IPCC; **p70**: C. J. Hapke and K. R. Green, U.S. Geological Survey; **p71(t)**: atthle / Shutterstock; **p71(b)**: Fulcanelli / Shutterstock; **p72(t)**: Rob Read / Alamy Stock Photo; **p72(b)**: Michael Hawkridge / Alamy Stock Photo; **p84**: Lake District Landscapes; **p85**: orxy / Shutterstock; **p93**: Crown Copyright; **p98(t)**: Jonatan Hedberg / Alamy Stock Photo; **p98(b)**: GRANGER - Historical Picture Archive / Alamy Stock Photo; **p99(t)**: All Canada Photos / Alamy Stock Photo; **p99(b)**: GFC Collection / Alamy Stock Photo; **p124**: Silent O / Shutterstock; **p130**: Coppola et al., (2020), HYPERLINK "http://www.mirovaweb.it"www.mirovaweb.it; **p144**: Belikova Oksana / Shutterstock; **p145**: B.G. Wilson Fire / Alamy Stock Photo; **p150**: Prasit Rodphan / Alamy Stock Photo; **p154**: Becky Stares / Shutterstock; **p158**: UK Parliament, Source: WWF 2020; **p174**: Mark Mason; **p177**: blurAZ / Shutterstock; **p180**: vadik4444 / Shutterstock; **p190**: alexsl / Getty Images; **p196(m)**: Crown Copyright; **p196(r)**: Judy Joel / Bridgeman Images; **p199**: Robert Clare / Alamy Stock Photo; **p200**: Richard W Miller / Shutterstock; **p201**: Neirfy / Shutterstock; **p204(t)**: Atomazul / Shutterstock; **p204(b)**: Richard Green / Alamy Stock Photo; **p205**: Jason Wells / Alamy Stock Photo; **p214(t)**: The Stapleton Collection / Bridgeman Images; **p214(b)**: Anthony Pemberton / Alamy Stock Photo; **p215(t)**: Crown Copyright; **p216**: NurPhoto / Getty Images; **p217**: Carmarthen County Council; **p228(t)**: Patrick Wang / Shutterstock; **p228(b)**: marchello74 / Shutterstock; **p244(t)**: NadyGinzburg / Shutterstock; **p244(m)**: Romanenkova / Shutterstock; **p244(b)**: philip openshaw / Shutterstock; **p245(t)**: Henry Franklin / Shutterstock; **p245(m)**: DedMityay / Shutterstock; **p245(b)**: nikonpete / Shutterstock; **p258**: Jasper Suijten / Shutterstock; **p277**: Feraru Nicolae / Shutterstock; **p282(b)**: Surveillance, Epidemiology, and End Results (SEER) Program, National Cancer Institute; **p318**: Photodiem / Shutterstock; **p319**: © Consumer Data Research Centre 2016 / © Crown Copyright & Database Right 2014-5.

Artwork by Adrian Smith, Aptara Inc., Barking Dog Art, David Russell Illustration, Dusan Lakicevic, Giorgio Bacchin, Jess McGeachin, Kamae Design, KJA Artists, Lovell Johns, Mark Duffin, Mike Connor, ODI, Q2A Media, Simon Tegg, Tracey Learoyd and Oxford University Press.

Ordnance Survey (OS) is the national mapping agency for Great Britain, and a world-leading geospatial data and technology organisation. As a reliable partner to government, business and citizens across Britain and the world, OS helps its customers in virtually all sectors improve quality of life.

**Text: p15**: Adapted from Figure 2.1 Panel a in IPCC, 2023: Climate Change 2023: Synthesis Report. Contribution of Working Groups I, II and III to the Sixth Assessment Report of the Intergovernmental Panel on Climate Change [Core Writing Team, H. Lee and J. Romero (eds.)]. IPCC, Geneva, Switzerland, pp. 35-115, doi: 10.59327/IPCC/AR6-9789291691647. With permission from IPCC; **p19**: Reprinted from Thomas, H. & Nisbet, T. R. (2016) 'Slowing the flow in Pickering: Quantifying the effect of catchment woodland and planting on flooding using the soil conservation service curve number method', International Journal of Safety and Security Engineering, Volume 6, Number 3, doi: 10.2495/SAFE-V6-N3-466-474, © 2016, with permission from WIT Press, Southampton, UK; **p21**: https://www.frontiersin.org/articles/10.3389/ffgc.2021.618401/full Covey, K., et al., Carbon and Beyond: The Biogeochemistry of Climate in a Rapidly Changing Amazon, Front. For. Glob. Change, 11 March 2021 Sec. Tropical Forests Volume 4 - 2021 | https://doi.org/10.3389/ffgc.2021.618401 Reproduced under the terms of the CC BY 4.0 Licence: <https://creativecommons.org/licenses/by/4.0>, via Wikimedia Commons; **p22**: Source: May 2023; page 11 – Hydrological Summary for the United Kingdom; National Hydrological Monitoring Programme (NHMP). Data from COSMOS-UK Soil Moisture Network (cosmos.ceh.ac.uk). Reproduced with permission from UK Centre for Ecology & Hydrology.; **p24**: OA in the surface ocean near Mauna Loa Observatory in Hawaii, USA.  Adapted from Dore et al. 2009. PNAS 106:12235-12240. © 2009. Freely available online through the PNAS open access option.; **p25**: Adapted from Figure 2.2 Panel b in IPCC, 2023: Climate Change 2023: Synthesis Report. Contribution of Working Groups I, II and III to the Sixth Assessment Report of the Intergovernmental Panel on Climate Change [Core Writing Team, H. Lee and J. Romero (eds.)]. IPCC, Geneva, Switzerland, pp. 35-115, doi: 10.59327/IPCC/AR6-9789291691647. With permission from IPCC.; **p27**: Adapted from Cherlet, M., et al. (Eds.), World Atlas of Desertification, Publication Office of the European Union, Luxembourg, 2018.  Data Source: Global Precipitation Climatology Centre and potential evapotranspiration data from the Climate Research Unit of the University of East Anglia (CRUTSv3.20), WAD3-JRC, modified from Spinoni J. 2015 [AP] Heat and cold waves trends in the Carpathian Region from 1961 to 2010 Spinoni J; Lakatos MSzentimrey T et al. International Journal of Climatology (2015) 35(14) 4197-4209; **p45**: Adapted with permission from S. Sherwood from A Drier Future? Global warming is likely to lead to overall drying of land surfaces. Steven Sherwood and Qiang Fu. Science, 14 Feb 2014, Vol 343, Issue 6172, pp. 737-739. DOI: 10.1126/science.1247620. Reprinted with permission from AAAS.; **p46**: Atlas of Namibia Project (2002) Directorate of Environmental Affairs, Ministry of Environment and Tourism © published edition, text and maps, Namibia Nature Foundation Open access: Licensed under the terms of the Creative Commons Attribution 4.0 International License (http://creativecommons.org/licenses/by/4.0/); **p69**: Figure 4.10 in Oppenheimer, M., et al., 2019: Sea Level Rise and Implications for Low-Lying Islands, Coasts and Communities. In: IPCC Special Report on the Ocean and Cryosphere in a Changing Climate [H.-O. Pörtner, et al. (eds.)]. Cambridge University Press, Cambridge, UK and New York, NY, USA, pp. 321–445. https://doi.org/10.1017/9781009157964.006.  With permission from IPCC.; **p130**: Reproduced with permission from the MIROVA Project. https://www.

Every effort has been made to contact copyright holders of material reproduced in this book. Any omissions will be rectified in subsequent printings if notice is given to the publisher.

The publisher would also like to thank Adam Robbins and Alice Griffiths for sharing their expertise and feedback in the development of this resource.

Links to third party websites are provided by Oxford in good faith and for information only. Oxford disclaims any responsibility for the materials contained in any third party website referenced in this work.